APPLIED REGRESSION
MODELING

APPLIED REGRESSION MODELING

Second Edition

IAIN PARDOE
Thompson Rivers University

A JOHN WILEY & SONS, INC., PUBLICATION

Published by John Wiley & Sons, Inc., Hoboken, New Jersey.
Published simultaneously in Canada.

For general information on our other products and services or for technical support, please contact our Customer Care Department within the United States at (800) 762-2974, outside the United States at (317) 572-3993 or fax (317) 572-4002.

Wiley also publishes its books in a variety of electronic formats. Some content that appears in print may not be available in electronic formats. For more information about Wiley products, visit our web site at www.wiley.com.

Library of Congress Cataloging-in-Publication Data:

Pardoe, Iain, 1970–
  Applied regression modeling [electronic resource] / Iain Pardoe. — 2nd ed.
    1 online resource.
  Includes index.
  Description based on print version record and CIP data provided by publisher; resource not viewed.
  ISBN 978-1-118-34502-3 (pdf) — ISBN 978-1-118-34503-0 (mobi) — ISBN 978-1-118-34504-7 (epub) — ISBN 978-1-118-09728-1 (hardback) (print)
1. Regression analysis. 2. Statistics. I. Title.
  QA278.2
  519.5'36—dc23   2012006617

Printed in the United States of America.

To Tanya, Bethany, and Sierra

CONTENTS

PREFACE

The first edition of this book was developed from class notes written for an applied regression course taken primarily by undergraduate business majors in their junior year at the University of Oregon. Since the regression methods and techniques covered in the book have broad application in many fields, not just business, this second edition widens its scope to reflect this. Details of the major changes for the second edition are included below.

The book is suitable for any undergraduate statistics course in which regression analysis is the main focus. A recommended prerequisite is an introductory probability and statistics course. It would also be suitable for use in an applied regression course for non-statistics major graduate students, including MBAs, and for vocational, professional, or other non-degree courses. Mathematical details have deliberately been kept to a minimum, and the book does not contain any calculus. Instead, emphasis is placed on applying regression analysis to data using statistical software, and understanding and interpreting results.

Chapter 1 reviews essential introductory statistics material, while Chapter 2 covers simple linear regression. Chapter 3 introduces multiple linear regression, while Chapters 4 and 5 provide guidance on building regression models, including transforming variables, using interactions, incorporating qualitative information, and using regression diagnostics. Each of these chapters includes homework problems, mostly based on analyzing real datasets provided with the book. Chapter 6 contains three in-depth case studies, while Chapter 7 introduces extensions to linear regression and outlines some related topics. The appendices contain a list of statistical software packages that can be used to carry out all the analyses covered in the book (each with detailed instructions available from the book website), a table of critical values for the t-distribution, notation and formulas used throughout the book, a glossary of important terms, a short mathematics refresher, and brief answers to selected homework problems.

The first five chapters of the book have been used successfully in quarter-length courses at a number of institutions. An alternative approach for a quarter-length course would be to skip some of the material in Chapters 4 and 5 and substitute one or more of the case studies in Chapter 6, or briefly introduce some of the topics in Chapter 7. A semester-length course could comfortably cover all the material in the book.

The website for the book, which can be found at www.iainpardoe.com/arm2e/, contains supplementary material designed to help both the instructor teaching from this book and the student learning from it. There you'll find all the datasets used for examples and homework problems in formats suitable for most statistical software packages, as well as detailed instructions for using the major packages, including SPSS, Minitab, SAS, JMP, Data Desk, EViews, Stata, Statistica, R, and S-PLUS. There is also some information on using the Microsoft Excel spreadsheet package for *some* of the analyses covered in the book (dedicated statistical software is necessary to carry out *all* of the analyses). The website also includes information on obtaining a solutions manual containing complete answers to all the homework problems, as well as further ideas for organizing class time around the material in the book.

The book contains the following stylistic conventions:

- When displaying calculated values, the general approach is to be as accurate as possible when it matters (such as in intermediate calculations for problems with many steps), but to round appropriately when convenient or when reporting final results for real-world questions. Displayed results from statistical software use the default rounding employed in R throughout.

- In the author's experience, many students find some traditional approaches to notation and terminology a barrier to learning and understanding. Thus, some traditions have been altered to improve ease of understanding. These include: using familiar Roman letters in place of unfamiliar Greek letters [e.g., $E(Y)$ rather than μ and b rather than β]; replacing the nonintuitive $\bar{Y}$ for the sample mean of Y with m_Y; using NH and AH for null hypothesis and alternative hypothesis, respectively, rather than the usual H_0 and H_a.

Major changes for the second edition

- The first edition of this book was used in the regression analysis course run by *Statistics.com* from 2008 to 2012. The lively discussion boards provided an invaluable source for suggestions for changes to the book. This edition clarifies and expands on concepts that students found challenging and addresses every question posed in those discussions.

- The foundational material on interval estimation has been rewritten to clarify the mathematics.

- There is new material on testing model assumptions, transformations, indicator variables, nonconstant variance, autocorrelation, power and sample size, model building, and model selection.

- As far as possible, I've replaced outdated data examples with more recent data, and also used more appropriate data examples for particular topics (e.g., autocorrelation). In total, about 40% of the data files have been replaced.

- Most of the data examples now use descriptive names for variables rather than generic letters such as Y and X.

- As in the first edition, this edition uses mathematics to explain methods and techniques only where necessary, and formulas are used within the text only when they are instructive. However, this edition also includes additional formulas in optional sections to aid those students who can benefit from more mathematical detail.

- I've added many more end-of-chapter problems. In total, the number of problems has increased by nearly 25%.

- I've updated and added new references, nearly doubling the total number of references.

- I've added a third case study to Chapter 6.

- The first edition included detailed computer software instructions for five major software packages (SPSS, Minitab, SAS Analyst, R/S-PLUS, and Excel) in an appendix. This appendix has been dropped from this edition; instead, instructions for newer software versions and other packages (e.g., JMP and Stata) are now just updated on the book website.

IAIN PARDOE

Nelson, British Columbia
April 2012

ACKNOWLEDGMENTS

I am grateful to a number of people who helped to make this book a reality. Dennis Cook and Sandy Weisberg first gave me the textbook-writing bug when they approached me to work with them on their classic applied regression book (Cook and Weisberg, 1999), and Dennis subsequently motivated me to transform my teaching class notes into my own applied regression book. People who provided data for examples used throughout the book include: Victoria Whitman for the house price examples; Wolfgang Jank for the autocorrelation example on beverage sales; Craig Allen for the case study on pharmaceutical patches; Cathy Durham for the Poisson regression example in the chapter on extensions. The multilevel and Bayesian modeling sections of the chapter on extensions are based on work by Andrew Gelman and Hal Stern. Gary A. Simon and a variety of anonymous reviewers provided extremely useful feedback on the first edition of the book, as did many of my students at the University of Oregon and Statistics.com. Finally, I'd like to thank my editor at Wiley, Steve Quigley, who encouraged me to prepare this second edition.

I. P.

INTRODUCTION

I.1 STATISTICS IN PRACTICE

Statistics is used in many fields of application since it provides an effective way to analyze quantitative information. Some examples include:

- A pharmaceutical company is developing a new drug for treating a particular disease more effectively. How might statistics help you decide whether the drug will be safe and effective if brought to market?

  Clinical trials involve large-scale statistical studies of people—usually both patients with the disease and healthy volunteers—who are assessed for their response to the drug. To determine that the drug is both safe and effective requires careful statistical analysis of the trial results, which can involve controlling for the personal characteristics of the people (e.g., age, gender, health history) and possible placebo effects, comparisons with alternative treatments, and so on.

- A manufacturing firm is not getting paid by its customers in a timely manner—this costs the firm money on lost interest. You've collected recent data for the customer accounts on amount owed, number of days since the customer was billed, and size of the customer (small, medium, large). How might statistics help you improve the on-time payment rate?

  You can use statistics to find out whether there is an association between the amount owed and the number of days and/or size. For example, there may be a positive association between amount owed and number of days for small and medium-sized customers but not for large-sized customers—thus it may be more profitable to focus collection efforts on small and medium-sized customers billed some time ago, rather than on large-sized customers or customers billed more recently.

- A firm makes scientific instruments and has been invited to make a sealed bid on a large government contract. You have cost estimates for preparing the bid and fulfilling the contract, as well as historical information on similar previous contracts on which the firm has bid (some successful, others not). How might statistics help you decide how to price the bid?

You can use statistics to model the association between the success/failure of past bids and variables such as bid cost, contract cost, bid price, and so on. If your model proves useful for predicting bid success, you could use it to set a maximum price at which the bid is likely to be successful.

- As an auditor, you'd like to determine the number of price errors in all of a company's invoices—this will help you detect whether there might be systematic fraud at the company. It is too time-consuming and costly to examine all of the company's invoices, so how might statistics help you determine an upper bound for the proportion of invoices with errors?

Statistics allows you to infer about a population from a relatively small random sample of that population. In this case, you could take a sample of 100 invoices, say, to find a proportion, p, such that you could be 95% confident that the population error rate is less than that quantity p.

- A firm manufactures automobile parts and the factory manager wants to get a better understanding of overhead costs. You believe two variables in particular might contribute to cost variation: machine hours used per month and separate production runs per month. How might statistics help you to quantify this information?

You can use statistics to build a multiple linear regression model that estimates an equation relating the variables to one another. Among other things you can use the model to determine how much cost variation can be attributed to the two cost drivers, their individual effects on cost, and predicted costs for particular values of the cost drivers.

- You work for a computer chip manufacturing firm and are responsible for forecasting future sales. How might statistics be used to improve the accuracy of your forecasts?

Statistics can be used to fit a number of different forecasting models to a time series of sales figures. Some models might just use past sales values and extrapolate into the future, while others might control for external variables such as economic indices. You can use statistics to assess the fit of the various models, and then use the best-fitting model, or perhaps an average of the few best-fitting models, to base your forecasts on.

- As a financial analyst, you review a variety of financial data, such as price/earnings ratios and dividend yields, to guide investment recommendations. How might statistics be used to help you make buy, sell, or hold recommendations for individual stocks?

By comparing statistical information for an individual stock with information about stock market sector averages, you can draw conclusions about whether the stock is overvalued or undervalued. Statistics is used for both "technical analysis" (which considers the trading patterns of stocks) and "quantitative analysis" (which studies economic or company-specific data that might be expected to affect the price or perceived value of a stock).

- You are a brand manager for a retailer and wish to gain a better understanding of the association between promotional activities and sales. How might statistics be used to help you obtain this information and use it to establish future marketing strategies for your brand?

  Electronic scanners at retail checkout counters and online retailer records can provide sales data and statistical summaries on promotional activities such as discount pricing and the use of in-store displays or e-commerce websites. Statistics can be used to model these data to discover which product features appeal to particular market segments and to predict market share for different marketing strategies.

- As a production manager for a manufacturer, you wish to improve the overall quality of your product by deciding when to make adjustments to the production process, for example, increasing or decreasing the speed of a machine. How might statistics be used to help you make those decisions?

  Statistical quality control charts can be used to monitor the output of the production process. Samples from previous runs can be used to determine when the process is "in control." Ongoing samples allow you to monitor when the process goes out of control, so that you can make the adjustments necessary to bring it back into control.

- As an economist, one of your responsibilities is providing forecasts about some aspect of the economy, for example, the inflation rate. How might statistics be used to estimate those forecasts optimally?

  Statistical information on various economic indicators can be entered into computerized forecasting models (also determined using statistical methods) to predict inflation rates. Examples of such indicators include the producer price index, the unemployment rate, and manufacturing capacity utilization.

- As general manager of a baseball team with limited financial resources, you'd like to obtain strong, yet undervalued players. How might statistics help you to do this?

  A wealth of statistical information on baseball player performance is available, and objective analysis of these data can reveal information on those players most likely to add value to the team (in terms of winning games) relative to a player's cost. This field of statistics even has its own name, sabermetrics.

I.2 LEARNING STATISTICS

- What is this book about?

  This book is about the application of statistical methods, primarily regression analysis and modeling, to enhance decision-making. Regression analysis is by far the most used statistical methodology in real-world applications. Furthermore, many other statistical techniques are variants or extensions of regression analysis, so once you have a firm foundation in this methodology, you can approach these other techniques without too much additional difficulty. In this book we show you how to apply and interpret regression models, rather than deriving results and formulas (there is no calculus in the book).

- Why are non-math major students required to study statistics?

  In many aspects of modern life, we have to make decisions based on incomplete information (e.g., health, climate, economics, business). This book will help you to understand, analyze, and interpret such data in order to make informed decisions in the face of uncertainty. Statistical theory allows a rigorous, quantifiable appraisal of this uncertainty.

- How is the book organized?

  Chapter 1 reviews the essential details of an introductory statistics course necessary for use in later chapters. Chapter 2 covers the simple linear regression model for analyzing the linear association between two variables (a "response" and a "predictor"). Chapter 3 extends the methods of Chapter 2 to multiple linear regression where there can be more than one predictor variable. Chapters 4 and 5 provide guidance on building regression models, including transforming variables, using interactions, incorporating qualitative information, and diagnosing problems. Chapter 6 contains three case studies that apply the linear regression modeling techniques considered in this book to examples on real estate prices, vehicle fuel efficiency, and pharmaceutical patches. Chapter 7 introduces some extensions to the multiple linear regression model and outlines some related topics. The appendices contain a list of statistical software that can be used to carry out all the analyses covered in the book, a t-table for use in calculating confidence intervals and conducting hypothesis tests, notation and formulas used throughout the book, a glossary of important terms, a short mathematics refresher, and brief answers to selected problems.

- What else do you need?

  The preferred calculation method for understanding the material and completing the problems is to use statistical software rather than a statistical calculator. It may be possible to apply many of the methods discussed using spreadsheet software (such as Microsoft Excel), although some of the graphical methods may be difficult to implement and statistical software will generally be easier to use. Although a statistical calculator is not recommended for use with this book, a traditional calculator capable of basic arithmetic (including taking logarithmic and exponential transformations) will be invaluable.

- What other resources are recommended?

  Good supplementary textbooks (some at a more advanced level) include Dielman (2004), Draper and Smith (1998), Kutner et al. (2004), Mendenhall and Sincich (2011), Ryan (2008), and Weisberg (2005).

CHAPTER 1

FOUNDATIONS

This chapter provides a brief refresher of the main statistical ideas that will be a useful foundation for the main focus of this book, regression analysis, covered in subsequent chapters. For more detailed discussion of this material, consult a good introductory statistics textbook such as Freedman et al. (2007) or Moore et al. (2011). To simplify matters at this stage, we consider *univariate* data, that is, datasets consisting of measurements of just a single variable on a sample of observations. By contrast, regression analysis concerns *multivariate* data where there are two or more variables measured on a sample of observations. Nevertheless, the statistical ideas for univariate data carry over readily to this more complex situation, so it helps to start as simply as possible and make things more complicated only as needed.

1.1 IDENTIFYING AND SUMMARIZING DATA

One way to think about *statistics* is as a collection of methods for using data to understand a problem quantitatively—we saw many examples of this in the introduction. This book is concerned primarily with analyzing data to obtain information that can be used to help make decisions in real-world contexts.

The process of framing a problem in such a way that it will be amenable to quantitative analysis is clearly an important step in the decision-making process, but this lies outside the scope of this book. Similarly, while data collection is also a necessary task—often the most time-consuming part of any analysis—we assume from this point on that we have

Applied Regression Modeling, Second Edition. By Iain Pardoe
Copyright © 2012 John Wiley & Sons, Inc.

already obtained some data relevant to the problem at hand. We will return to the issue of the manner in which these data have been collected—namely, whether the sample data can be considered to be representative of some larger population that we wish to make statistical inferences for—in Section 1.3.

For now, we consider identifying and summarizing the data at hand. For example, suppose that we have moved to a new city and wish to buy a home. In deciding on a suitable home, we would probably consider a variety of factors, such as size, location, amenities, and price. For the sake of illustration we focus on price and, in particular, see if we can understand the way in which sale prices vary in a specific housing market. This example will run through the rest of the chapter, and, while no one would probably ever obsess over this problem to this degree in real life, it provides a useful, intuitive application for the statistical ideas that we use in the rest of the book in more complex problems.

For this example, identifying the data is straightforward: the *units of observation* are a random *sample* of size $n = 30$ single-family homes in our particular housing market, and we have a single measurement for each observation, the sale price in thousands of dollars (\$), represented using the notation $Y = Price$. Here, Y is the generic letter used for any univariate data variable, while *Price* is the specific variable name for this dataset. These data, obtained from Victoria Whitman, a realtor in Eugene, Oregon, are available in the **HOMES1** data file—they represent sale prices of 30 homes in south Eugene during 2005. This represents a subset of a larger file containing more extensive information on 76 homes, which is analyzed as a case study in Section 6.1.

The particular sample in the **HOMES1** data file is *random* because the 30 homes have been selected randomly somehow from the population of all single-family homes in this housing market. For example, consider a list of homes currently for sale, which are considered to be representative of this population. A random number generator—commonly available in spreadsheet or statistical software—can be used to pick out 30 of these. Alternative selection methods may or may not lead to a random sample. For example, picking the first 30 homes on the list would not lead to a random sample if the list were ordered by the size of the sale price.

We can simply list small datasets such as this. The values of *Price* in this case are:

155.5	195.0	197.0	207.0	214.9	230.0	239.5	242.0	252.5	255.0
259.9	259.9	269.9	270.0	274.9	283.0	285.0	285.0	299.0	299.9
319.0	319.9	324.5	330.0	336.0	339.0	340.0	355.0	359.9	359.9

However, even for these data, it can be helpful to summarize the numbers with a small number of *sample statistics* (such as the sample mean and standard deviation), or with a graph that can effectively convey the manner in which the numbers vary. A particularly effective graph is a *stem-and-leaf plot*, which places the numbers along the vertical axis of the plot, with numbers that are close together in magnitude next to one another on the plot. For example, a stem-and-leaf plot for the 30 sample prices looks like the following:

```
1 | 6
2 | 0011344
2 | 5666777899
3 | 002223444
3 | 666
```

In this plot, the decimal point is two digits to the right of the stem. So, the "1" in the stem and the "6" in the leaf represents 160, or, because of rounding, any number between 155 and

164.9. In particular, it represents the lowest price in the dataset of 155.5 (thousand dollars). The next part of the graph shows two prices between 195 and 204.9, two prices between 205 and 214.9, one price between 225 and 234.9, two prices between 235 and 244.9, and so on. A stem-and-leaf plot can easily be constructed by hand for small datasets such as this, or it can be constructed automatically using statistical software. The appearance of the plot can depend on the type of statistical software used—this particular plot was constructed using R statistical software (as are all the plots in this book). Instructions for constructing stem-and-leaf plots are available as computer help #13 in the software information files available from the book website at www.iainpardoe.com/arm2e/software.htm.

The overall impression from this graph is that the sample prices range from the mid-150s to the mid-350s, with some suggestion of clustering around the high 200s. Perhaps the sample represents quite a range of moderately priced homes, but with no very cheap or very expensive homes. This type of observation often arises throughout a data analysis—the data begin to tell a story and suggest possible explanations. A good analysis is usually not the end of the story since it will frequently lead to other analyses and investigations. For example, in this case, we might surmise that we would probably be unlikely to find a home priced at much less than $150,000 in this market, but perhaps a realtor might know of a nearby market with more affordable housing.

A few modifications to a stem-and-leaf plot produce a *histogram*—the value axis is now horizontal rather than vertical, and the counts of observations within adjoining data intervals (called "bins") are displayed in bars (with the counts, or frequency, shown on the vertical axis) rather than by displaying individual values with digits. Figure 1.1 shows a histogram for the home prices data generated by statistical software (see computer help #14).

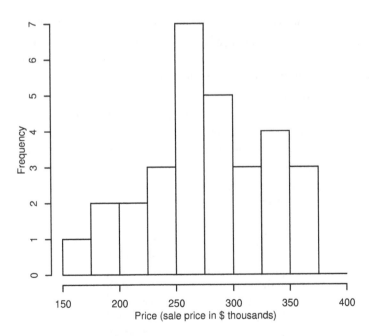

Figure 1.1   Histogram for home prices example.

Histograms can convey very different impressions depending on the bin width, start point, and so on. Ideally, we want a large enough bin size to avoid excessive sampling "noise" (a histogram with many bins that looks very wiggly), but not so large that it is hard to see the underlying distribution (a histogram with few bins that looks too blocky). A reasonable pragmatic approach is to use the default settings in whichever software package we are using, and then perhaps to create a few more histograms with different settings to check that we're not missing anything. There are more sophisticated methods, but for the purposes of the methods in this book, this should suffice.

In addition to graphical summaries such as the stem-and-leaf plot and histogram, sample statistics can summarize data numerically. For example:

- The *sample mean*, m_Y, is a measure of the "central tendency" of the data Y-values.
- The *sample standard deviation*, s_Y, is a measure of the spread or variation in the data Y-values.

We won't bother here with the formulas for these sample statistics. Since almost all of the calculations necessary for learning the material covered by this book will be performed by statistical software, the book only contains formulas when they are helpful in understanding a particular concept or provide additional insight to interested readers.

We can calculate *sample standardized Z-values* from the data Y-values:

$$Z = \frac{Y - m_Y}{s_Y}.$$

Sometimes, it is useful to work with sample standardized Z-values rather than the original data Y-values since sample standardized Z-values have a sample mean of 0 and a sample standard deviation of 1. Try using statistical software to calculate sample standardized Z-values for the home prices data, and then check that the mean and standard deviation of the Z-values are 0 and 1, respectively.

Statistical software can also calculate additional sample statistics, such as:

- the *median* (another measure of central tendency, but which is less sensitive than the sample mean to very small or very large values in the data)—half the dataset values are smaller than this quantity and half are larger;
- the minimum and maximum;
- *percentiles* or *quantiles* such as the 25th percentile—this is the smallest value that is larger than 25% of the values in the dataset (i.e., 25% of the dataset values are smaller than the 25th percentile, while 75% of the dataset values are larger).

Here are the values obtained by statistical software for the home prices example (see computer help #10):

Sample size, n	Valid	30
	Missing	0
Mean		278.6033
Median		278.9500
Std. Deviation		53.8656
Minimum		155.5000
Maximum		359.9000
Percentiles	25	241.3750
	50	278.9500
	75	325.8750

There are many other methods—numerical and graphical—for summarizing data. For example, another popular graph besides the histogram is the *boxplot*; see Chapter 6 for some examples of boxplots used in case studies.

1.2 POPULATION DISTRIBUTIONS

While the methods of the preceding section are useful for describing and displaying sample data, the real power of statistics is revealed when we use samples to give us information about *populations*. In this context, a population is the entire collection of objects of interest, for example, the sale prices for all single-family homes in the housing market represented by our dataset. We'd like to know more about this population to help us make a decision about which home to buy, but the only data we have is a random sample of 30 sale prices.

Nevertheless, we can employ "statistical thinking" to draw inferences about the population of interest by analyzing the sample data. In particular, we use the notion of a *model*—a mathematical abstraction of the real world—which we fit to the sample data. If this model provides a reasonable fit to the data, that is, if it can approximate the manner in which the data vary, then we assume that it can also approximate the behavior of the population. The model then provides the basis for making decisions about the population, by, for example, identifying patterns, explaining variation, and predicting future values. Of course, this process can work only if the sample data can be considered representative of the population. One way to address this is to randomly select the sample from the population. There are other more complex sampling methods that are used to select representative samples, and there are also ways to make adjustments to models to account for known nonrandom sampling. However, we do not consider these here—any good sampling textbook should cover these issues.

Sometimes, even when we know that a sample has not been selected randomly, we can still model it. Then, we may not be able to formally infer about a population from the sample, but we can still model the underlying structure of the sample. One example would be a *convenience sample*—a sample selected more for reasons of convenience than for its statistical properties. When modeling such samples, any results should be reported with a caution about restricting any conclusions to objects similar to those in the sample. Another kind of example is when the sample comprises the whole population. For example, we could model data for all 50 states of the United States of America to better understand any patterns or systematic associations among the states.

Since the real world can be extremely complicated (in the way that data values vary or interact together), models are useful because they simplify problems so that we can better understand them (and then make more effective decisions). On the one hand, we therefore need models to be simple enough that we can easily use them to make decisions, but on the other hand, we need models that are flexible enough to provide good approximations to complex situations. Fortunately, many statistical models have been developed over the years that provide an effective balance between these two criteria. One such model, which provides a good starting point for the more complicated models we consider later, is the *normal distribution*.

From a statistical perspective, a *distribution* (strictly speaking, a *probability distribution*) is a theoretical model that describes how a *random variable* varies. For our purposes, a random variable represents the data values of interest in the population, for example, the

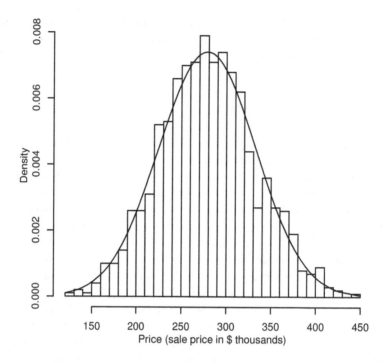

Figure 1.2 Histogram for a simulated population of 1,000 sale prices, together with a normal density curve.

sale prices of all single-family homes in our housing market. One way to represent the population distribution of data values is in a histogram, as described in Section 1.1. The difference now is that the histogram displays the whole population rather than just the sample. Since the population is so much larger than the sample, the bins of the histogram (the consecutive ranges of the data that comprise the horizontal intervals for the bars) can be much smaller than in Figure 1.1. For example, Figure 1.2 shows a histogram for a simulated population of 1,000 sale prices. The scale of the vertical axis now represents proportions (density) rather than the counts (frequency) of Figure 1.1.

As the population size gets larger, we can imagine the histogram bars getting thinner and more numerous, until the histogram resembles a smooth curve rather than a series of steps. This smooth curve is called a *density curve* and can be thought of as the theoretical version of the population histogram. Density curves also provide a way to visualize probability distributions such as the normal distribution. A normal density curve is superimposed on Figure 1.2. The simulated population histogram follows the curve quite closely, which suggests that this simulated population distribution is quite close to normal.

To see how a theoretical distribution can prove useful for making statistical inferences about populations such as that in our home prices example, we need to look more closely at the normal distribution. To begin, we consider a particular version of the normal distribution, the *standard normal*, as represented by the density curve in Figure 1.3. Random variables that follow a standard normal distribution have a mean of 0 (represented in Figure 1.3 by the curve being symmetric about 0, which is under the highest point of the curve) and a standard deviation of 1 (represented in Figure 1.3 by the curve having a point

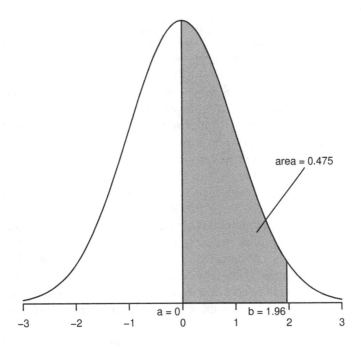

Figure 1.3   Standard normal density curve together with a shaded area of 0.475 between $a=0$ and $b=1.96$, which represents the probability that a standard normal random variable lies between 0 and 1.96.

of inflection—where the curve bends first one way and then the other—at $+1$ and -1). The normal density curve is sometimes called the "bell curve" since its shape resembles that of a bell. It is a slightly odd bell, however, since its sides never quite reach the ground (although the ends of the curve in Figure 1.3 are quite close to zero on the vertical axis, they would never actually quite reach there, even if the graph were extended a very long way on either side).

The key feature of the normal density curve that allows us to make statistical inferences is that areas under the curve represent probabilities. The entire area under the curve is one, while the area under the curve between one point on the horizontal axis (a, say) and another point (b, say) represents the probability that a random variable that follows a standard normal distribution is between a and b. So, for example, Figure 1.3 shows that the probability is 0.475 that a standard normal random variable lies between $a=0$ and $b=1.96$, since the area under the curve between $a=0$ and $b=1.96$ is 0.475.

We can obtain values for these areas or probabilities from a variety of sources: tables of numbers, calculators, spreadsheet or statistical software, Internet websites, and so on. In this book we print only a few select values since most of the later calculations use a generalization of the normal distribution called the "t-distribution." Also, rather than areas such as that shaded in Figure 1.3, it will become more useful to consider "tail areas" (e.g., to the right of point b), and so for consistency with later tables of numbers, the following table allows calculation of such tail areas:

Upper-tail area	0.1	0.05	0.025	0.01	0.005	0.001
Horizontal axis value	1.282	1.645	1.960	2.326	2.576	3.090
Two-tail area	0.2	0.1	0.05	0.02	0.01	0.002

In particular, the upper-tail area to the right of 1.96 is 0.025; this is equivalent to saying that the area between 0 and 1.96 is 0.475 (since the entire area under the curve is 1 and the area to the right of 0 is 0.5). Similarly, the two-tail area, which is the sum of the areas to the right of 1.96 and to the left of -1.96, is two times 0.025, or 0.05.

How does all this help us to make statistical inferences about populations such as that in our home prices example? The essential idea is that we fit a normal distribution model to our sample data and then use this model to make inferences about the corresponding population. For example, we can use probability calculations for a normal distribution (as shown in Figure 1.3) to make probability statements about a population modeled using that normal distribution—we'll show exactly how to do this in Section 1.3. Before we do that, however, we pause to consider an aspect of this inferential sequence that can make or break the process. Does the model provide a close enough approximation to the pattern of sample values that we can be confident the model adequately represents the population values? The better the approximation, the more reliable our inferential statements will be.

We saw in Figure 1.2 how a density curve can be thought of as a histogram with a very large sample size. So one way to assess whether our population follows a normal distribution model is to construct a histogram from our sample data and visually determine whether it "looks normal," that is, approximately symmetric and bell-shaped. This is a somewhat subjective decision, but with experience you should find that it becomes easier to discern clearly nonnormal histograms from those that are reasonably normal. For example, while the histogram in Figure 1.2 clearly looks like a normal density curve, the normality of the histogram of 30 sample sale prices in Figure 1.1 is less certain. A reasonable conclusion in this case would be that while this sample histogram isn't perfectly symmetric and bell-shaped, it is close enough that the corresponding (hypothetical) population histogram could well be normal.

An alternative way to assess normality is to construct a *QQ-plot* (quantile–quantile plot), also known as a *normal probability plot*, as shown in Figure 1.4 (see computer help #22 in the software information files available from the book website). If the points in the QQ-plot lie close to the diagonal line, then the corresponding population values could well be normal. If the points generally lie far from the line, then normality is in question. Again, this is a somewhat subjective decision that becomes easier to make with experience. In this case, given the fairly small sample size, the points are probably close enough to the line that it is reasonable to conclude that the population values could be normal.

There are also a variety of quantitative methods for asessing normality—brief details and references are provided on page 124.

Optional—technical details of QQ-plots. For the purposes of this book, the technical details of QQ-plots are not too important. For those that are curious, however, a brief description follows. First, calculate a set of n equally spaced percentiles (quantiles) from a standard normal distribution. For example, if the sample size, n, is 9, then the calculated percentiles would be the 10th, 20th, ..., 90th. Then construct a scatterplot with the n observed data values ordered from low to high on the vertical axis and the calculated

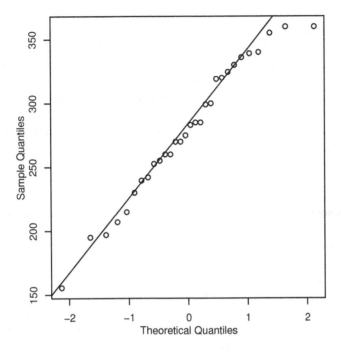

Figure 1.4   QQ-plot for the home prices example.

percentiles on the horizontal axis. If the two sets of values are similar (i.e., if the sample values closely follow a normal distribution), then the points will lie roughly along a straight line. To facilitate this assessment, a diagonal line that passes through the first and third quartiles is often added to the plot. The exact details of how a QQ-plot is drawn can differ depending on the statistical software used (e.g., sometimes the axes are switched or the diagonal line is constructed differently).

1.3   SELECTING INDIVIDUALS AT RANDOM—PROBABILITY

Having assessed the normality of our population of sale prices by looking at the histogram and QQ-plot of sample sale prices, we now return to the task of making probability statements about the population. The crucial question at this point is whether the sample data are representative of the population for which we wish to make statistical inferences. One way to increase the chance of this being true is to select the sample values from the population at random—we discussed this in the context of our home prices example on page 2. We can then make reliable statistical inferences about the population by considering properties of a model fit to the sample data—provided the model fits reasonably well.

We saw in Section 1.2 that a normal distribution model fits the home prices example reasonably well. However, we can see from Figure 1.1 that a *standard* normal distribution is inappropriate here, because a standard normal distribution has a mean of 0 and a standard deviation of 1, whereas our sample data have a mean of 278.6033 and a standard deviation of 53.8656. We therefore need to consider more general normal distributions with a mean that can take any value and a standard deviation that can take any positive value (standard deviations cannot be negative).

Let Y represent the population values (sale prices in our example) and suppose that Y is normally distributed with mean (or *expected value*), $E(Y)$, and standard deviation, $SD(Y)$. This textbook uses this notation with familiar Roman letters in place of the traditional Greek letters, μ and σ, which, in the author's experience, are unfamiliar and awkward for many students. We can abbreviate this normal distribution as $Normal(E(Y), SD(Y)^2)$, where the first number is the mean and the second number is the square of the standard deviation (also known as the *variance*). Then the *population standardized Z-value*,

$$Z = \frac{Y - E(Y)}{SD(Y)},$$

has a standard normal distribution with mean 0 and standard deviation 1. In symbols,

$$Y \sim Normal(E(Y), SD(Y)^2) \quad \Longleftrightarrow \quad Z = \frac{Y - E(Y)}{SD(Y)} \sim Normal(0, 1^2).$$

We are now ready to make a probability statement for the home prices example. Suppose that we would consider a home as being too expensive to buy if its sale price is higher than \$380,000. What is the probability of finding such an expensive home in our housing market? In other words, if we were to randomly select one home from the population of all homes, what is the probability that it has a sale price higher than \$380,000? To answer this question we need to make a number of assumptions. We've already decided that it is probably safe to assume that the population of sale prices (*Price*) could be normal, but we don't know the mean, $E(Price)$, or the standard deviation, $SD(Price)$, of the population of home prices. For now, let's assume that $E(Price) = 280$ and $SD(Price) = 50$ (fairly close to the sample mean of 278.6033 and sample standard deviation of 53.8656). (We'll be able to relax these assumptions later in this chapter.) From the theoretical result above, $Z = (Price - 280)/50$ has a standard normal distribution with mean 0 and standard deviation 1.

Next, to find the probability that a randomly selected *Price* is greater than 380, we perform some standard algebra on probability statements. In particular, if we write "the probability that a is bigger than b" as "$Pr(a > b)$," then we can make changes to a (such as adding, subtracting, multiplying, and dividing other quantities) as long as we do the same thing to b. It is perhaps easier to see how this works by example:

$$Pr(Price > 380) = Pr\left(\frac{Price - 280}{50} > \frac{380 - 280}{50}\right)$$
$$= Pr(Z > 2.00).$$

The second equality follows since $(Price - 280)/50$ is defined to be Z, which is a standard normal random variable with mean 0 and standard deviation 1. From the table on page 8, the probability that a standard normal random variable is greater than 1.96 is 0.025. Thus, $Pr(Z > 2.00)$ is slightly less than 0.025 (draw a picture of a normal density curve with 1.96 and 2.00 marked on the horizontal axis to convince yourself of this fact). In other words, there is slightly less than a 2.5% chance of finding an expensive home ($> \$380,000$) in our housing market, under the assumption that $Price \sim Normal(280, 50^2)$.

For further practice of this kind of calculation, suppose that we have a budget of \$215,000. What is the probability of finding such an affordable home in our housing market? (You should find it is slightly less than a 10% chance; see Problem 1.6 at the end of this chapter.)

We can also turn these calculations around. For example, which value of *Price* has a probability of 0.025 to the right of it? To answer this, consider the following calculation:

$$\Pr(Z > 1.96) = 0.025$$
$$\Pr\left(\frac{Price - 280}{50} > 1.96\right) = 0.025$$
$$\Pr(Price > 1.96(50) + 280) = 0.025$$
$$\Pr(Price > 378) = 0.025.$$

So, the value 378 has a probability of 0.025 to the right of it. Another way of expressing this is that "the 97.5th percentile of the variable *Price* is $378,000."

1.4 RANDOM SAMPLING

In the preceding section we had to make some pretty restrictive assumptions (normality, known mean, known variance) in order to make statistical inferences. We now explore the connection between samples and populations a little more closely so that we can draw conclusions using fewer assumptions.

Recall that the population is the entire collection of objects under consideration, while the sample is a (random) subset of the population. Sometimes we may have a complete listing of the population (a census), but most of the time a census is too expensive and time consuming to collect. Moreover, it is seldom necessary to consider an entire population in order to make some fairly strong statistical inferences about it using just a random sample.

We are particularly interested in making statistical inferences not only about values in the population, denoted Y, but also about numerical summary measures such as the population mean, denoted $E(Y)$—these population summary measures are called *parameters*. While population parameters are unknown (in the sense that we do not have all the individual population values and so cannot calculate them), we can calculate similar quantities in the sample, such as the sample mean—these sample summary measures are called *statistics*. (Note the dual use of the term "statistics." Up until now it has represented the notion of a general methodology for analyzing data based on probability theory, and just now it was used to represent a collection of summary measures calculated from sample data.)

Next we'll see how statistical inference essentially involves estimating population parameters (and assessing the precision of those estimates) using sample statistics. When our sample data is a subset of the population that has been selected randomly, statistics calculated from the sample can tell us a great deal about corresponding population parameters. For example, a sample mean tends to be a good estimate of the population mean, in the following sense. *If* we were to take random samples over and over again, each time calculating a sample mean, then the mean of all these sample means would be equal to the population mean. (There may seem to be a surfeit of "means" in that last sentence, but if you read it slowly enough it will make sense.) Such an estimate is called *unbiased* since on average it estimates the correct value. It is not actually necessary to take random samples over and over again to show this—probability theory (beyond the scope of this book) allows us to prove such theorems without expending the time and expense of administering a large number of samples.

However, it is not enough to just have sample statistics (such as the sample mean) that average out (over a large number of hypothetical samples) to the correct target (i.e., the population mean). We would also like sample statistics that would have "low" variability from one hypothetical sample to another. At the very least we need to be able to quantify this variability, known as sampling uncertainty. One way to do this is to consider the *sampling distribution* of a statistic, that is, the distribution of values of a statistic under repeated (hypothetical) samples. Again, we can use results from probability theory to tell us what these sampling distributions are. So, all we need to do is take a single random sample, calculate a statistic, and we'll know the theoretical sampling distribution of that statistic (i.e., we'll know what the statistic should average out to over repeated samples, and how much the statistic should vary over repeated samples).

1.4.1   Central limit theorem—normal version

Suppose that a random sample of n data values, represented by $Y_1, Y_2, \ldots, Y_n$, comes from a population that has a mean of $E(Y)$ and a standard deviation of $SD(Y)$. The sample mean, m_Y, is a pretty good estimate of the population mean, $E(Y)$. This textbook uses m_Y for the sample mean of Y rather than the traditional $\bar{y}$ ("y-bar"), which, in the author's experience, is unfamiliar and awkward for many students. The very famous sampling distribution of this statistic derives from the *central limit theorem*. This theorem states that under very general conditions, the sample mean has an approximate normal distribution with mean $E(Y)$ and standard deviation $SD(Y)/\sqrt{n}$ (under repeated sampling). In other words, if we were to take a large number of random samples of n data values and calculate the mean for each sample, the distribution of these sample means would be a normal distribution with mean $E(Y)$ and standard deviation $SD(Y)/\sqrt{n}$. Since the mean of this sampling distribution is $E(Y)$, m_Y is an unbiased estimate of $E(Y)$.

An amazing fact about the central limit theorem is that there is no need for the population itself to be normal (remember that we had to assume this for the calculations in Section 1.3). However, the more symmetric the distribution of the population, the better is the normal approximation for the sampling distribution of the sample mean. Also, the approximation tends to be better the larger the sample size n.

So, how can we use this information? Well, the central limit theorem by itself won't help us to draw statistical inferences about the population without still having to make some restrictive assumptions. However, it is certainly a step in the right direction, so let's see what kind of calculations we can now make for the home prices example. As in Section 1.3, we'll assume that $E(Price) = 280$ and $SD(Price) = 50$, but now we no longer need to assume that the population is normal. Imagine taking a large number of random samples of size 30 from this population and calculating the mean sale price for each sample. To get a better handle on the sampling distribution of these mean sale prices, we'll find the 90th percentile of this sampling distribution. Let's do the calculation first, and then see why this might be a useful number to know.

First, we need to get some notation straight. In this section we're *not* thinking about the specific sample mean we got for our actual sample of 30 sale prices, $m_Y = 278.6033$. Rather we're imagining a list of potential sample means from a population distribution with mean 280 and standard deviation 50—we'll call a potential sample mean in this list M_Y. From the central limit theorem, the sampling distribution of M_Y is normal with mean 280 and standard deviation $50/\sqrt{30} = 9.129$. Then the standardized Z-value from M_Y,

$$Z = \frac{M_Y - \mathrm{E}(Y)}{\mathrm{SD}(Y)/\sqrt{n}} = \frac{M_Y - 280}{9.129},$$

is standard normal with mean 0 and standard deviation 1. From the table on page 8, the 90th percentile of a standard normal random variable is 1.282 (since the horizontal axis value of 1.282 corresponds to an upper-tail area of 0.1). Then

$$\mathrm{Pr}\,(Z < 1.282) = 0.90$$
$$\mathrm{Pr}\left(\frac{M_Y - 280}{9.129} < 1.282\right) = 0.90$$
$$\mathrm{Pr}\,(M_Y < 1.282(9.129) + 280) = 0.90$$
$$\mathrm{Pr}\,(M_Y < 291.703) = 0.90.$$

Thus, the 90th percentile of the sampling distribution of M_Y is \$292,000. In other words, under repeated sampling, M_Y has a distribution with an area of 0.90 to the left of \$292,000 (and an area of 0.10 to the right of \$292,000). This illustrates a crucial distinction between the distribution of population Y-values and the sampling distribution of M_Y—the latter is much less spread out. For example, suppose for the sake of argument that the population distribution of Y is normal (although this is not actually required for the central limit theorem to work). Then we can do a similar calculation to the one above to find the 90th percentile of this distribution (normal with mean 280 and standard deviation 50). In particular,

$$\mathrm{Pr}\,(Z < 1.282) = 0.90$$
$$\mathrm{Pr}\left(\frac{Y - 280}{50} < 1.282\right) = 0.90$$
$$\mathrm{Pr}\,(Y < 1.282(50) + 280) = 0.90$$
$$\mathrm{Pr}\,(Y < 344.100) = 0.90.$$

Thus, the 90th percentile of the population distribution of Y is \$344,000. This is much larger than the value we got above for the 90th percentile of the sampling distribution of M_Y (\$292,000). This is because the sampling distribution of M_Y is less spread out than the population distribution of Y—the standard deviations for our example are 9.129 for the former and 50 for the latter. Figure 1.5 illustrates this point.

We can again turn these calculations around. For example, what is the probability that M_Y is greater than 291.703? To answer this, consider the following calculation:

$$\mathrm{Pr}\,(M_Y > 291.703) = \mathrm{Pr}\left(\frac{M_Y - 280}{9.129} > \frac{291.703 - 280}{9.129}\right)$$
$$= \mathrm{Pr}\,(Z > 1.282)$$
$$= 0.10.$$

So, the probability that M_Y is greater than 291.703 is 0.10.

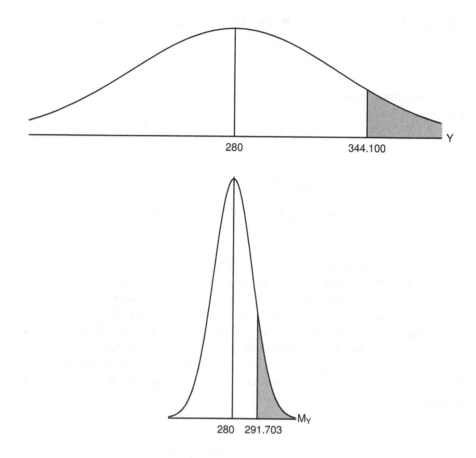

Figure 1.5   The central limit theorem in action. The upper density curve shows a normal population distribution for Y with mean 280 and standard deviation 50: the shaded area is 0.10, which lies to the right of the 90th percentile, 344.100. The lower density curve shows a normal sampling distribution for M_Y with mean 280 and standard deviation 9.129: the shaded area is also 0.10, which lies to the right of the 90th percentile, 291.703. It is not necessary for the population distribution of Y to be normal for the central limit theorem to work—we have used a normal population distribution here just for the sake of illustration.

1.4.2 Central limit theorem—t-version

One major drawback to the normal version of the central limit theorem is that to use it we have to assume that we know the value of the population standard deviation, SD(Y). A generalization of the standard normal distribution called *Student's t-distribution* solves this problem. The density curve for a t-distribution looks very similar to a normal density curve, but the tails tend to be a little "thicker," that is, t-distributions are a little more spread out than the normal distribution. This "extra variability" is controlled by an integer number called the *degrees of freedom*. The smaller this number, the more spread out the t-distribution density curve (conversely, the higher the degrees of freedom, the more like a normal density curve it looks).

For example, the following table shows tail areas for a t-distribution with 29 degrees of freedom:

Upper-tail area	0.1	0.05	0.025	0.01	0.005	0.001
Critical value of t_{29}	1.311	1.699	2.045	2.462	2.756	3.396
Two-tail area	0.2	0.1	0.05	0.02	0.01	0.002

Compared with the corresponding table for the normal distribution on page 8, the critical values (i.e., horizontal axis values or percentiles) are slightly larger in this table.

We will use the t-distribution from this point on because it will allow us to use an estimate of the population standard deviation (rather than having to assume this value). A reasonable estimate to use is the sample standard deviation, s_Y. Since we will be using an estimate of the population standard deviation, we will be a little less certain about our probability calculations—this is why the t-distribution needs to be a little more spread out than the normal distribution, to adjust for this extra uncertainty. This extra uncertainty will be of particular concern when we're not too sure if our sample standard deviation is a good estimate of the population standard deviation (i.e., in small samples). So, it makes sense that the degrees of freedom increases as the sample size increases. In this particular application, we will use the t-distribution with $n-1$ degrees of freedom in place of a standard normal distribution in the following t-version of the central limit theorem.

Suppose that a random sample of n data values, represented by $Y_1, Y_2, \ldots, Y_n$, comes from a population that has a mean of $E(Y)$. Imagine taking a large number of random samples of n data values and calculating the mean and standard deviation for each sample. As before, we'll let M_Y represent the imagined list of repeated sample means, and similarly, we'll let S_Y represent the imagined list of repeated sample standard deviations. Define

$$t = \frac{M_Y - E(Y)}{S_Y/\sqrt{n}}.$$

Under very general conditions, t has an approximate t-distribution with $n-1$ degrees of freedom. The two differences from the normal version of the central limit theorem that we used before are that the repeated sample standard deviations, S_Y, replace an assumed population standard deviation, $SD(Y)$, and that the resulting sampling distribution is a t-distribution (not a normal distribution).

So far, we have focused on the sampling distribution of sample means, M_Y, but what we would really like to do is infer what the observed sample mean, m_Y, tells us about the population mean, $E(Y)$. Thus, while the calculations in Section 1.4.1 have been useful for building up intuition about sampling distributions and manipulating probability statements, their main purpose has been to prepare the ground for the next two sections, which cover how to make statistical inferences about the population mean, $E(Y)$.

1.5 INTERVAL ESTIMATION

We have already seen that the sample mean, m_Y, is a good *point estimate* of the population mean, $E(Y)$ (in the sense that it is unbiased—see page 11). It is also helpful to know how reliable this estimate is, that is, how much sampling uncertainty is associated with it.

A useful way to express this uncertainty is to calculate an *interval estimate* or *confidence interval* for the population mean, $E(Y)$. The interval should be centered at the point estimate (in this case, m_Y), and since we are probably equally uncertain that the population mean could be lower or higher than this estimate, it should have the same amount of uncertainty either side of the point estimate. In other words, the confidence interval is of the form "point estimate ± uncertainty" or "(point estimate − uncertainty, point estimate + uncertainty)."

We can obtain the exact form of the confidence interval from the t-version of the central limit theorem, where $t = (M_Y - E(Y))/(S_Y/\sqrt{n})$ has an approximate t-distribution with $n-1$ degrees of freedom. In particular, suppose that we want to calculate a 95% confidence interval for the population mean, $E(Y)$, for the home prices example—in other words, an interval such that there will be an area of 0.95 between the two endpoints of the interval (and an area of 0.025 to the left of the interval in the lower tail, and an area of 0.025 to the right of the interval in the upper tail). Let's consider just one side of the interval first. Since 2.045 is the 97.5th percentile of the t-distribution with 29 degrees of freedom (see the table on page 15), then

$$\Pr(t_{29} < 2.045) = 0.975$$
$$\Pr\left(\frac{M_Y - E(Y)}{S_Y/\sqrt{n}} < 2.045\right) = 0.975$$
$$\Pr(M_Y - E(Y) < 2.045(S_Y/\sqrt{n})) = 0.975$$
$$\Pr(-E(Y) < -M_Y + 2.045(S_Y/\sqrt{n})) = 0.975$$
$$\Pr(E(Y) > M_Y - 2.045(S_Y/\sqrt{n})) = 0.975.$$

The difference from earlier calculations is that this time $E(Y)$ is the focus of inference, so we have not assumed that we know its value. One consequence for the probability calculation is that in the fourth line we have "$-E(Y)$." To change this to "$+E(Y)$" in the fifth line, we multiply each side of the inequality sign by "-1" (this also has the effect of changing the direction of the inequality sign).

This probability statement must be true for all potential values of M_Y and S_Y. In particular, it must be true for our observed sample statistics, $m_Y = 278.6033$ and $s_Y = 53.8656$. Thus, to find the values of $E(Y)$ that satisfy the probability statement, we plug in our sample statistics to find

$$M_Y - 2.045(S_Y/\sqrt{n}) = 278.6033 - 2.045(53.8656/\sqrt{30}) = 258.492.$$

This shows that a population mean greater than \$258,492 would satisfy the expression $\Pr(t_{29} < 2.045) = 0.975$. In other words, we have found that the lower bound of our confidence interval is \$258,492, or approximately \$258,000.

To find the upper bound we perform a similar calculation:

$$\Pr(t_{29} > -2.045) = 0.975$$
$$\Pr\left(\frac{M_Y - E(Y)}{S_Y/\sqrt{n}} > -2.045\right) = 0.975$$
$$\Pr(M_Y - E(Y) > -2.045(S_Y/\sqrt{n})) = 0.975$$
$$\Pr(-E(Y) > -M_Y - 2.045(S_Y/\sqrt{n})) = 0.975$$
$$\Pr(E(Y) < M_Y + 2.045(S_Y/\sqrt{n})) = 0.975.$$

To find the values of $E(Y)$ that satisfy this expression, we plug in our sample statistics to find

$$M_Y + 2.045(S_Y/\sqrt{n}) = 278.6033 + 2.045(53.8656/\sqrt{30}) = 298.715.$$

This shows that a population mean less than \$298,715 would satisfy the expression $\Pr(t_{29} < 2.045) = 0.975$. In other words, we have found that the upper bound of our confidence interval is \$298,715, or approximately \$299,000.

We can write these two calculations a little more concisely as

$$\Pr(-2.045 < t_{29} < 2.045) = 0.95$$
$$\Pr\left(-2.045 < \frac{M_Y - E(Y)}{S_Y/\sqrt{n}} < 2.045\right) = 0.95$$
$$\Pr(-2.045(S_Y/\sqrt{n}) < M_Y - E(Y) < 2.045(S_Y/\sqrt{n})) = 0.95$$
$$\Pr(-M_Y - 2.045(S_Y/\sqrt{n}) < -E(Y) < -M_Y + 2.045(S_Y/\sqrt{n})) = 0.95$$
$$\Pr(M_Y + 2.045(S_Y/\sqrt{n}) > E(Y) > M_Y - 2.045(S_Y/\sqrt{n})) = 0.95.$$

As before, we plug in our sample statistics to find the values of $E(Y)$ that satisfy this expression:

$$M_Y - 2.045(S_Y/\sqrt{n}) = 278.6033 - 2.045(53.8656/\sqrt{30}) = 258.492$$
$$M_Y + 2.045(S_Y/\sqrt{n}) = 278.6033 + 2.045(53.8656/\sqrt{30}) = 298.715.$$

This shows that a population mean between \$258,492 and \$298,715 would satisfy the expression $\Pr(-2.045 < t_{29} < 2.045) = 0.95$. In other words, we have found that a 95% confidence interval for $E(Y)$ for this example is (\$258,492, \$298,715), or approximately (\$258,000, \$299,000). It is traditional to write confidence intervals with the lower number on the left.

More generally, using symbols, a 95% confidence interval for a univariate population mean, $E(Y)$, results from the following:

$$\Pr(-97.5\text{th percentile} < t_{n-1} < 97.5\text{th percentile}) = 0.95$$
$$\Pr\left(-97.5\text{th percentile} < \frac{M_Y - E(Y)}{S_Y/\sqrt{n}} < 97.5\text{th percentile}\right) = 0.95$$
$$\Pr(M_Y - 97.5\text{th percentile}(S_Y/\sqrt{n}) < E(Y) < M_Y + 97.5\text{th percentile}(S_Y/\sqrt{n})) = 0.95,$$

where the 97.5th percentile comes from the t-distribution with $n-1$ degrees of freedom. In other words, plugging in our observed sample statistics, m_Y and s_Y, we can write the 95% confidence interval as $m_Y \pm 97.5\text{th percentile}(s_Y/\sqrt{n})$.

For a lower or higher level of confidence than 95%, the percentile used in the calculation must be changed as appropriate. For example, for a 90% interval (i.e., with 5% in each tail), the 95th percentile would be needed, whereas for a 99% interval (i.e., with 0.5% in each tail), the 99.5th percentile would be needed. These percentiles can be obtained from Table B.1 on page 291 (which is an expanded version of the table on page 15). Instructions for using the table can be found in Appendix B.

Thus, in general, we can write a confidence interval for a univariate mean, $E(Y)$, as

$$m_Y \pm \text{t-percentile}(s_Y/\sqrt{n}),$$

where the t-percentile comes from a t-distribution with $n-1$ degrees of freedom. The example above thus becomes

$$m_Y \pm (\text{97.5th percentile from } t_{29}) \left(s_Y/\sqrt{n}\right) = 278.6033 \pm 2.045 \left(53.8656/\sqrt{30}\right)$$
$$= 278.6033 \pm 20.111$$
$$= (258.492, 298.715).$$

Computer help #23 in the software information files available from the book website shows how to use statistical software to calculate confidence intervals for the population mean. As further practice, calculate a 90% confidence interval for the population mean for the home prices example (see Problem 1.6 at the end of this chapter)—you should find that it is ($262,000, $295,000).

Now that we've calculated a confidence interval, what exactly does it tell us? Well, for the home prices example, *loosely speaking*, we can say that "we're 95% confident that the mean single-family home sale price in this housing market is between $258,000 and $299,000." This will get you by among friends (as long as none of your friends happen to be expert statisticians). But to provide a more precise interpretation we have to revisit the notion of hypothetical repeated samples. If we were to take a large number of random samples of size 30 from our population of sale prices and calculate a 95% confidence interval for each, then 95% of those confidence intervals would contain the (unknown) population mean. We do not know (nor will we ever know) whether the 95% confidence interval for our particular sample contains the population mean—thus, strictly speaking, we *cannot* say "the probability that the population mean is in our interval is 0.95." All we know is that the procedure that we have used to calculate the 95% confidence interval tends to produce intervals that under repeated sampling contain the population mean 95% of the time. Stick with the phrase "95% confident" and avoid using the word "probability" and chances are that no one (not even expert statisticians) will be too offended.

Before moving on to Section 1.6, which describes another way to make statistical inferences about population means—hypothesis testing—let us consider whether we can now forget the normal distribution. The calculations in this section are based on the central limit theorem, which does not require the population to be normal. We have also seen that t-distributions are more useful than normal distributions for calculating confidence intervals. For large samples, it doesn't make much difference (note how the percentiles for t-distributions get closer to the percentiles for the standard normal distribution as the degrees of freedom get larger in Table B.1), but for smaller samples it can make a large difference. So for this type of calculation we always use a t-distribution from now on. However, we can't completely forget about the normal distribution yet; it will come into play again in a different context in later chapters.

When using a t-distribution, how do we know how many degrees of freedom to use? One way to think about degrees of freedom is in terms of the information provided by the data we are analyzing. Roughly speaking, each data observation provides one degree of freedom (this is where the n in the degrees of freedom formula comes in), but we lose a degree of freedom for each population parameter that we have to estimate. So, in this chapter, when we are estimating the population mean, the degrees of freedom formula is $n-1$. In Chapter 2, when we will be estimating two population parameters (the intercept and the slope of a regression line), the degrees of freedom formula will be $n-2$. For the remainder of the book, the general formula for the degrees of freedom in a multiple linear regression model will be $n-(k+1)$ or $n-k-1$, where k is the number of predictor variables in the model. Note that this general formula actually also works for Chapter 2 (where $k=1$) and even this chapter (where $k=0$, since a linear regression model with zero predictors is equivalent to estimating the population mean for a univariate dataset).

1.6 HYPOTHESIS TESTING

Another way to make statistical inferences about a population parameter such as the mean is to use *hypothesis testing* to make decisions about the parameter's value. Suppose that we are interested in a particular value of the mean single-family home sale price, for example, a claim from a realtor that the mean sale price in this market is $255,000. Does the information in our sample support this claim, or does it favor an alternative claim?

1.6.1 The rejection region method

To decide between two competing claims, we can conduct a hypothesis test as follows.

- Express the claim about a specific value for the population parameter of interest as a *null hypothesis*, denoted NH. In this textbook, we use this notation in place of the traditional H_0, which, in the author's experience, is unfamiliar and awkward for many students. The null hypothesis needs to be in the form "parameter = some hypothesized value," for example, NH: $E(Y)=255$. A frequently used legal analogy is that the null hypothesis is equivalent to a presumption of innocence in a trial before any evidence has been presented.

- Express the alternative claim as an *alternative hypothesis*, denoted AH. Again, in this book, we use this notation in place of the traditional H_a. The alternative hypothesis can be in a lower-tail form, for example, AH: $E(Y)<255$, or an upper-tail form, for example, AH: $E(Y)>255$, or a two-tail form, for example, AH: $E(Y)\neq255$. The alternative hypothesis, also sometimes called the *research hypothesis*, is what we would like to demonstrate to be the case, and needs to be stated *before looking at the data*. To continue the legal analogy, the alternative hypothesis is guilt, and we will only reject the null hypothesis (innocence) if we favor the alternative hypothesis (guilt) beyond a reasonable doubt. To illustrate, we will presume for the home prices example that we have some reason to suspect that the mean sale price is higher than claimed by the realtor (perhaps a political organization is campaigning on the issue of high housing costs and has employed us to investigate whether sale prices are "too high" in this housing market). Thus, our alternative hypothesis is AH: $E(Y) > 255$.

- Calculate a test statistic based on the assumption that the null hypothesis is true. For hypothesis tests for a univariate population mean the relevant test statistic is

$$\text{t-statistic} = \frac{m_Y - E(Y)}{s_Y / \sqrt{n}},$$

where m_Y is the sample mean, $E(Y)$ is the value of the population mean in the null hypothesis, s_Y is the sample standard deviation, and n is the sample size.

- Under the assumption that the null hypothesis is true, this test statistic will have a particular probability distribution. For testing a univariate population mean, this t-statistic has a t-distribution with $n-1$ degrees of freedom. We would therefore expect it to be "close" to zero (if the null hypothesis is true). Conversely, if it is far from zero, then we might begin to doubt the null hypothesis:

- For an upper-tail test, a t-statistic that is positive and far from zero would then lead us to favor the alternative hypothesis (a t-statistic that was far from zero but negative would favor neither hypothesis and the test would be inconclusive).

- For a lower-tail test, a t-statistic that is negative and far from zero would then lead us to favor the alternative hypothesis (a t-statistic that was far from zero but positive would favor neither hypothesis and the test would be inconclusive).

- For a two-tail test, any t-statistic that is far from zero (positive or negative) would lead us to favor the alternative hypothesis.

- To decide how far from zero a t-statistic would have to be before we reject the null hypothesis in favor of the alternative, recall the legal analogy. To deliver a guilty verdict (the alternative hypothesis), the jury must establish guilt beyond a reasonable doubt. In other words, a jury rejects the presumption of innocence (the null hypothesis) only if there is compelling evidence of guilt. In statistical terms, compelling evidence of guilt is found only in the tails of the t-distribution density curve. For example, in conducting an upper-tail test, if the t-statistic is way out in the upper tail, then it seems unlikely that the null hypothesis could have been true—so we reject it in favor of the alternative. Otherwise, the t-statistic could well have arisen while the null hypothesis held true—so we do not reject it in favor of the alternative. How far out in the tail does the t-statistic have to be to favor the alternative hypothesis rather than the null? Here we must make a decision about how much evidence we will require before rejecting a null hypothesis. There is always a chance that we might mistakenly reject a null hypothesis when it is actually true (the equivalent of pronouncing an innocent defendant guilty). Often, this chance—called the *significance level*—will be set at 5%, but more stringent tests (such as in clinical trials of new pharmaceutical drugs) might set this at 1%, while less stringent tests (such as in sociological studies) might set this at 10%. For the sake of argument, we use 5% as a default value for hypothesis tests in this book (unless stated otherwise).

- The significance level dictates the *critical value(s)* for the test, beyond which an observed t-statistic leads to rejection of the null hypothesis in favor of the alternative. This region, which leads to rejection of the null hypothesis, is called the *rejection region*. For example, for a significance level of 5%:

- For an upper-tail test, the critical value is the 95th percentile of the t-distribution with $n-1$ degrees of freedom; reject the null in favor of the alternative if the t-statistic is greater than this.

- For a lower-tail test, the critical value is the 5th percentile of the t-distribution with $n-1$ degrees of freedom; reject the null in favor of the alternative if the t-statistic is less than this.

- For a two-tail test, the *two* critical values are the 2.5th and the 97.5th percentiles of the t-distribution with $n-1$ degrees of freedom; reject the null in favor of the alternative if the t-statistic is less than the 2.5th percentile or greater than the 97.5th percentile.

As previously, it is perhaps easier to see how all this works by example. It is best to lay out hypothesis tests in a series of steps (see computer help #24 in the software information files available from the book website):

- State null hypothesis: NH: $E(Y) = 255$.

- State alternative hypothesis: AH: $E(Y) > 255$.

- Calculate test statistic: t-statistic $= \dfrac{m_Y - E(Y)}{s_Y/\sqrt{n}} = \dfrac{278.6033 - 255}{53.8656/\sqrt{30}} = 2.40$.

- Set significance level: 5%.

- Look up critical value: The 95th percentile of the t-distribution with 29 degrees of freedom is 1.699 (from Table B.1); the rejection region is therefore any t-statistic greater than 1.699.

- Make decision: Since the t-statistic of 2.40 falls in the rejection region, we reject the null hypothesis in favor of the alternative.

- Interpret in the context of the situation: The 30 sample sale prices suggest that a population mean of \$255,000 seems implausible—the sample data favor a value greater than this (at a significance level of 5%).

1.6.2 The p-value method

An alternative way to conduct a hypothesis test is to again assume initially that the null hypothesis is true, but then to calculate the probability of observing a t-statistic as extreme as the one observed or even more extreme (in the direction that favors the alternative hypothesis). This is known as the *p-value* (sometimes also called the *observed significance level*):

- For an upper-tail test, the p-value is the area under the curve of the t-distribution (with $n-1$ degrees of freedom) to the right of the observed t-statistic.

- For a lower-tail test, the p-value is the area under the curve of the t-distribution (with $n-1$ degrees of freedom) to the left of the observed t-statistic.

- For a two-tail test, the p-value is the sum of the areas under the curve of the t-distribution (with $n-1$ degrees of freedom) beyond both the observed t-statistic and the negative of the observed t-statistic.

If the p-value is too "small," then this suggests that it seems unlikely that the null hypothesis could have been true—so we reject it in favor of the alternative. Otherwise, the t-statistic could well have arisen while the null hypothesis held true—so we do not reject it in favor of the alternative. Again, the significance level chosen tells us how small is small: If the p-value is less than the significance level, then reject the null in favor of the alternative; otherwise, do not reject it. For the home prices example (see computer help #24 in the software information files available from the book website):

- State null hypothesis: NH: $E(Y) = 255$.

- State alternative hypothesis: AH: $E(Y) > 255$.

- Calculate test statistic: t-statistic $= \dfrac{m_Y - E(Y)}{s_Y/\sqrt{n}} = \dfrac{278.6033 - 255}{53.8656/\sqrt{30}} = 2.40$.

- Set significance level: 5%.

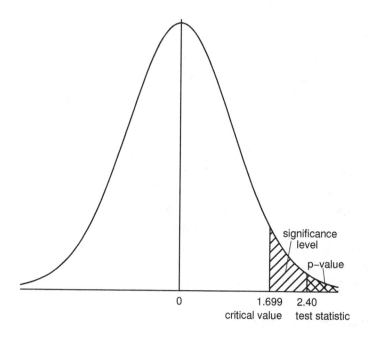

Figure 1.6   Home prices example—density curve for the t-distribution with 29 degrees of freedom, together with the critical value of 1.699 corresponding to a significance level of 0.05, as well as the test statistic of 2.40 corresponding to a p-value less than 0.05.

- Look up p-value: The area to the right of the t-statistic (2.40) for the t-distribution with 29 degrees of freedom is less than 0.025 but greater than 0.01 (since from Table B.1 the 97.5th percentile of this t-distribution is 2.045 and the 99th percentile is 2.462); thus the upper-tail p-value is between 0.01 and 0.025.

- Make decision: Since the p-value is between 0.01 and 0.025, it must be less than the significance level (0.05), so we reject the null hypothesis in favor of the alternative.

- Interpret in the context of the situation: The 30 sample sale prices suggest that a population mean of $255,000 seems implausible—the sample data favor a value greater than this (at a significance level of 5%).

Figure 1.6 shows why the rejection region method and the p-value method will always lead to the same decision (since if the t-statistic is in the rejection region, then the p-value must be smaller than the significance level, and vice versa). Why do we need two methods if they will always lead to the same decision? Well, when learning about hypothesis tests and becoming comfortable with their logic, many people find the rejection region method a little easier to apply. However, when we start to rely on statistical software for conducting hypothesis tests in later chapters of the book, we will find the p-value method easier to use. At this stage, when doing hypothesis test calculations by hand, it is helpful to use both the rejection region method and the p-value method to reinforce learning of the general concepts. This also provides a useful way to check our calculations since if we reach a different conclusion with each method we will know that we have made a mistake.

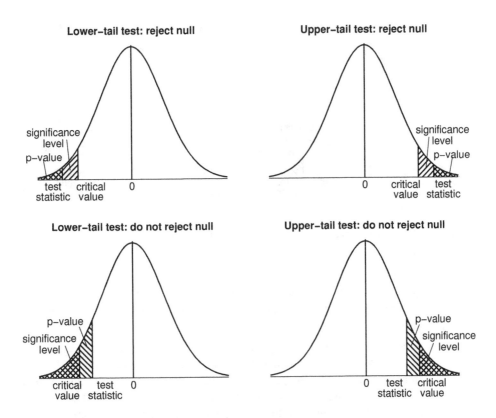

Figure 1.7 Relationships between critical values, significance levels, test statistics, and p-values for one-tail hypothesis tests.

Lower-tail tests work in a similar way to upper-tail tests, but all the calculations are performed in the negative (left-hand) tail of the t-distribution density curve; Figure 1.7 illustrates. A lower-tail test would result in an inconclusive result for the home prices example (since the large, positive t-statistic means that the data favor neither the null hypothesis, NH: $E(Y) = 255$, nor the alternative hypothesis, AH: $E(Y) < 255$).

Two-tail tests work similarly, but we have to be careful to work with both tails of the t-distribution; Figure 1.8 illustrates. For the home prices example, we might want to do a two-tail hypothesis test if we had no prior expectation about how large or small sale prices are, but just wanted to see whether or not the realtor's claim of $255,000 was plausible. The steps involved are as follows (see computer help #24).

- State null hypothesis: NH: $E(Y) = 255$.
- State alternative hypothesis: AH: $E(Y) \neq 255$.
- Calculate test statistic: t-statistic $= \dfrac{m_Y - E(Y)}{s_Y/\sqrt{n}} = \dfrac{278.6033 - 255}{53.8656/\sqrt{30}} = 2.40$.
- Set significance level: 5%.
- Look up t-table:
    - critical value: The 97.5th percentile of the t-distribution with 29 degrees of freedom is 2.045 (from Table B.1); the rejection region is therefore any t-statistic

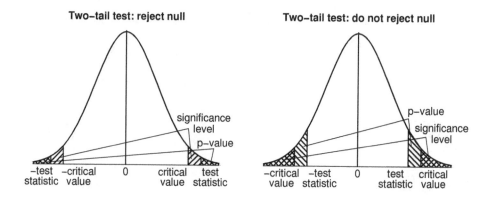

Figure 1.8 Relationships between critical values, significance levels, test statistics, and p-values for two-tail hypothesis tests.

greater than 2.045 or less than −2.045 (we need the 97.5th percentile in this case because this is a two-tail test, so we need half the significance level in each tail).

- p-value: The area to the right of the t-statistic (2.40) for the t-distribution with 29 degrees of freedom is less than 0.025 but greater than 0.01 (since from Table B.1 the 97.5th percentile of this t-distribution is 2.045 and the 99th percentile is 2.462); thus the upper-tail area is between 0.01 and 0.025 and the two-tail p-value is twice as big as this, that is, between 0.02 and 0.05.

- Make decision:

  - Since the t-statistic of 2.40 falls in the rejection region, we reject the null hypothesis in favor of the alternative.

  - Since the p-value is between 0.02 and 0.05, it must be less than the significance level (0.05), so we reject the null hypothesis in favor of the alternative.

- Interpret in the context of the situation: The 30 sample sale prices suggest that a population mean of $255,000 seems implausible—the sample data favor a value different from this (at a significance level of 5%).

1.6.3 Hypothesis test errors

When we introduced the significance level on page 20, we saw that the person conducting the hypothesis test gets to choose this value. We now explore this notion a little more fully.

Whenever we conduct a hypothesis test, either we reject the null hypothesis in favor of the alternative or we do not reject the null hypothesis. "Not rejecting" a null hypothesis isn't quite the same as "accepting" it. All we can say in such a situation is that we do not have enough evidence to reject the null—recall the legal analogy where defendants are not found "innocent" but rather are found "not guilty." Anyway, regardless of the precise terminology we use, we hope to reject the null when it really is false and to "fail to reject it" when it really is true. Anything else will result in a *hypothesis test error*. There are two types of error that can occur, as illustrated in the following table:

		Decision	
		Do not reject NH in favor of AH	Reject NH in favor of AH
Reality	NH true	Correct decision	Type 1 error
	NH false	Type 2 error	Correct decision

A type 1 error can occur if we reject the null hypothesis when it is really true—the probability of this happening is precisely the significance level. If we set the significance level lower, then we lessen the chance of a type 1 error occurring. Unfortunately, lowering the significance level *increases* the chance of a type 2 error occurring—when we fail to reject the null hypothesis but we should have rejected it because it was false. Thus, we need to make a trade-off and set the significance level low enough that type 1 errors have a low chance of happening, but not so low that we greatly increase the chance of a type 2 error happening. The default value of 5% tends to work reasonably well in many applications at balancing both goals. However, other factors also affect the chance of a type 2 error happening for a specific significance level. For example, the chance of a type 2 error tends to decrease the greater the sample size.

1.7 RANDOM ERRORS AND PREDICTION

So far, we have focused on estimating a univariate population mean, $E(Y)$, and quantifying our uncertainty about the estimate via confidence intervals or hypothesis tests. In this section, we consider a different problem, that of "prediction." In particular, rather than estimating the mean of a population of Y-values based on a sample, $Y_1, \dots, Y_n$, consider predicting an individual Y-value picked at random from the population.

Intuitively, this sounds like a more difficult problem. Imagine that rather than just estimating the mean sale price of single-family homes in the housing market based on our sample of 30 homes, we have to predict the sale price of an individual single-family home that has just come onto the market. Presumably, we'll be less certain about our prediction than we were about our estimate of the population mean (since it seems likely that we could be farther from the truth with our prediction than when we estimated the mean—for example, there is a chance that the new home could be a real bargain or totally overpriced). Statistically speaking, Figure 1.5 illustrates this "extra uncertainty" that arises with prediction—the population distribution of data values, Y (more relevant to prediction problems), is much more variable than the sampling distribution of sample means, M_Y (more relevant to mean estimation problems).

We can tackle prediction problems with a similar process to that of using a confidence interval to tackle estimating a population mean. In particular, we can calculate a *prediction interval* of the form "point estimate $\pm$ uncertainty" or "(point estimate $-$ uncertainty, point estimate $+$ uncertainty)." The point estimate is the same one that we used for estimating the population mean, that is, the observed sample mean, m_Y. This is because m_Y is an unbiased estimate of the population mean, $E(Y)$, and we assume that the individual Y-value we are predicting is a member of this population. As discussed in the preceding paragraph, however, the "uncertainty" is larger for prediction intervals than for confidence intervals. To see how much larger, we need to return to the notion of a model that we introduced on page 5.

We can express the model we've been using to estimate the population mean, $E(Y)$, as

$$Y\text{-value} = \text{deterministic part} + \text{random error}$$
$$Y_i = E(Y) + e_i \quad (i = 1, \ldots, n).$$

In other words, each sample Y_i-value (the index i keeps track of the sample observations) can be decomposed into two pieces, a *deterministic* part that is the same for all values, and a *random error* part that varies from observation to observation. A convenient choice for the deterministic part is the population mean, $E(Y)$, since then the random errors have a (population) mean of zero. Since $E(Y)$ is the same for all Y-values, the random errors, e, have the same standard deviation as the Y-values themselves, that is, $SD(Y)$. We can use this decomposition to derive the confidence interval and hypothesis test results of Sections 1.5 and 1.6 (although it would take more mathematics than we really need for our purposes in this book). Moreover, we can also use this decomposition to motivate the precise form of the uncertainty needed for prediction intervals (without having to get into too much mathematical detail).

In particular, write the Y-value to be predicted as Y^*, and decompose this into two pieces as above:

$$Y^* = E(Y) + e^*.$$

Then subtract M_Y, which represents potential values of repeated sample means, from both sides of this equation:

$$Y^* - M_Y = (E(Y) - M_Y) + e^*$$
$$\text{prediction error} = \text{estimation error} + \text{random error.} \tag{1.1}$$

Thus, whereas in estimating the population mean the only error we have to worry about is estimation error, in predicting an individual Y-value we have to worry about both estimation error and random error.

Recall from page 17 that the form of a confidence interval for the population mean is

$$m_Y \pm \text{t-percentile}(s_Y/\sqrt{n}).$$

The term $s_Y/\sqrt{n}$ in this formula is an estimate of the standard deviation of the sampling distribution of sample means, M_Y, and is called the *standard error of estimation*. The square of this quantity, s_Y^2/n, is the estimated variance of the sampling distribution of sample means, M_Y. Then, thinking of $E(Y)$ as some fixed, unknown constant, s_Y^2/n is also the estimated variance of the estimation error, $E(Y) - M_Y$, in expression (1.1).

The estimated variance of the random error, e^*, in expression (1.1) is s_Y^2. It can then be shown that the estimated variance of the prediction error, $Y^* - M_Y$, in expression (1.1) is $s_Y^2/n + s_Y^2 = s_Y^2(1/n+1) = s_Y^2(1+1/n)$. Then $s_Y\sqrt{1+1/n}$ is called the *standard error of prediction* and leads to the formula for a prediction interval for an individual Y-value as

$$m_Y \pm \text{t-percentile}\left(s_Y\sqrt{1+1/n}\right).$$

As with confidence intervals for the mean, the t-percentile used in the calculation comes from a t-distribution with $n-1$ degrees of freedom. For example, for a 95% interval (i.e., with 2.5% in each tail), the 97.5th percentile would be needed, whereas for a 90% interval (i.e., with 5% in each tail), the 95th percentile would be needed. These percentiles can be obtained from Table B.1. For example, the 95% prediction interval for an individual value of *Price* picked at random from the population of single-family home sale prices is calculated as

$$m_Y \pm (\text{97.5th percentile from } t_{29}) \left(s_Y \sqrt{1+1/n} \right)$$
$$= 278.6033 \pm 2.045 \left(53.8656 \sqrt{1+1/30} \right)$$
$$= 278.6033 \pm 111.976$$
$$= (166.627, 390.579).$$

What about the interpretation of a prediction interval? Well, for the home prices example, *loosely speaking*, we can say that "we're 95% confident that the sale price for an individual home picked at random from all single-family homes in this housing market will be between $167,000 and $391,000." More precisely, if we were to take a large number of random samples of size 30 from our population of sale prices and calculate a 95% prediction interval for each, then 95% of those prediction intervals would contain the (unknown) sale price for an individual home picked at random from the population.

As discussed at the beginning of this section on page 25, this interval is much wider than the 95% confidence interval for the population mean single-family home sale price, which was calculated as

$$m_Y \pm (\text{97.5th percentile from } t_{29}) \left(s_Y / \sqrt{n} \right)$$
$$= 278.6033 \pm 2.045 \left(53.8656 / \sqrt{30} \right)$$
$$= 278.6033 \pm 20.111$$
$$= (258.492, 298.715).$$

Unlike for confidence intervals for the population mean, statistical software does not generally provide an automated method to calculate prediction intervals for an individual Y-value. Thus they have to be calculated by hand using the sample statistics, m_Y and s_Y. However, there is a trick that can get around this (although it makes use of simple linear regression, which we cover in Chapter 2). First, create a variable that consists only of the value 1 for all observations. Then, fit a simple linear regression model using this variable as the predictor variable and Y as the response variable, and restrict the model to fit *without an intercept* (see computer help #25 in the software information files available from the book website). The estimated regression line for this model will be a horizontal line at a value equal to the sample mean of the response variable. Prediction intervals for this model will be the same for each value of the predictor variable (see computer help #30), and will be the same as a prediction interval for an individual Y-value. As further practice, calculate a 90% prediction interval for an individual sale price (see Problem 1.6 at the end of this chapter). Calculate it by hand or using the trick just described. You should find that the interval is ($186,000, $372,000).

We derived the formula for a confidence interval for a univariate population mean from the t-version of the central limit theorem, which does not require the data Y-values to be normally distributed. However, the formula for a prediction interval for an individual univariate Y-value tends to work better for datasets in which the Y-values are at least approximately normally distributed—see Problem 1.8 at the end of this chapter.

1.8 CHAPTER SUMMARY

We spent some time in this chapter coming to grips with summarizing data (graphically and numerically) and understanding sampling distributions, but the four major concepts that will carry us through the rest of the book are as follows:

Statistical thinking is the process of analyzing quantitative information about a random sample of n observations and drawing conclusions (statistical inferences) about the population from which the sample was drawn. An example is using a univariate sample mean, m_Y, as an estimate of the corresponding population mean and calculating the sample standard deviation, s_Y, to evaluate the precision of this estimate.

Confidence intervals are one method for calculating the sample estimate of a parameter (such as the population mean) and its associated uncertainty. An example is the confidence interval for a univariate population mean, which takes the form

$$m_Y \pm \text{t-percentile}(s_Y/\sqrt{n}).$$

Hypothesis testing provides another means of making decisions about the likely values of a population parameter. An example is hypothesis testing for a univariate population mean, whereby the magnitude of a calculated sample test statistic,

$$\text{t-statistic} = \frac{m_Y - \text{E}(Y)}{s_Y/\sqrt{n}},$$

indicates which of two hypotheses (about likely values for the population mean) we should favor.

Prediction intervals, while similar in spirit to confidence intervals, tackle the different problem of predicting the value of an individual observation picked at random from the population. An example is the prediction interval for an individual univariate Y-value, which takes the form

$$m_Y \pm \text{t-percentile}\left(s_Y\sqrt{1+1/n}\right).$$

PROBLEMS

- "Computer help" refers to the numbered items in the software information files available from the book website.

- There are *brief* answers to the even-numbered problems in Appendix E.

1.1 The **NBASALARY** data file contains salary information for 214 guards in the National Basketball Association (NBA) for 2009–2010 (obtained from the online USA Today NBA Salaries Database).

   (a) Construct a histogram of the *Salary* variable, representing 2009–2010 salaries in thousands of dollars [computer help #14].

   (b) What would we expect the histogram to look like if the data were normal?

   (c) Construct a QQ-plot of the *Salary* variable [computer help #22].

   (d) What would we expect the QQ-plot to look like if the data were normal?

   (e) Compute the natural logarithm of guard salaries (call this variable *Logsal*) [computer help #6], and construct a histogram of this *Logsal* variable [computer help #14].
   Hint: The "natural logarithm" transformation (also known as "log to base-e," or by the symbols $\log_e$ or ln) is a way to transform (rescale) skewed data to make them more symmetric and normal.

   (f) Construct a QQ-plot of the *Logsal* variable [computer help #22].

   (g) Based on the plots in parts (a), (c), (e), and (f), say whether salaries or log-salaries more closely follow a normal curve, and justify your response.

1.2 Assume that final scores in a statistics course follow a normal distribution with mean $E(Y) = 70$ and standard deviation $SD(Y) = 10$. Find the scores, Y, that correspond to the:

   (a) 90th percentile;

   (b) 99th percentile;

   (c) 5th percentile.

1.3 Consider the data on 2009–2010 salaries of 214 NBA guards from Problem 1.1.

   (a) Calculate a 95% confidence interval for the population mean *Salary* in thousands of dollars [computer help #23].
   Hint: Calculate by hand (using the fact that the sample mean of Salary is 3980.318, the sample standard deviation is 4525.378, and the 97.5th percentile of the t-distribution with 213 degrees of freedom is approximately 1.971) and check your answer using statistical software.

   (b) Consider *Logsal*, the natural logarithms of the salaries. The sample mean of *Logsal* is 7.664386. Reexpress this number in thousands of dollars (the original units of salary).
   Hint: To back-transform a number in natural logarithms to its original scale, use the "exponentiation" function on a calculator [denoted exp(X) or e^X, where X is the variable expressed in natural logarithms]. This is because $\exp(\log_e(Y)) = Y$.

   (c) Compute a 95% confidence interval for the population mean *Logsal* in natural logarithms of thousands of dollars [computer help #23].
   Hint: Calculate by hand (using the fact that the sample mean of Logsal is 7.664386, the sample standard deviation of Logsal is 1.197118, and the 97.5th

percentile of the t-distribution with 213 degrees of freedom is approximately 1.971) and check your answer using statistical software.

(d) Reexpress each interval endpoint of your 95% confidence interval computed in part (c) in thousands of dollars and say what this interval means in words.

(e) The confidence interval computed in part (a) is *exactly* symmetric about the sample mean of *Salary*. Is the confidence interval computed in part (d) *exactly* symmetric about the sample mean of *Logsal* back-transformed to thousands of dollars that you computed in part (b)? How does this relate to quantifying our uncertainty about the population mean salary?

Hint: Looking at the histogram from Problem 1.1 part (a), if someone asked you to give lower and upper bounds on the population mean salary using your intuition rather than statistics, would you give a symmetric or an asymmetric interval?

1.4 A company's pension plan includes 50 mutual funds, with each fund expected to earn a mean, $E(Y)$, of 3% over the risk-free rate with a standard deviation of $SD(Y)=10\%$. Based on the assumption that the funds are randomly selected from a population of funds with normally distributed returns in excess of the risk-free rate, find the probability that an individual fund's return in excess of the risk-free rate is, respectively, greater than 34.1%, greater than 15.7%, or less than -13.3%. In other words, if Y represents potential values of individual fund returns, find:

(a) $Pr(Y > 34.1\%)$;

(b) $Pr(Y > 15.7\%)$;

(c) $Pr(Y < -13.3\%)$.

Use the normal version of the central limit theorem to approximate the probability that the pension plan's overall mean return in excess of the risk-free rate is, respectively, greater than 7.4%, greater than 4.8%, or less than 0.7%. In other words, if M_Y represents potential values of repeated sample means, find:

(d) $Pr(M_Y > 7.4\%)$;

(e) $Pr(M_Y > 4.8\%)$;

(f) $Pr(M_Y < 0.7\%)$.

1.5 *Gapminder* is a "non-profit venture promoting sustainable global development and achievement of the United Nations Millennium Development Goals." It provides related time series data for all countries in the world at the website www.gapminder.org. For example, the **COUNTRIES** data file contains the 2010 population count (variable *Pop* in millions) of the 55 most populous countries together with 2010 life expectancy at birth (variable *Life* in years).

(a) Calculate the sample mean and sample standard deviation of *Pop* [computer help #10].

(b) Briefly say why calculating a confidence interval for the population mean *Pop* would *not* be useful for understanding mean population counts for all countries in the world.

(c) Consider the variable *Life*, which represents the average number of years a newborn child would live if current mortality patterns were to stay the same. Suppose that for *this* variable, these 55 countries *could* be considered a random sample from the population of all countries in the world. Calculate a 95% confidence interval for the population mean of *Life* [computer help #23].

Hint: Calculate by hand (using the fact that the sample mean of Life is 69.787, the sample standard deviation is 9.2504, and the 97.5th percentile of the t-distribution with 54 degrees of freedom is approximately 2.005) and check your answer using statistical software.

1.6 Consider the housing market represented by the sale prices in the **HOMES1** data file.

(a) As suggested on page 11, calculate the probability of finding an affordable home (less than $215,000) in this housing market. Assume that the population of sale prices (*Price*) is normal, with mean $E(Price) = 280$ and standard deviation $SD(Price) = 50$.

(b) As suggested on page 18, calculate a 90% confidence interval for the population mean in this housing market. Recall that the sample mean $m_Y = 278.6033$, the sample standard deviation $s_Y = 53.8656$, and the sample size $n = 30$. Check your answer using statistical software [computer help #23].

(c) Practice the mechanics of hypothesis tests by conducting the following tests using a significance level of 5%.
(i) NH: $E(Y) = 265$ versus AH: $E(Y) > 265$;
(ii) NH: $E(Y) = 300$ versus AH: $E(Y) < 300$;
(iii) NH: $E(Y) = 290$ versus AH: $E(Y) < 290$;
(iv) NH: $E(Y) = 265$ versus AH: $E(Y) \neq 265$.

(d) As suggested on page 27, calculate a 90% prediction interval for an individual sale price in this market.

1.7 Consider the **COUNTRIES** data file from Problem 1.5.

(a) A journalist speculates that the population mean of *Life* is at least 68 years. Based on the sample of 55 countries, a smart statistics student thinks that there is insufficient evidence to conclude this. Do a hypothesis test to show who is correct based on a significance level of 5% [computer help #24].

Hint: Make sure that you lay out all the steps involved—as on page 21—and include a short sentence summarizing your conclusion; that is, who do you think is correct, the journalist or the student?

(b) Calculate a 95% prediction interval for the variable *Life*. Discuss why this interval is so much wider than the confidence interval calculated in Problem 1.5 part (c).

Hint: Calculate by hand (using the fact that the sample mean of Life is 69.787, the sample standard deviation is 9.2504, and the 97.5th percentile of the t-distribution with 54 degrees of freedom is approximately 2.005) and check your answer using statistical software (if possible—see page 27).

1.8 This problem is adapted from one in Frees (1995). The **HOSP** data file contains data on charges for patients at a Wisconsin hospital in 1989, as analyzed by Frees (1994). Managers wish to estimate health care costs and to measure how reliable their estimates are. Suppose that a risk manager for a large corporation is trying to understand the cost of one aspect of health care, hospital costs for a small, homogeneous group of claims, *Charge* = the charges (in thousands of dollars) for $n = 33$ female patients aged 30–49 who were admitted to the hospital for circulatory disorders.

(a) Calculate a 95% confidence interval for the population mean, E(*Charge*). Use the following in your calculation: the sample mean, m_Y, is 2.9554, the sample standard deviation, s_Y, is 1.48104, and the 97.5th percentile of the t-distribution with 32 degrees of freedom is 2.037. Check your answer using statistical software [computer help #23].

(b) Also calculate a 95% prediction interval for an individual claim, *Charge**. Does this interval seem reasonable given the range of values in the data?

(c) Transform the data by taking the reciprocal of the claim values (i.e., 1/*Charge*). Calculate a 95% confidence interval for the population mean of the reciprocal-transformed claims. Use the following sample statistics: the sample mean of 1/*Charge* is 0.3956 and the sample standard deviation of 1/*Charge* is 0.12764. Check your answer using statistical software [computer help #23].

(d) Back-transform the endpoints of the interval you just calculated into the original units of *Charge* (thousands of dollars).

(e) Do the same for a 95% prediction interval—that is, calculate the reciprocal-transformed interval and back-transform to the original units. Does this interval seem reasonable given the range of values in the data? If so, why did transforming the data help here?

1.9   The following questions allow you to practice important concepts from Chapter 1 without having to use a computer.

(a) In the construction of confidence intervals, will an increase in the sample size lead to a wider or narrower interval (if all other quantities are unchanged)?

(b) Suppose that a 95% confidence interval for the population mean, E(Y), turns out to be (50, 105). Give a definition of what it means to be "95% confident" here.

(c) A government department is supposed to respond to requests for information within 5 days of receiving the request. Studies show a mean time to respond of 5.28 days and a standard deviation of 0.40 day for a sample of $n = 9$ requests. Construct a 90% confidence interval for the mean time to respond. Then do an appropriate hypothesis test at significance level 5% to determine if the mean time to respond exceeds 5 days. (You may find some of the following information useful in answering these questions: The 90th percentile of the t-distribution with 8 degrees of freedom is 1.397; the 95th percentile of the t-distribution with 8 degrees of freedom is 1.860.)

(d) Students have claimed that the average number of classes missed per student during a quarter is 2. College professors dispute this claim and believe that the average is more than this. They sample $n = 16$ students and find that the sample mean is 2.3 and the sample standard deviation is 0.6. State the null and alternative hypotheses that the professors wish to test. Then calculate the test statistic for this test and, using a 5% significance level, determine who appears to be correct, the students or the professors. (You may find some of the following information useful: The 95th percentile of the t-distribution with 15 degrees of freedom is 1.753; the 97.5th percentile of the t-distribution with 15 degrees of freedom is 2.131.)

(e) Consider the following computer output:

Sample size, n	150
Mean	2.94
Std. Deviation	0.50

Suppose that we desire a two-tail test of the null hypothesis that the population mean is equal to 3 versus the alternative hypothesis that the population mean is not equal to 3. Find upper and lower limits for the p-value for the test. (You may find some of the following information useful: The 90th percentile of the t-distribution with 149 degrees of freedom is 1.287; the 95th percentile of the t-distribution with 149 degrees of freedom is 1.655.)

(f) A developer would like to see if the average sale price of condominiums in a particular locality has changed in the last 12 months. A study conducted 12 months ago indicated that the average sale price of condominiums in this locality was $280,000. Data on recent sales were as follows:

Sample size, n	28
Mean	289,280
Std. Deviation	21,728

Write down the null and alternative hypotheses for this problem. Then specify the rejection region for conducting a two-tail test at significance level 5%. Based on the computer output, would you reject or fail to reject the null hypothesis for this test? (You may find some of the following information useful: The 95th percentile of the t-distribution with 27 degrees of freedom is 1.703; the 97.5th percentile of the t-distribution with 27 degrees of freedom is 2.052.)

(g) In a hypothesis test, is it true that the smaller the p-value, the less likely you are to reject the null hypothesis? Explain.

CHAPTER 2

SIMPLE LINEAR REGRESSION

In the preceding chapter, we considered univariate data, that is, datasets consisting of measurements of just a single variable on a sample of observations. In this chapter we consider two variables measured on a sample of observations, that is, *bivariate data*. In particular, we will learn about *simple linear regression*, a technique for analyzing bivariate data which can help us to understand the linear association between the two variables, to see how a change in one of the variables is associated with a change in the other variable, and to estimate or predict the value of one of the variables knowing the value of the other variable. We will use the statistical thinking concepts from Chapter 1 to accomplish these goals.

2.1 PROBABILITY MODEL FOR X AND Y

As in Chapter 1, we will quantify our understanding of a particular problem—now involving two variables—using a model. The *simple linear regression model* is represented mathematically as an algebraic relationship between two quantifiable phenomena.

- Y is the *response* variable, which can also go by the name dependent, outcome, or output variable. This variable should be *quantitative*, having meaningful numerical values where the numbers represent actual quantities of time, money, weight, and so on. Chapter 7 introduces some extensions to qualitative (categorical) response variables.

- X is the *predictor* variable, which can also go by the name independent or input variable, or covariate. For the purposes of this chapter, this variable should also be quantitative. We will see in Section 4.3 how to incorporate qualitative information in a predictor variable.

We have a sample of n pairs of (X, Y) values, which we denote (X_i, Y_i) for $i = 1, 2, \ldots, n$ (the index i keeps track of the sample observations). For any particular problem, it is often clear which variable is which. As suggested by its name, the response variable, Y, often "responds" in some way (not necessarily causally, however—see later) to a change in the value of the predictor variable, X. Similarly, if our model provides a useful approximation to the association between Y and X, then knowing a value for the predictor variable can help us to "predict" a corresponding value for the response variable.

The two variables therefore take on very different roles in a simple linear regression analysis, so it is important to identify for any particular problem with bivariate data which is the response variable and which is the predictor variable. Recall the home prices example from Chapter 1, in which we analyzed sale prices for single-family homes in Eugene, Oregon. We could extend this analysis to see if there are any features related to the homes that could help us to understand variation in sale prices. For example, it is likely that there is an association between the sale price (in thousands of dollars) for a particular single-family home and the corresponding floor size (in thousands of square feet). It makes more sense to think of sale price as "responding" to floor size rather than the other way around, so we should set the response variable, Y, as sale price, and the predictor variable, X, as floor size. Furthermore, we are more likely to be interested in predicting sale price for different values of floor size rather than the other way around.

In some problems it may be less clear which is the response variable and which is the predictor variable, but usually the nature of the questions that we wish to answer will dictate which is which. For example, a question such as "How much do we expect variable A to change by when we change the value of variable B?" requires variable A to be the response and variable B to be the predictor. Similarly, a question such as "What do we expect the value of variable C to be when we set the value of variable D at 100?" requires variable C to be the response and variable D to be the predictor.

The simple linear regression model does not require there to be a *causal* link between Y and X (e.g., such that "X causes Y"). In fact, it is quite difficult to establish a causal link between two variables using *observational data* (when data are observed naturally in society or a business or a trading market, say, and the units of observation can potentially differ not only on the measured variables, but also on unmeasured characteristics). It is easier to establish causal links between variables using *experimental data* (when data are collected in a carefully controlled experiment in which the units of observation differ only on the measured variables). The regression modeling techniques described in this book can really only be used to quantify associations between variables and to identify whether a change in one variable is associated with a change in another variable, not to establish whether changing one variable "causes" another to change. For further discussion of causality in the context of regression, see Gelman and Hill (2006).

Once the response and predictor variables have been identified, it is important to define them carefully. For example, returning to the home prices–floor size dataset: Does floor size include just living space or also areas such as unfinished basements? Are measurements in square feet or square meters? If square feet, is the unit of measurement square feet or thousands of square feet? Suppose that we have answered these questions and defined

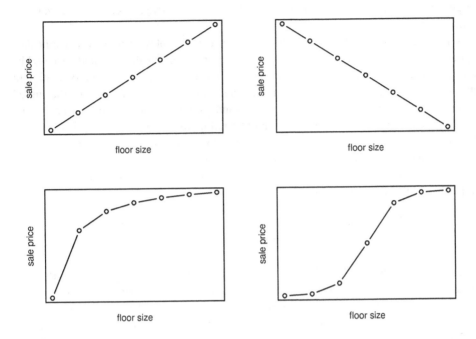

Figure 2.1   Different kinds of association between sale price and floor size. Which factors might lead to the different associations?

Price = sale price in thousands of dollars and *Floor* = floor size in thousands of square feet. Now, we wish to quantify the linear association between *Price* and *Floor* in order to find out how changes in *Floor* are associated with changes in *Price*, and to estimate or predict values for *Price* from particular values of *Floor*. To do this, we will take a random sample of *n* pairs of values of (*Floor*, *Price*) and use the observed linear association between *Price* and *Floor* in this sample to make statistical inferences about the corresponding population association. [We might think of the population in this case as a probability distribution of possible (*Floor*, *Price*) pairs for the particular housing market that we are considering.]

Before we specify the model algebraically, consider the kind of association between sale price and floor size that we might expect. It is convenient to conceptualize possible associations in a *scatterplot* with the response variable, *Price*, plotted on the vertical axis, and the predictor variable, *Floor*, plotted on the horizontal axis. Figure 2.1 displays four simulated datasets in this form, with the plotted points joined by line segments to focus the eye on the associations. Consider the factors that might lead to these different associations. Which association is probably most realistic for typical housing markets? It is often useful to think about these matters before analyzing the data. Often, expert knowledge can be tapped to find expected associations between variables. For example, there may be business or economic theories as to why certain variables tend to have particular associations, or previous research or "common sense" may suggest how certain variables tend to be associated with one another. In this case, common sense suggests some kind of positive association between sale price and floor size, but an experienced realtor might need to be consulted to reason which of the three positive associations in Figure 2.1 might be the most appropriate.

It also makes sense to look at a scatterplot of the data so that we can see whether the data appear to conform to expectations. This can also help us see whether the model we will be using to analyze the data is capable of providing a useful mathematical approximation to the actual association we can observe in the scatterplot. In particular, although the curved associations in the lower two plots of Figure 2.1 may be more likely for a particular housing market, the simple linear regression model covered in this chapter is appropriate only for linear associations such as those in the upper two plots. We will see how to analyze certain types of curved (or nonlinear) associations in Chapter 4.

We can express the simple linear regression model for associations like those in the upper two plots of Figure 2.1 as

$$Y\text{-value} \mid X\text{-value} = \text{deterministic part} + \text{random error}$$
$$Y_i \mid X_i = \text{E}(Y \mid X_i) + e_i \quad (i = 1, \ldots, n),$$

where the vertical bar symbol, "$\mid$," stands for the word "given," so that, for example, $\text{E}(Y \mid X_i)$ means "the expected value of Y given that X is equal to X_i." In other words, each sample Y_i-value can be decomposed into two pieces, a deterministic part that depends on the value of X_i and a random error part that varies from observation to observation. The difference between this model and the model used in Chapter 1 (see page 26) is that the deterministic part now varies with the value of X_i and so is not the same for all observations (except for those observations that have the same value of X).

As an example, suppose that for this housing market, sale price is (on average) $190,300 plus 40.8 times floor size (i.e., homes have a "base price" of $190,300 with an additional $40,800 for each thousand square feet): We will see where these numbers come from in Section 2.2. If we let *Price* represent sale price and *Floor* represent floor size, the deterministic part of the model for such a situation is thus

$$\text{E}(Price \mid Floor_i) = 190.3 + 40.8\,Floor_i \quad (i = 1, \ldots, n).$$

The whole model including the random error part is

$$Price_i \mid Floor_i = 190.3 + 40.8\,Floor_i + e_i \quad (i = 1, \ldots, n),$$

although typically only the deterministic part of the model is reported in a regression analysis, usually with the index i suppressed. The random error part of this model is the difference between the value of *Price* actually observed with a particular observed *Floor*-value, and what we expect *Price* to be on average for that particular observed *Floor*-value: that is, $e_i = Price_i - \text{E}(Price \mid Floor_i) = Price_i - 190.3 - 40.8\,Floor_i$. This random error is not an error in the sense of a "mistake," but rather, represents variation in *Price* due to factors other than *Floor* that we haven't measured. In this example, these might be factors related to lot size (property land area), numbers of bedrooms/bathrooms, property age, garage size, nearby schools, and so on.

Let us now look at the observed data for this example more closely. To keep things simple to aid understanding of the concepts in this chapter, the data for this example consist of a subset of a larger data file containing more extensive information on 76 homes, which is analyzed as a case study in Section 6.1. This subset consists of $n = 5$ of these homes and

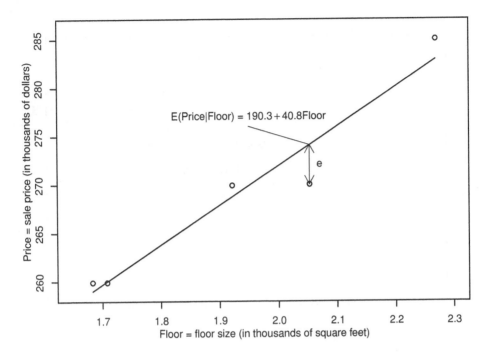

Figure 2.2   Scatterplot showing the simple linear regression model for the home prices–floor size example.

is available in the **HOMES2** data file:

$Price$ = Sale price in thousands of dollars	259.9	259.9	269.9	270.0	285.0
$Floor$ = Floor size in thousands of square feet	1.683	1.708	1.922	2.053	2.269

Figure 2.2 displays a scatterplot of the data points, together with a straight line, called the *regression line*, going through the points (see computer help #15 and #26 in the software information files available from the book website). The regression line represents the deterministic part of the model, $\text{E}(Price \mid Floor) = 190.3 + 40.8\,Floor$. For example, for the fourth observation in the dataset, $Floor = 2.053$, so $\text{E}(Price \mid Floor) = 190.3 + 40.8(2.053) = 274.1$. The observed value of *Price* when $Floor = 2.053$ is 270.0, so the random error for the fourth observation (shown in the figure) is $e = 270.0 - 274.1 = -4.1$. Perhaps that particular home has some unappealing features (that haven't been accounted for in this analysis), so it has a slightly lower sale price than might be expected based on floor size alone. For further practice of these concepts, calculate the expected *Price*-values and random errors for each observation in this dataset, and make sure that the answers you obtain correspond to what you can see in the scatterplot.

Up until now, we have used the deterministic part of the model, $\text{E}(Price \mid Floor) = 190.3 + 40.8\,Floor$, but we haven't said where this equation comes from. Before we see how to obtain the values "190.3" and "40.8," we need to formalize the representation of a linear association in an algebraic expression. In general, we represent the linear association in a simple linear regression model with response variable, Y, and predictor variable, X, as

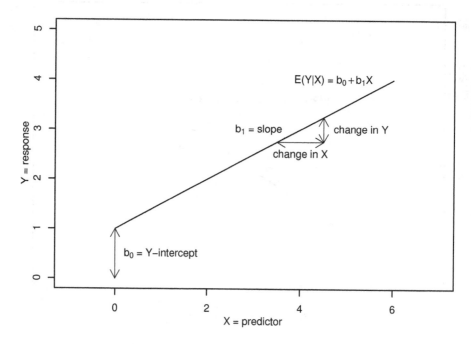

Figure 2.3   Linear equation for the simple linear regression model.

a deterministic linear association plus some random error:

$$Y_i \mid X_i = E(Y \mid X_i) + e_i \quad (i = 1, \ldots, n),$$
$$\text{where } E(Y \mid X_i) = b_0 + b_1 X_i \quad (i = 1, \ldots, n).$$

We usually write this more compactly without the i indices as

$$Y \mid X = E(Y \mid X) + e,$$
$$\text{where } E(Y \mid X) = b_0 + b_1 X.$$

As mentioned earlier, typically only the latter expression—the deterministic part of the model—is reported in a regression analysis. The "b's" in this expression are called *regression parameters* (or *regression coefficients*). Graphically, the parameter b_0 ("b-zero") is the vertical axis intercept (the value of Y when $X = 0$), while the parameter b_1 ("b-one") is the slope (the change in Y for a 1-unit change in X). The expression $E(Y \mid X) = b_0 + b_1 X$ is called the *regression equation* and is an algebraic representation of the regression line. Figure 2.3 displays a hypothetical example where b_0 is 1 and b_1 is 0.5. Having set up our notation, we are now ready to find a systematic, easily calculated way to obtain particular values (estimates) for b_0 and b_1—we explore this in more detail in the next section.

2.2 LEAST SQUARES CRITERION

Figure 2.3 represents a hypothetical population linear association between Y and X. However, as in Chapter 1, we don't get to observe all the values in the population. We just

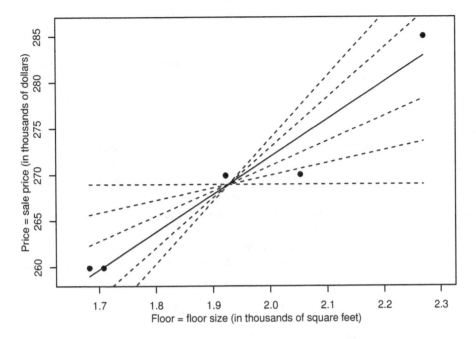

Figure 2.4   Illustration of the least squares criterion for the simple linear regression model.

get to observe the ones in the sample, in this case the n pairs of (X, Y) values. If we can estimate a "best fit" regression line to go through our sample (X, Y) values, then we can use probability theory results to make inferences about the corresponding regression line for the population—we will see how to do that in Section 2.3. In the meantime, how can we estimate a "best fit" regression line for our sample?

Figure 2.4 displays the sample data for the home prices–floor size example, together with six potential regression lines. The dashed horizontal line represents a model in which we would expect the same *Price*-value whatever the value of *Floor*—in this case, that value would be the sample mean of the *Price*-values. However, since there is a clear positive association between *Price* and *Floor* for the five sample values represented by the data points (filled circles) in the scatterplot, it would make sense for our model to expect higher *Price*-values to be associated with higher *Floor*-values. In other words, a regression line with a positive slope makes more sense than a horizontal line (which has a slope of zero). The dashed lines in Figure 2.4 gradually get more and more steeply sloped in an attempt to match the pattern of the data points better, until they finally seem to overshoot and become too steep. Which positively sloped line do you think fits the data the best?

Most people would probably agree that the solid line shows the best fit to the data, and in fact the human eye is very good at finding this "best fit" line for bivariate data displayed in a scatterplot. While good for building intuition about estimating a regression line for bivariate data, this method will not be very good for finding the specific values for the regression parameter estimates (the intercept of "190.3" and the slope of "40.8"). We need a calculation method that is clear and unambiguous and that will always give the same answer no matter who applies the method.

It turns out, however, that the process the human eye goes through to fit a straight line to bivariate data can lead us to a workable calculation method. The eye looks for a straight line that is as close to the data points as possible. One way to do this is to make the vertical distances between the data points and the line as small as possible: these vertical distances are the random errors in the model, $e_i = Price_i - \mathrm{E}(Price | Floor_i) = Price_i - b_0 - b_1 Floor_i$. Since some random errors are positive (corresponding to data points above the line) and some are negative (corresponding to data points below the line), a mathematical way to make the "magnitudes" of the random errors as small as possible is to square them, add them up, and then minimize the resulting *error sum of squares*.

To express this in more formal terms, we need to clarify the distinction between sample and population once more. Since we only observe the sample (and not the population), we can only find an *estimated* regression line for the sample. To make this clear in the text from now on, we will use "hat" notation, such that estimated quantities in the sample have "hats." For example, $\hat{e}$ ("*e*-hat") represents an estimated random error in the sample, called a *residual*. We will also drop the "| X" (given X) notation, so that from this point on this concept will be implicit in all expressions relating Y and X.

In particular, we write the estimated simple linear regression model as an estimated deterministic linear association plus some estimated random error:

$$Y_i = \hat{Y}_i + \hat{e}_i \quad (i = 1, \ldots, n),$$
$$\text{where } \hat{Y}_i = \hat{b}_0 + \hat{b}_1 X_i \quad (i = 1, \ldots, n).$$

Again, we usually write this more compactly without the i indices as

$$Y = \hat{Y} + \hat{e},$$
$$\text{where } \hat{Y} = \hat{b}_0 + \hat{b}_1 X.$$

$\hat{Y}$ ("*Y*-hat") represents an estimated expected value of Y, also known as a *fitted (or predicted) value* of Y. The expression $\hat{Y} = \hat{b}_0 + \hat{b}_1 X$ is called the *estimated (or sample) regression equation* and is an algebraic representation of the estimated (or sample) regression line. To find values for the particular estimates (point estimates), $\hat{b}_0$ and $\hat{b}_1$, we minimize the residual sum of squares (RSS):

$$\begin{aligned} \text{RSS} &= \sum_{i=1}^{n} \hat{e}_i^2 \\ &= \sum_{i=1}^{n} (Y_i - \hat{Y}_i)^2 \\ &= \sum_{i=1}^{n} (Y_i - \hat{b}_0 - \hat{b}_1 X_i)^2. \end{aligned}$$

This is known as the method of *least squares*. Mathematically, to minimize a function such as this, we can set the partial derivatives with respect to b_0 and b_1 equal to zero and then solve for b_0 and b_1, which leads to relatively simple expressions for $\hat{b}_0$ and $\hat{b}_1$. Since we will be using statistical software to estimate $\hat{b}_0$ and $\hat{b}_1$, these expressions are not provided here; also, the intuition behind the expressions is more useful than the formulas themselves. Since the regression line is estimated by using the method of least squares, it is also known as the *least squares line*. Figure 2.5 illustrates this process for the home prices–floor size

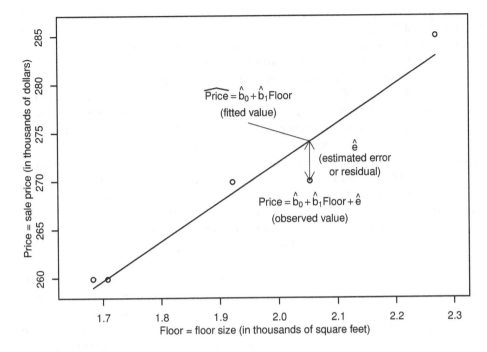

Figure 2.5   Simple linear regression model fitted to sample data for the home prices–floor size example.

example. The values of b_0 and b_1 that minimize RSS are $\hat{b}_0 = 190.3$ and $\hat{b}_1 = 40.8$; this minimum value of RSS is 23.3. For further practice of this concept, confirm for yourself that this is the minimum value of RSS that can be obtained. One way to approach this is to change $b_0 = 190.3$ and $b_1 = 40.8$ slightly (to 191 and 41, say) and calculate RSS (this should be straightforward to set up using spreadsheet software such as Microsoft Excel). You should find that whatever values you try, you cannot get lower than 23.3 (e.g., $b_0 = 191$ and $b_1 = 41$ lead to an RSS of 29.0).

In practice, we use statistical software to find the regression parameter estimates, $\hat{b}_0$ and $\hat{b}_1$. Here is part of the output produced by statistical software when a simple linear regression model is fit to the home prices–floor size example (see computer help #25 in the software information files available from the book website):

Parameters[a]

Model		Estimate	Std. Error	t-stat	$\Pr(>|t|)$
1	(Intercept)	190.318	11.023	17.266	0.000
	Floor	40.800	5.684	7.179	0.006

[a] Response variable: *Price*.

The estimates $\hat{b}_0$ and $\hat{b}_1$ are in the column headed "Estimate" with the intercept in the row labeled "(Intercept)" and the slope in the row labeled with the name of the predictor, "*Floor*" in this case. We discuss the other numbers in the output later in the book.

Having obtained these estimates, how can we best interpret these numbers? Overall, we have found that *if* we were to model *Price* = sale price and *Floor* = floor size for a housing market population represented by this sample with a linear association, then the best-fitting line is $\widehat{Price} = 190.3 + 40.8\,Floor$. This association holds only over the range of the sample *Floor*-values, that is, from 1,683 to 2,269 square feet. In the next section, we will formally discuss whether a simple linear regression model is appropriate for this example, but Figure 2.5 indicates informally that a simple linear regression model is probably reasonable here (the data points are scattered fairly close to the line with no obvious curvature to the dominant pattern in the points).

How should we interpret the particular values for $\hat{b}_0$ and $\hat{b}_1$? Since $\hat{b}_0$ is the estimated response value when the predictor variable is zero, it makes sense to interpret this estimate only *if* a predictor value of zero makes sense for the particular situation being considered, *and* if we have some data with predictor values close to zero. In this example, it does *not* make sense to estimate sale price when floor size is zero. Also, we don't have sample data anywhere close to *Floor* = 0. So, in this case it is not appropriate to interpret $\hat{b}_0 = 190.3$ in any practical terms.

The estimate $\hat{b}_1 = 40.8$ does have a straightforward practical interpretation. It represents the slope of the linear association, that is, the expected change in the response variable for each 1-unit increase in the predictor variable. In particular, we can say that we expect *Price* to increase by 40.8 for each one-unit increase in *Floor*. In other words, bearing in mind that *Price* is measured in thousands of dollars and *Floor* is measured in thousands of square feet, we expect sale price to increase by \$40,800 for each 1,000-square foot increase in floor size. It is always important to state the units of measurement for the response and predictor variables like this when interpreting the slope in a simple linear regression model. Again, this interpretation is valid only over the range of the sample *Floor*-values, that is, from *Floor* = 1,683 to 2,269 square feet.

In some circumstances, it might make more sense to interpret the slope estimate in more practically meaningful units. For example, a 1,000-square foot increase in floor size is fairly large given the overall range in sample *Floor*-values. What about the effect on sale price that corresponds to an increase of 100 square feet? Since we have modeled a linear association between sale price and floor size, if we expect sale price to increase by \$40,800 for each 1,000-square foot increase in floor size, then we should expect sale price to increase by \$4,080 for each 100-square foot increase.

Remember that $\hat{b}_1 = 40.8$ cannot be given a *causal* interpretation. The regression modeling described in this book can really only be used to quantify linear associations and to identify whether a change in one variable is associated with a change in another variable, not to establish whether changing one variable "causes" another to change.

Some statistical software will calculate *standardized* regression parameter (or coefficient) estimates in addition to the (unstandardized) estimates considered here. These are the parameter estimates that would result if the response variable and predictor variable were first standardized to have mean 0 and standard deviation 1. Their interpretations are then in terms of "standard deviations." For example, a standardized slope estimate in a simple linear regression model would represent the number of standard deviations of change in the response variable for each 1-standard deviation increase in the predictor variable. We stick to unstandardized estimates in this book, for reasons discussed in Section 3.2.

The next section covers methods for evaluating whether a linear association—in other words, a simple linear regression model—is appropriate for a particular bivariate dataset.

2.3 MODEL EVALUATION

Before making the kinds of interpretations discussed at the end of the preceding section, we need to be reasonably confident that our simple linear regression model provides a useful approximation to the actual association between response variable, Y, and predictor variable, X. All we have to base that decision on are the visual impression of the regression line going through the sample data in a scatterplot of Y versus X, and the results of the fit of the model to the sample data. With experience, a data analyst can often tell from looking at the data scatterplot whether a simple linear regression model is adequate to explain the association between Y and X (the data points should be scattered fairly close to the line with no obvious curved pattern to the points).

For example, consider the four simulated bivariate datasets represented by the scatterplots and simple linear regression lines in Figure 2.6. The strength of the linear association clearly increases from one scatterplot to the next. In the upper-left scatterplot there is no apparent linear association. In the upper-right scatterplot there is an increasing pattern to the data points overall, but some points are scattered quite far from the regression line. In the lower-left scatterplot the increasing pattern is more pronounced, and nearly all the points are quite close to the regression line. In the lower-right scatterplot the points all lie exactly on the regression line, and the regression line represents the association between Y and X perfectly. In terms of how well a simple linear regression model fits each of the four bivariate datasets, we can say in very general terms that the fit is poor for the upper-left dataset, reasonable for the upper-right dataset, very good for the lower-left dataset, and perfect for the lower-right dataset.

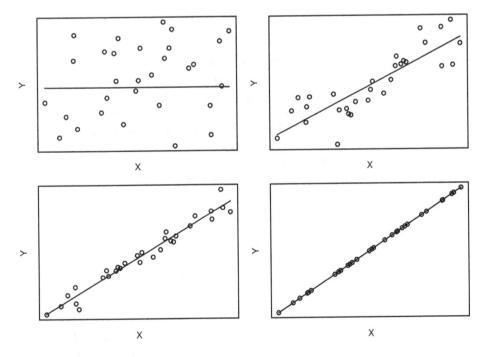

Figure 2.6   How well does the model fit each dataset?

Nevertheless, it is important to be able to present unambiguous numerical justification for whether or not the model provides a good fit. There are three standard methods for evaluating numerically how well a simple linear regression model fits some sample data. These methods can be categorized by the type of question they are designed to answer:

- How close are the actual observed Y-values to the model-based fitted values, $\hat{Y}$? The *regression standard error*, discussed in Section 2.3.1, quantifies this concept.
- How much of the variability in Y have we been able to explain with our model? The *coefficient of determination* or R^2, discussed in Section 2.3.2, quantifies this concept.
- How strong is the evidence of a linear association between Y and X? Estimating and testing the *slope parameter*, b_1, discussed in Section 2.3.3, addresses this concept.

Each of these methods produces a number that can confirm the visual impression of scatterplots such as those in Figure 2.6 while strengthening those impressions with a meaningful numerical value.

2.3.1   Regression standard error

Recall the least squares method used for estimating the regression parameters, b_0 and b_1. The estimates $\hat{b}_0$ and $\hat{b}_1$ are the values that minimize the residual sum of squares,

$$\text{RSS} = \sum_{i=1}^{n} \hat{e}_i^2 = \sum_{i=1}^{n} (Y_i - \hat{Y}_i)^2 = \sum_{i=1}^{n} (Y_i - \hat{b}_0 - \hat{b}_1 X_i)^2.$$

We can use this minimum value of RSS to say how far (on average) the actual observed response values, Y_i, are from the model-based fitted values, $\hat{Y}_i$, by calculating the regression standard error, s:

$$s = \sqrt{\frac{\text{RSS}}{n-2}},$$

which is an estimate of the standard deviation of the random errors in the simple linear regression model. The unit of measurement for s is the same as the unit of measurement for Y. For example, the value of the regression standard error for the home prices–floor size dataset is $s = 2.79$ (see the statistical software output below). In other words, loosely speaking, the actual observed *Price*-values are, on average, a distance of \$2,790 from the model-based fitted values, $\widehat{Price}$.

Here is the output produced by statistical software that displays the value of s for the simple linear regression model fit to the home prices–floor size example (see computer help #25 in the software information files available from the book website):

Model Summary

Model	Sample Size	Multiple R Squared	Adjusted R Squared	Regression Std. Error
1[a]	5	0.945	0.927	2.786

[a] Predictors: (Intercept), *Floor*.

The value of the regression standard error, s, is in the column headed "Regression Std. Error." This can go by a different name depending on the statistical software used. For example, in SPSS it is called the "standard error of the estimate," while in SAS it is

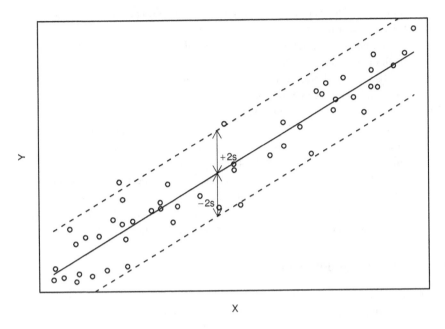

Figure 2.7   Interpretation of the regression standard error for simple linear regression models. Approximately 95% of the observed Y-values lie within $\pm 2s$ of the regression line.

called the "root mean squared error," and in R it is called the "residual standard error." We discuss the other numbers in the output later in the book.

A simple linear regression model is more effective the closer the observed Y-values are to the fitted $\hat{Y}$-values. Thus, for a particular dataset, we would prefer a *small* value of s to a large one. How small is small depends on the scale of measurement of Y (since Y and s have the same unit of measurement). Thus, s is most useful for comparing one model to another *for the same response variable Y*. For example, suppose that we have an alternative possible predictor to use instead of floor size, say, lot size (property land area). We might fit a simple linear regression model with *Price* = sale price in thousands of dollars and *Lot* = lot size and find that the value of the regression standard error for this model is $s = 7.72$. Thus, the observed *Price*-values are farther away (on average) from the fitted $\widehat{Price}$-values for this model than they were for the home prices–floor size model (which had $s = 2.79$). In other words, the random errors tend to be larger, and consequently the deterministic part of the home prices–lot size model must be less well estimated on average: We cannot determine a linear association between sale price and lot size as accurately as we can a linear association between sale price and floor size.

Another way of interpreting s is to multiply its value by 2 to provide an approximate range of "prediction uncertainty." In particular, approximately 95% of the observed Y-values lie within plus or minus $2s$ of their fitted $\hat{Y}$-values. In other words, if we use a simple linear regression model to predict unobserved Y-values from potential X-values, we can expect to be accurate to within approximately $\pm 2s$ (at a 95% confidence level). Figure 2.7 illustrates this interpretation for a hypothetical simple linear regression model. We can see that "most" (on average, 95%) of the data points observed in the scatterplot lie within a parallel band that lies a (vertical) distance of $\pm 2s$ from the regression line of fitted $\hat{Y}$-values. We might expect that unobserved data points (i.e., those in the population but not in the sample) would also mostly lie within this band. In particular, an unobserved Y-value that we might try to predict from a potential X-value will most likely lie within this band. Returning to the home prices–floor size example, $2s = 5.57$, so if we use a simple linear regression model to predict an unobserved sale price value from a particular floor size value, we can expect to be accurate to within approximately $\pm\$5,570$ (at a 95% confidence level).

Where does this result come from? Well, it's an approximation derived from the central limit theorem that we covered in Section 1.4. It can be shown that under repeated sampling, the vertical distances between the observed data points and the fitted regression line have an approximate normal distribution with mean equal to zero and standard deviation equal to s. Since the area under a standard normal density curve between -1.96 and $+1.96$ is 0.95—see the table on page 8—we can make the approximation that about 95% of the observed Y-values lie within $\pm 2s$ of their fitted $\hat{Y}$-values. We will refine this approximation with a more accurate method for finding prediction intervals in Section 2.6.2.

2.3.2   Coefficient of determination—R^2

Another way to evaluate the fit of a simple linear regression model is to contrast the model with a hypothetical situation in which the predictor X is not available. If there is no predictor, then all we have is a list of Y-values, that is, we are in the situation we found ourselves in for Chapter 1. Then, when predicting an individual Y-value, we found that the sample mean, m_Y, is the best point estimate in terms of having no bias and relatively small sampling variability. One way to summarize how well this univariate model fits the sample data is to compute the sum of squares of the differences between the Y_i-values and this point estimate m_Y; this is known as the *total sum of squares* (TSS):

$$\text{TSS} = \sum_{i=1}^{n} (Y_i - m_Y)^2.$$

This is similar to the residual sum of squares (RSS) on page 42, but it measures how far off our observed Y-values are from predictions, m_Y, which *ignore* the predictor X.

For the simple linear regression model, knowing the value of the predictor, X, should allow us to predict an individual Y-value more accurately, particularly if there is a strong linear association between Y and X. To see how much more accurately X helps us to predict an individual Y-value, we can see how much we can reduce the random errors between the observed Y-values and our new predictions, the fitted $\hat{Y}$-values. Recall from page 42 that RSS for the simple linear regression model is

$$\text{RSS} = \sum_{i=1}^{n} (Y_i - \hat{Y}_i)^2.$$

To quantify how much smaller RSS is than TSS, we can calculate the proportional reduction from TSS to RSS, known as the coefficient of determination or R^2 ("R-squared"):

$$R^2 = \frac{\text{TSS} - \text{RSS}}{\text{TSS}} = 1 - \frac{\text{RSS}}{\text{TSS}} = 1 - \frac{\sum_{i=1}^{n}(Y_i - \hat{Y}_i)^2}{\sum_{i=1}^{n}(Y_i - m_Y)^2}.$$

To fully understand what R^2 measures, think of simple linear regression as a method for using a predictor variable, X, to help explain the variation in a response variable, Y.

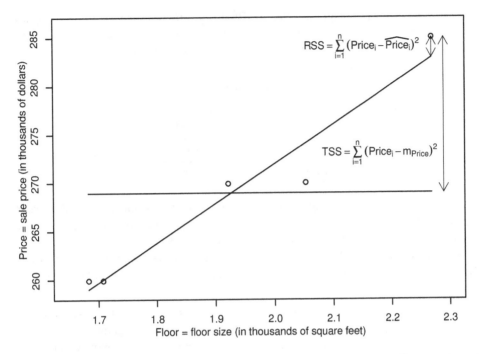

$$RSS = \sum_{i=1}^{n} (Price_i - \widehat{Price_i})^2$$

$$TSS = \sum_{i=1}^{n} (Price_i - m_{Price})^2$$

Figure 2.8 Measures of variation used to derive the coefficient of determination, R^2. The horizontal line represents the sample mean, m_{Price}, while the positively sloped line represents the estimated regression line, $\widehat{Price} = \hat{b}_0 + \hat{b}_1 Floor$.

The "total variation" in Y is measured by TSS (which ignores X) and considers how far the Y-values are from their sample mean, m_Y. If instead we use X in a simple linear regression model, this can help us to predict Y (through the estimated regression equation $\hat{Y} = \hat{b}_0 + \hat{b}_1 X$). Any differences between observed Y-values and fitted $\hat{Y}$-values remains "unexplained" and is measured by RSS. The quantity TSS − RSS therefore represents the variation in Y-values (about their sample mean) that has been "explained" by the simple linear regression model.

In other words, R^2 is the proportion of variation in Y (about its mean) explained by a linear association between Y and X. Figure 2.8 illustrates this interpretation for the simple linear regression model fit to the home prices–floor size dataset. For this example, TSS is the sum of squares of the vertical distances between the observed *Price*-values and the sample mean, m_{Price}, and comes to 423.4 (confirm for yourself that this is the value of TSS by using the fact that $m_{Price} = 268.94$). RSS is the sum of squares of the vertical distances between the observed *Price*-values and the fitted $\widehat{Price}$-values and comes to 23.3 (we saw this earlier on page 43). Therefore,

$$R^2 = \frac{TSS - RSS}{TSS} = \frac{423.4 - 23.3}{423.4} = 0.945.$$

To interpret this number, it is standard practice to report the value as a percentage. In this case, we would conclude that 94.5% of the variation in sale price (about its mean) can be explained by a linear association between sale price and floor size.

In practice, we can obtain the value for R^2 directly from statistical software for any particular simple linear regression model. For example, here is the output that displays the value of R^2 for the home prices–floor size dataset (see computer help #25 in the software information files available from the book website):

Model Summary

Model	Sample Size	Multiple R Squared	Adjusted R Squared	Regression Std. Error
1 [a]	5	0.945	0.927	2.786

[a] Predictors: (Intercept), *Floor*.

The value of the coefficient of determination, R^2, is in the column headed "Multiple R Squared." We discuss the other numbers in the output later in the book (although we have already discussed the "Regression Std. Error" in Section 2.3.1).

Since TSS and RSS are both nonnegative (i.e., each is greater than or equal to zero), and TSS is always greater than or equal to RSS, the value of R^2 must always be between 0 and 1. Consider three possibilities:

- If TSS is equal to RSS, then R^2 is 0. In this case, using X to predict Y hasn't helped and we might as well use the sample mean, m_Y, to predict Y regardless of the value of X. We saw an example of this situation in the upper-left scatterplot in Figure 2.6.

- If RSS is equal to 0, then R^2 is 1. In this case, using X allows us to predict Y perfectly (with no random errors). We saw an example of this situation in the lower-right scatterplot in Figure 2.6.

- These two extremes rarely happen in practice; in all other situations, R^2 lies between 0 and 1, with higher values of R^2 corresponding to better-fitting simple linear regression models.

Figure 2.9 contains scatterplots containing some simulated bivariate datasets with simple linear regression lines and corresponding values of R^2 displayed. These scatterplots can help to build intuition about the connection between a calculated value of R^2 and the visual impression conveyed by a scatterplot.

Correlation   A concept related to R^2 in simple linear regression is *correlation*. The correlation coefficient, r, sometimes known as the Pearson product moment correlation coefficient, measures the strength of linear association between two variables, Y and X:

$$r = \frac{\sum_{i=1}^{n}(Y_i - m_Y)(X_i - m_X)}{\sqrt{\sum_{i=1}^{n}(Y_i - m_Y)^2}\sqrt{\sum_{i=1}^{n}(X_i - m_X)^2}},$$

where m_Y and m_X are the sample means of Y and X, respectively.

The correlation coefficient not only indicates the strength of any linear association between Y and X, but also whether the association is positive (i.e., increasing X-values tend to be associated with increasing Y-values) or negative (i.e., increasing X-values tend to be associated with decreasing Y-values). In particular, correlation coefficients are always between -1 and $+1$, with values close to -1 indicating a negative linear association, values close to $+1$ indicating a positive linear association, and values close to zero suggesting that there is little evidence of any linear association.

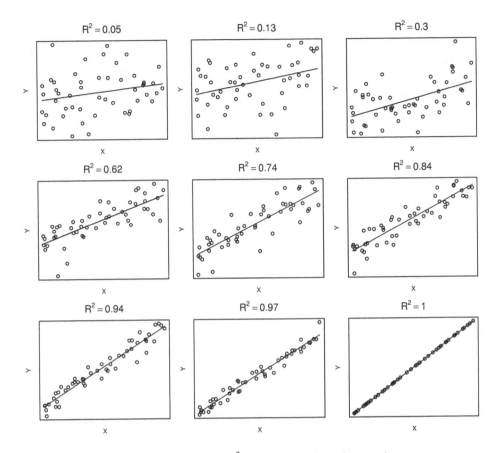

Figure 2.9 Examples of R^2 values for a variety of scatterplots.

For simple linear regression, there is an algebraic relationship between the correlation coefficient, r, and the coefficient of determination, R^2:

$$\text{absolute value of } r = \sqrt{R^2}.$$

However, both measures have their uses. While the correlation coefficient can tell us the strength *and direction* of any linear association between Y and X, R^2 is a more general concept. It has a distinct interpretation for simple linear regression (it represents the proportion of variation in Y explained by a linear association between Y and X), and we will also see in Chapter 3 that this interpretation also carries over to multiple linear regression models with more than one predictor variable. In fact, we will also see that while correlation is an intuitive concept for bivariate data, that intuition can break down when there is more than one predictor variable.

Thus, in this book we tend to prefer the use of R^2 rather than r when looking at linear associations between two variables. Nevertheless, correlation coefficients are popular in real-life applications, so it is helpful to build up some intuition about the connection between calculated values of r and R^2 and the visual impression conveyed by a scatterplot; Figure 2.10 provides some examples.

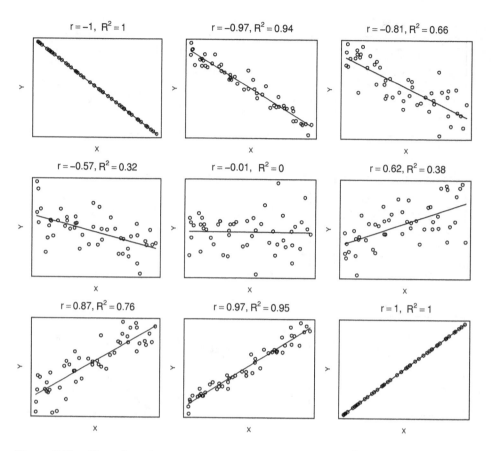

Figure 2.10   Examples of correlation values and corresponding R^2 values for a variety of scatterplots.

2.3.3   Slope parameter

The third method for evaluating the fit of a simple linear regression model that we discuss concerns the slope parameter in the model, b_1. Recall that we have taken a random sample of n pairs of values of (X, Y) and modeled the linear association between Y and X in this sample. However, we are really interested in drawing conclusions about the corresponding *population* association. [We might think of the population in this case as a probability distribution of possible (X, Y) pairs for the particular dataset that we are considering.] Therefore, we need to apply statistical thinking to make statistical inferences about the population. For example, what does the estimated sample slope, $\hat{b}_1$, tell us about likely values for the population slope, b_1?

Figure 2.11 reproduces Figure 2.5, which showed the simple linear regression model fitted to the home prices–floor size sample data, but adds some hypothetical population data to the scatterplot. The scatterplot also displays a dashed hypothetical population regression line, $\mathrm{E}(Y) = b_0 + b_1 X$, and a solid estimated sample regression line, $\hat{Y} = \hat{b}_0 + \hat{b}_1 X$. Since we assume that the sample has been randomly selected from the population, under repeated sampling we would expect the sample and population regression lines to match *on average*, but for any particular sample the two lines will probably differ (as they do here).

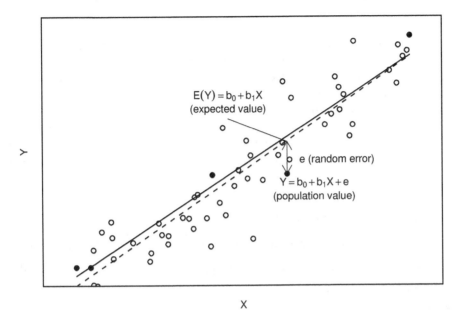

Figure 2.11 Simple linear regression model fitted to hypothetical population data. The dashed line represents the population regression line, $\mathrm{E}(Y) = b_0 + b_1 X$. The filled circles are the sample data from Figure 2.5, while the solid line is the estimated (sample) regression line, $\hat{Y} = \hat{b}_0 + \hat{b}_1 X$.

We cannot be sure how much they will differ, but we can quantify this uncertainty using the sampling distribution of the estimated slope parameter, $\hat{b}_1$. This is where the intuition we built up regarding sampling distributions in Section 1.4 will come in handy. Recall that when analyzing univariate data, we calculated a test statistic:

$$\text{univariate t-statistic} = \frac{m_Y - \mathrm{E}(Y)}{\text{standard error of estimation}},$$

where the standard error of estimation is $s_Y/\sqrt{n}$, and s_Y is the sample standard deviation of $Y_1, Y_2, \ldots, Y_n$. Under very general conditions, this univariate t-statistic has an approximate t-distribution with $n-1$ degrees of freedom. We used this result to conduct hypothesis tests and construct confidence intervals for the univariate population mean, $\mathrm{E}(Y)$.

We can use a similar result to conduct hypothesis tests and construct confidence intervals for the population slope parameter, b_1, in simple linear regression. In particular, we can calculate the following test statistic:

$$\text{slope t-statistic} = \frac{\hat{b}_1 - b_1}{s_{\hat{b}_1}},$$

where $s_{\hat{b}_1}$ is the standard error of the slope estimate:

$$s_{\hat{b}_1} = \frac{s}{\sqrt{\sum_{i=1}^{n}(X_i - m_X)^2}},$$

and s is the regression standard error (see page 46). Under very general conditions, this slope t-statistic has an approximate t-distribution with $n-2$ degrees of freedom.

Slope hypothesis test Suppose that we are interested in a particular value of the population slope for a simple linear regression model. Usually, "zero" is an interesting value for the slope since a zero slope is equivalent to there being no linear association between Y and X. Does the information in our sample support a population slope of zero, or does it favor some alternative value? For the home prices–floor size example, *before* looking at the sample data we might have reason to believe that sale price and floor size have a positive linear association. To see whether the sample data provide compelling evidence that this is the case, we should conduct an upper-tail hypothesis test for the population slope.

- State null hypothesis: NH: $b_1 = 0$.
- State alternative hypothesis: AH: $b_1 > 0$.
- Calculate test statistic: t-statistic $= \frac{\hat{b}_1 - b_1}{s_{\hat{b}_1}} = \frac{40.8 - 0}{5.684} = 7.18$ ($\hat{b}_1$ and $s_{\hat{b}_1}$ can be obtained using statistical software—see output below—while b_1 is the value in NH).
- Set significance level: 5%.
- Look up t-table:
  - critical value: The 95th percentile of the t-distribution with $n-2=3$ degrees of freedom is 2.353 (from Table B.1); the rejection region is therefore any t-statistic greater than 2.353.
  - p-value: The area to the right of the t-statistic (7.18) for the t-distribution with 3 degrees of freedom is less than 0.005 but greater than 0.001 (since from Table B.1 the 99.5th percentile of this t-distribution is 5.841 and the 99.9th percentile is 10.215); thus the upper-tail p-value is between 0.001 and 0.005.
- Make decision:
  - Since the t-statistic of 7.18 falls in the rejection region, we reject the null hypothesis in favor of the alternative.
  - Since the p-value is between 0.001 and 0.005, it must be less than the significance level (0.05), so we reject the null hypothesis in favor of the alternative.
- Interpret in the context of the situation: The five sample observations suggest that a population slope of zero seems implausible—the sample data favor a positive slope (at a significance level of 5%).

A lower-tail test works in a similar way, except that all the calculations are performed in the negative (left-hand) tail of the t-distribution density curve. A lower-tail test would result in an inconclusive result for the home prices–floor size example (since the large, positive t-statistic means that the data favor neither the null hypothesis, NH: $b_1 = 0$, nor the alternative hypothesis, AH: $b_1 < 0$).

On the other hand, before looking at the sample data, we might believe that sale price and floor size have a linear association but not anticipate the direction of the association. To see whether the sample data provide compelling evidence that this is the case, we should conduct a two-tail hypothesis test for the population slope.

- State null hypothesis: NH: $b_1 = 0$.

- State alternative hypothesis: AH: $b_1 \neq 0$.
- Calculate test statistic: t-statistic $= \dfrac{\hat{b}_1 - b_1}{s_{\hat{b}_1}} = \dfrac{40.8 - 0}{5.684} = 7.18$ (the same as for the upper-tail test above).
- Set significance level: 5%.
- Look up t-table:
  - critical value: The 97.5th percentile of the t-distribution with $n-2=3$ degrees of freedom is 3.182 (from Table B.1—we use the 97.5th percentile because this is a two-tail test and we need half the significance level in each tail); the rejection region is therefore any t-statistic greater than 3.182 or less than -3.182.
  - p-value: The area to the right of the t-statistic (7.18) for the t-distribution with 3 degrees of freedom is less than 0.005 but greater than 0.001 (since from Table B.1 the 99.5th percentile of this t-distribution is 5.841 and the 99.9th percentile is 10.215); thus the upper-tail area is between 0.001 and 0.005 and the two-tail p-value is twice this, that is, between 0.002 and 0.01.
- Make decision:
  - Since the t-statistic of 7.18 falls in the rejection region, we reject the null hypothesis in favor of the alternative.
  - Since the p-value is between 0.002 and 0.01, it must be less than the significance level (0.05), and so we reject the null hypothesis in favor of the alternative.
- Interpret in the context of the situation: The five sample observations suggest that a population slope of zero seems implausible—the sample data favor a nonzero slope (at a significance level of 5%).

Figure 2.12 illustrates why the sample data strongly favor the alternative hypothesis of a nonzero slope for this example. The light gray lines represent hypothetical sample regression lines from a population in which the regression slope really was zero. The actual sample data and estimated regression line shown in black do not conform to the pattern of the light gray lines. The visual impression of the graph suggests that it is unlikely that our actual sample could have come from a population with a zero slope. The hypothesis test just described shows this formally.

In practice, we can conduct a hypothesis test for the population slope in a simple linear regression model directly using statistical software. For example, here is the output that displays the information needed for the home prices–floor size dataset (see computer help #25 in the software information files available from the book website):

Parameters[a]

Model		Estimate	Std. Error	t-stat	$\Pr(>|t|)$
1	(Intercept)	190.318	11.023	17.266	0.000
	Floor	40.800	5.684	7.179	0.006

[a] Response variable: *Price*.

The estimate $\hat{b}_1$ is in the column headed "Estimate" and the row labeled with the name of the predictor, "X" in this case. The standard error of the slope estimate, $s_{\hat{b}_1}$, is in the column headed "Std. Error," while the t-statistic is in the column headed "t-stat," and

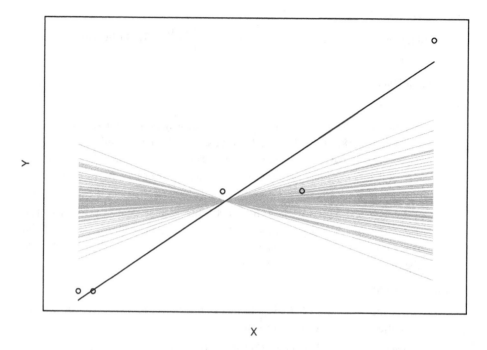

Figure 2.12   Illustration of the sampling distribution of the slope for the simple linear regression model.

the two-tail p-value is in the column headed "$\Pr(>|t|)$" (meaning "the probability that a t random variable with $n-2$ degrees of freedom could be larger than the absolute value of the t-statistic or smaller than the negative of the absolute value of the t-statistic"). This is why we mentioned on page 22 that when we start to rely on statistical software for conducting hypothesis tests, we will find the p-value method easier to use. In general:

- To carry out a two-tail hypothesis test for a zero population slope in simple linear regression, all we have to do is decide on our significance level (e.g., 5%), and check to see whether the two-tail p-value ("$\Pr(>|t|)$" in the statistical software output) is smaller than this significance level. If it is, we reject NH: $b_1 = 0$ in favor of AH: $b_1 \neq 0$ and conclude that the sample data favor a nonzero slope (at the chosen significance level). Otherwise, there is insufficient evidence to reject NH: $b_1 = 0$ in favor of AH: $b_1 \neq 0$, and we conclude that a zero population slope cannot be ruled out (at the chosen significance level).

- To carry out an upper-tail hypothesis test for a zero population slope in simple linear regression is only marginally more difficult. All we have to do is decide on our significance level (e.g., 5%), and calculate the upper-tail p-value. Remember that the upper-tail p-value is the area to the right of the t-statistic under the appropriate t-distribution density curve. If the t-statistic is positive, this area is equal to the two-tail p-value divided by 2. Then, if the upper-tail p-value is smaller than the chosen significance level, we reject NH: $b_1 = 0$ in favor of AH: $b_1 > 0$ and conclude that the sample data favor a positive slope (at the chosen significance level). Otherwise, there is insufficient evidence to reject NH: $b_1 = 0$ in favor of AH: $b_1 > 0$, and we conclude that a zero population slope cannot be ruled out (at the chosen significance level).

However, be careful when conducting an upper-tail hypothesis test when the t-statistic is negative. In such a situation the upper-tail p-value must be at least 0.5 (draw a picture to convince yourself of this). So, in this case, the upper-tail p-value is also going to be larger than any reasonable significance level we might have picked, and we won't be able to reject NH: $b_1 = 0$ in favor of AH: $b_1 > 0$.

- To carry out a lower-tail hypothesis test for a zero population slope is much like the upper-tail test. Now, the lower-tail p-value is the area to the left of the t-statistic under the appropriate t-distribution density curve. If the t-statistic is negative, then this area is equal to the two-tail p-value divided by 2. Then, if the lower-tail p-value is smaller than the chosen significance level, we reject NH: $b_1 = 0$ in favor of AH: $b_1 < 0$ and conclude that the sample data favor a negative slope (at the chosen significance level). Otherwise, there is insufficient evidence to reject NH: $b_1 = 0$ in favor of AH: $b_1 < 0$, and we conclude that a zero population slope cannot be ruled out (at the chosen significance level).

Again, be careful when conducting a lower-tail hypothesis test when the t-statistic is positive. In such a situation the lower-tail p-value must be at least 0.5, so it is also going to be larger than any reasonable significance level we might have picked, and we won't be able to reject NH: $b_1 = 0$ in favor of AH: $b_1 < 0$.

For the home prices–floor size example, since the two-tail p-value is 0.006, we reject NH: $b_1 = 0$ in favor of AH: $b_1 \neq 0$ and conclude that the sample data favor a nonzero slope (at a significance level of 5%). For an upper-tail test, since the t-statistic is positive, the upper-tail p-value is 0.003 and we reject NH: $b_1 = 0$ in favor of AH: $b_1 > 0$ and conclude that the sample data favor a positive slope (at a significance level of 5%). For a lower-tail test, the positive t-statistic means that the lower-tail p-value is at least 0.5, so we cannot reject NH: $b_1 = 0$ in favor of AH: $b_1 < 0$.

We present all three flavors of hypothesis test here (two-tail, upper-tail, and lower-tail), but in real-life applications we would usually only conduct one—selected *before* looking at the data.

It is also possible to conduct a hypothesis test for the intercept parameter, b_0; the procedure is exactly the same as for the slope parameter, b_1. For example, if the intercept p-value is greater than a significance level of 5%, then we cannot reject the null hypothesis that the intercept (b_0) is zero (at that significance level). In other words, a zero intercept is quite plausible. Whether this makes sense depends on the practical context. For example, in some physical systems we might know that when the predictor is zero, the response must necessarily be zero. In contexts such as this it might make sense to drop the intercept from the model and fit what is known as *regression through the origin*. In most practical applications, however, testing a zero intercept (and dropping the intercept from the model if the p-value is quite high) is relatively rare (it generally occurs only in contexts where an intercept exactly equal to zero is expected). Thus we won't dwell any further on this.

Recall from page 44 that it really only makes sense to interpret the estimated intercept, $\hat{b}_0$, if a predictor value of zero makes sense for the particular situation being considered and if we have some data with predictor values close to zero. However, if this is not the case, don't confuse being unable to interpret $\hat{b}_0$ meaningfully with dropping the intercept from the model. For example, consider the following dataset for the length of a baby boy from 15 to 24 months:

Age (months)	Length (cm)
15	79.2
16	79.9
17	80.8
18	82.1
19	82.8
20	83.5
21	84.2
22	86.3
23	87.1
24	88.5

If we fit a simple linear regression model to predict *Length* from *Age*, we obtain the estimated regression equation, $\widehat{Length} = 63.5 + 1.0 Age$. Both intercept and slope are highly significant and the points lie pretty close to the regression line in a scatterplot of *Length* versus *Age* (i.e., it's a pretty good model). As discussed on page 44, we cannot conclude from the estimated regression equation that boys with an age of 0 (i.e., newborn babies) are 63.5 centimeters long on average (the actual average birth length for male babies is closer to 50 cm). But that does not mean that we should drop the intercept term from the model. If we do drop the intercept term from the model, we obtain the estimated regression through the origin, $\widehat{Length} = 4.2 Age$, which is a lousy model (the regression line is much too steep, predicting way too short for low values of *Age* and way too long for high values of *Age*). The important point to keep in mind is that the original model (which included the intercept) is a great model for *Age* between about 15 and 24 months, but it has little of relevance to tell us about *Age* much younger or older than this. Furthermore, the estimated intercept parameter, $\hat{b}_0 = 63.5$, has no practical interpretation but is very necessary to the model.

In principle, we can also test values other than zero for the population slope, b_1—just plug the appropriate value for b_1 into the t-statistic formula on page 53 and proceed as usual (see Problem 2.4 at the end of this chapter). This is quite rare in practice since testing whether the population slope could be zero is usually of most interest. Finally, we can also test values other than zero for b_0—again, this is quite rare in practice.

Slope confidence interval   Another way to express our level of uncertainty about the population slope, b_1, is with a confidence interval. For example, a 95% confidence interval for b_1 results from the following:

$$\text{Pr}\left(-97.5\text{th percentile} < t_{n-2} < 97.5\text{th percentile}\right) = 0.95$$

$$\text{Pr}\left(-97.5\text{th percentile} < \frac{\hat{b}_1 - b_1}{s_{\hat{b}_1}} < 97.5\text{th percentile}\right) = 0.95$$

$$\text{Pr}\left(\hat{b}_1 - 97.5\text{th percentile}\,(s_{\hat{b}_1}) < b_1 < \hat{b}_1 + 97.5\text{th percentile}\,(s_{\hat{b}_1})\right) = 0.95,$$

where the 97.5th percentile comes from the t-distribution with $n-2$ degrees of freedom. In other words, the 95% confidence interval can be written $\hat{b}_1 \pm 97.5\text{th percentile}\,(s_{\hat{b}_1})$.

For example, the 95% confidence interval for b_1 in the home prices–floor size dataset is

$$\hat{b}_1 \pm 97.5\text{th percentile}\,(s_{\hat{b}_1}) = 40.8 \pm 3.182 \times 5.684$$
$$= 40.8 \pm 18.1$$
$$= (22.7, 58.9).$$

Here is the output produced by statistical software that displays the 95% confidence interval for the slope, b_1, in the simple linear regression model fit to the home prices–floor size example (see computer help #27 in the software information files available from the book website):

		Parameters[a]				95% Confidence Interval	
Model		Estimate	Std. Error	t-stat	Pr($>$|t|)	Lower Bound	Upper Bound
1	(Intercept)	190.318	11.023	17.266	0.000	155.238	225.398
	Floor	40.800	5.684	7.179	0.006	22.712	58.888

[a] Response variable: *Price*.

The confidence interval is in the columns headed "95% Confidence Interval" and the row labeled with the name of the predictor, "*Floor*" in this case.

Now that we've calculated a confidence interval, what exactly does it tell us? Well, for this home prices–floor size example, *loosely speaking*, we can say that "we're 95% confident that the population slope, b_1, is between 22.7 and 58.9." In other words, "we're 95% confident that sale price increases by between \$22,700 and \$58,900 for each 1,000-square foot increase in floor size." Again, a more practically useful interpretation in this case might be to say that "we're 95% confident that sale price increases by between \$2,270 and \$5,890 for each 100-square foot increase in floor size." To provide a more precise interpretation we would have to say something like: "If we were to take a large number of random samples of size 5 from our population of home prices–floor size numbers and calculate a 95% confidence interval for b_1 in each, then 95% of those confidence intervals would contain the true (unknown) population slope."

For a lower or higher level of confidence than 95%, the percentile used in the calculation must be changed as appropriate. For example, for a 90% interval (i.e., with 5% in each tail), the 95th percentile would be needed, whereas for a 99% interval (i.e., with 0.5% in each tail), the 99.5th percentile would be needed. These percentiles can be obtained from Table B.1. As further practice, calculate a 90% confidence interval for the population slope for the home prices–floor size example (see Problem 2.4 at the end of this chapter)—you should find that it is (27.4, 54.2). Some statistical software can automatically calculate a 95% interval only for this type of confidence interval, but it is easy to calculate any size interval by hand using $\hat{b}_1$, $s_{\hat{b}_1}$, and the appropriate t-percentile.

2.4   MODEL ASSUMPTIONS

The simple linear regression model relies on a number of assumptions being satisfied in order for it to provide a reliable approximation to a linear association between two variables. These assumptions describe the probability distributions of the random errors in the model:

$$\text{random error} = e = Y - E(Y) = Y - b_0 - b_1 X.$$

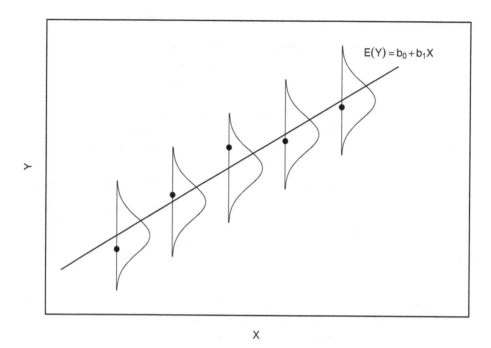

Figure 2.13 Scatterplot illustrating random error probability distributions.

In particular, there are four assumptions about these random errors, e:

- The probability distribution of e at each X-value has a **mean of zero** (in other words, the data points in a scatterplot balance along both sides of the regression line so that the random errors average out to zero as we move across the plot from left to right).

- The probability distribution of e at each X-value has **constant variance**, sometimes called *homoscedasticity* [in other words, the data points in a scatterplot spread out evenly around the regression line so that the (vertical) variation of the random errors is similar as we move across the plot from left to right].

- The probability distribution of e at each X-value is **normal** (in other words, the data points in a scatterplot are more likely to be closer to the regression line than farther away and have a gradually decreasing chance of being farther away).

- The value of e for one observation is **independent** of the value of e for any other observation (in other words, knowing the value of one random error gives us no information about the value of another).

Figure 2.13 illustrates the first three of these assumptions for a hypothetical population. The normal density bell curves show the probability of each data point being a particular distance from the regression line. Each curve is centered over the regression line (the "zero mean" assumption), each curve has the same spread (the "constant variance" assumption), and each curve indicates that each data point is most likely to be close to the line and has a gradually decreasing chance of being farther away (the "normality" assumption). It is difficult to illustrate the final "independence" assumption, but essentially this says that the data points should be randomly scattered about the line and exhibit no systematic patterns.

2.4.1 Checking the model assumptions

The model assumptions relate to the random errors in the population, so we are left with the usual statistical problem. Is there information in the sample data that we can use to ascertain what is likely to be going on in the population? One way to address this is to consider the estimated random errors from the simple linear regression model fit to the sample data. We can calculate these estimated errors or *residuals* as:

$$\text{residual} = \hat{e} = Y - \hat{Y} = Y - \hat{b}_0 - \hat{b}_1 X.$$

This is a list of numbers that represent the vertical distances between the sample Y-values in a scatterplot and the fitted $\hat{Y}$-values along the corresponding regression line. We can therefore construct scatterplots of the residuals themselves, called *residual plots*; these have $\hat{e}$ along the vertical axis and X along the horizontal axis.

We can assess these plots by eye to see whether it seems plausible that the four model assumptions described on page 60 could hold *in the population*. Since this involves somewhat subjective decision-making, to help build intuition Figure 2.14 displays residual plots generated from simulated populations in which the four model assumptions hold, while Figure 2.15 displays simulated residual plots in which the four model assumptions fail. We can use a residual plot to assess the assumptions as follows.

- To assess the **zero mean** assumption, visually divide the residual plot into five or six vertical slices and consider the approximate average value of the residuals in each slice; within-slice averages should each be "close" to zero. There is no need to formally calculate the average of the residuals in each slice. Rather, we can mentally estimate these averages to quickly assess this assumption. We'll see later on page 120 an automated way to carry out this slicing and averaging method using "loess smoothing," but for now we'll just do this by eye. In Figure 2.14 some within-slice averages vary from zero (shown by the horizontal lines), but the variation is small relative to the overall variation of the individual residuals. We should seriously question the zero mean assumption only if there are clear differences from zero for some of the within-slice averages. For example, the upper left residual plot in Figure 2.15 has within-slice averages well above zero at extreme values of X, but well below zero at medium values, suggesting violation of the zero mean assumption. There are also clear violations for the other residual plots in the upper row of Figure 2.15.

- To assess the **constant variance** assumption, again visually divide the residual plot into five or six vertical slices, but now consider the spread of the residuals in each slice; variation should be approximately the same within each slice. For the residual plots in Figure 2.14 there are small changes in variation as we move across each plot from left to right, but nothing that really jumps out. We should seriously question the constant variance assumption only if there are clear changes in variation between some of the slices. For example, the left-hand residual plot in the second row of Figure 2.15 shows a clearly increasing trend in variation from low values of X to high values, suggesting violation of the constant variance assumption. There are also clear violations for the other residual plots in the second row of Figure 2.15.

- The **normality** assumption is difficult to check with the "slicing" technique since there are usually too few residuals within each slice to assess their normality. Instead, we can visually assess whether residuals seem to be approximately normally

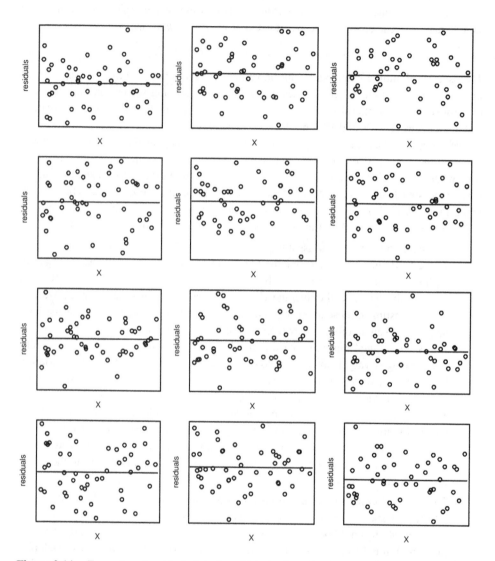

Figure 2.14   Examples of residual plots for which the four simple linear regression assumptions hold. Moving across each plot from left to right, the residuals appear to average close to zero and remain equally variable. Across the whole plot, they seem to be approximately normally distributed, and there are no clear nonrandom patterns.

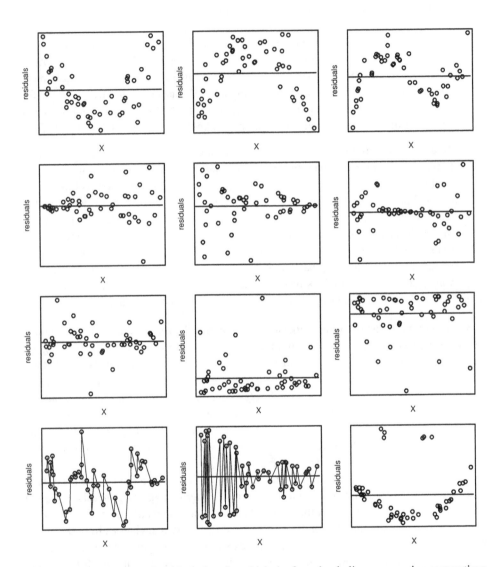

Figure 2.15 Examples of residual plots for which the four simple linear regression assumptions fail. The upper row of three plots fail the zero mean assumption, the second row of three plots fail the constant variance assumption, the third row of three plots fail the normality assumption, and the lower row of three plots fail the independence assumption.

distributed over the entire residual plot. This does seem to be the case for the plots in Figure 2.14, so the normality assumption seems reasonable. We should seriously question the normality assumption only if the distribution of residuals is clearly different from normal. For example, if we contrast the left-hand residual plot in the third row of Figure 2.15 with the residual plots in Figure 2.14, the residuals seem closer to the zero line on average and there are a couple of residuals that are very far away (this example was actually constructed using errors that have a t-distribution with 2 degrees of freedom rather than a normal distribution). The other two plots in the third row of Figure 2.15 show highly skewed distributions. Each of these plots suggests violation of the normality assumption. We discuss alternative methods for checking the normality assumption below: these use histograms and QQ-plots and can be a little easier to use than residual plots like those in Figures 2.14 and 2.15.

- To assess the **independence** assumption, take a final quick look at the residual plot to see if any nonrandom patterns jump out to suggest that this assumption may be in doubt. Otherwise, the independence assumption is probably satisfied, as it is for the residual plots in Figure 2.14. Detecting nonrandom patterns can be difficult, but the residual plots in the lower row of Figure 2.15 provide some examples. By joining up the residuals in the left-hand plot we can see that each residual seems strongly related to its neighbors, suggesting a lack of independence. By contrast, the residuals in the middle plot seem to jump back and forth from positive to negative in a systematic way, again suggesting a lack of independence. Finally, there is a curious pattern to the residuals in the right-hand plot, which seems seriously nonrandom.

While looking at residual plots works well for assessing the zero mean, constant variance, and independence assumptions, histograms and QQ-plots are more useful for assessing the normality assumption. Figure 2.16 displays some histograms of residuals which look sufficiently normal in the upper row of plots, but which suggest violation of the normality assumption in the lower row. The three lower histograms use the same sets of residuals displayed in the third row of Figure 2.15. Figure 2.17 displays some QQ-plots of residuals, which again look sufficiently normal in the upper row of plots, but suggest violation of the normality assumption in the lower row. The three lower QQ-plots again use the same sets of residuals displayed in the third row of Figure 2.15.

The residuals in the left-hand plot in the third row of Figure 2.15 (and in the lower left-hand plots of Figures 2.16 and 2.17) have a specific pattern, suggesting that there are too many values that are too far from the mean relative to a normal distribution. For example, there appear to be more than 5% of the values beyond 2 standard deviations from the mean. Since these residuals farthest from zero are more extreme than might be expected based on a normal distribution, this throws some doubt on the normality assumption. It is possible to modify the standard regression model in such cases to one that uses a t-distribution for the errors rather than a normal distribution—this is an example of *robust regression*, which lies beyond the scope of this book.

Checking the four simple linear regression assumptions using residual plots, histograms, and QQ-plots is a tricky, somewhat subjective technique that becomes easier with experience. We will see in Section 3.4 how to apply a graphical tool in statistical software to make the process a little easier. One important thing to remember is that simple linear regression is reasonably robust to mild violations of the four model assumptions. We really need to worry only when there is a *clear* violation of one or more of the assumptions.

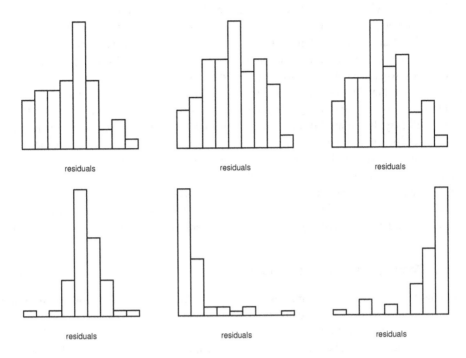

Figure 2.16 Examples of histograms of residuals for which the normality regression assumption holds for the upper row of three plots but fails for the lower row of three plots.

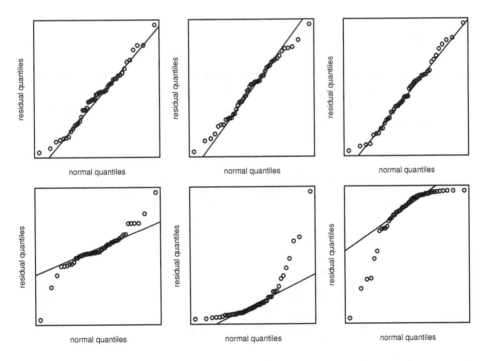

Figure 2.17 Examples of QQ-plots of residuals for which the normality regression assumption holds for the upper row of three plots but fails for the lower row of three.

What should we do if we find a clear violation of one or more of the assumptions? At this point in the book we have no remedies—we'll have to wait until Chapters 3 and 4 to learn about some strategies for dealing with such a situation (see also the suggested remedies on page 190). For now, all we can say is that we should rely on simple linear regression model results only when we can be reasonably confident that the model assumptions check out. If they do not, then the results are questionable and should probably not be used.

One final note of caution regarding residual plots relates to the sample size, n. If n is very small (say, less than 20) there may be insufficient data points to assess the patterns associated with the assumptions. If n is very large (say, in the thousands or more), then there may be substantial overplotting (multiple points plotted in the same position on a graph), which also makes it challenging to assess the patterns associated with the assumptions. One possible approach for the latter case is to take an appropriate random sample of, say, 100 residuals and then to use the methods of this section as normal. More comprehensive still would be to take a few such random samples and assess the assumptions for each sample.

2.4.2 Testing the model assumptions

In many cases, visually assessing residual plots, histograms, and QQ-plots to check the four simple linear regression assumptions is sufficient. However, in ambiguous cases or when we wish to complement our visual conclusions with quantitative evidence, there are a number of test statistics available that address each of the assumptions. A full discussion of these tests lies beyond the scope of this book, but there are brief details and references in Sections 3.4.2, 5.2.1, and 5.2.2.

2.5 MODEL INTERPRETATION

Once we are satisfied that the four simple linear regression assumptions seem plausible, we can interpret the model results. This requires relating the numerical information from the statistical software output back to the subject matter. For example, in the home prices–floor size dataset, the relevant output is:

Model Summary

Model	Sample Size	Multiple R Squared	Adjusted R Squared	Regression Std. Error
1 [a]	5	0.945	0.927	2.786

[a] Predictors: (Intercept), *Floor*.

| | Parameters [a] | | | | 95% Confidence Interval | |
Model	Estimate	Std. Error	t-stat	$\Pr(>|t|)$	Lower Bound	Upper Bound
1 (Intercept)	190.318	11.023	17.266	0.000	155.238	225.398
Floor	40.800	5.684	7.179	0.006	22.712	58.888

[a] Response variable: *Price*.

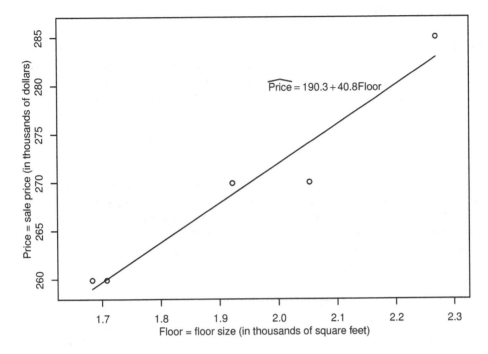

Figure 2.18 Simple linear regression model for the home prices–floor size example.

The corresponding practical interpretations of the results are the following:

- We found a linear association between *Price* = sale price in thousands of dollars and *Floor* = floor size in thousands of square feet that is statistically significant at the 5% significance level. (*Hypothesis test for the slope parameter, b_1.*)

- We expect sale price to increase by \$4,080 for each 100-square foot increase in floor size, for homes from 1,683 to 2,269 square feet. To express our uncertainty due to sampling variation, we could say that we're 95% confident that sale price increases by between \$2,270 and \$5,890 for each 100-square foot increase in floor size. (*Point estimate and confidence interval for the slope parameter, b_1.*)

- If we use a simple linear regression model to predict an unobserved sale price value from a particular floor size value, we can expect to be accurate to within approximately $\pm$\$5,570 (at a 95% confidence level). (*Regression standard error, s.*)

- 94.5% of the variation in sale price (about its mean) can be explained by a linear association between sale price and floor size. (*Coefficient of determination, R^2.*)

A graphical display of the sample data in a scatterplot is also very useful here because all the interpretations above relate directly to such a graph. For example, Figure 2.18 shows the simple linear regression model fit to the home prices–floor size data (see computer help #15 and #26 in the software information files available from the book website). To reinforce understanding of the interpretations discussed above, make sure that you can relate each of them to Figure 2.18.

2.6   ESTIMATION AND PREDICTION

Just as we made a distinction between a confidence interval for the population mean, E(Y), in Section 1.5, and a prediction interval for an individual Y-value in Section 1.7, we make the same distinction for simple linear regression models. The only difference now is that confidence and prediction intervals are for particular values of the predictor variable, X.

2.6.1   Confidence interval for the population mean, E(Y)

Consider estimating the mean (or expected) value of Y at a particular X-value, X_p, say, based on a linear association between Y and X. Since we have estimated the association to be $\hat{Y} = \hat{b}_0 + \hat{b}_1 X$, our best point estimate is $\hat{Y} = \hat{b}_0 + \hat{b}_1 X_p$. For example, if for the home prices–floor size dataset we would like to estimate the average sale price of 2,000-square foot homes, our best point estimate is $\widehat{Price} = 190.3 + 40.8 \times 2 = \$271{,}900$.

How sure are we about this answer? One way to express our uncertainty is with a confidence interval. For example, a 95% confidence interval for E(Y) results from

$$\Pr\left(-97.5\text{th percentile} < t_{n-2} < 97.5\text{th percentile}\right) = 0.95$$
$$\Pr\left(-97.5\text{th percentile} < (\hat{Y} - E(Y))/s_{\hat{Y}} < 97.5\text{th percentile}\right) = 0.95$$
$$\Pr\left(\hat{Y} - 97.5\text{th percentile}\,(s_{\hat{Y}}) < E(Y) < \hat{Y} + 97.5\text{th percentile}\,(s_{\hat{Y}})\right) = 0.95,$$

where $s_{\hat{Y}}$ is the *standard error of estimation* for the simple linear regression mean, and the 97.5th percentile comes from the t-distribution with $n-2$ degrees of freedom. In other words, the 95% confidence interval for E(Y) can be written $\hat{Y} \pm 97.5$th percentile $(s_{\hat{Y}})$.

While we usually rely on statistical software to calculate $s_{\hat{Y}}$ for a particular value of X that we might be interested in, the formula for it is instructive:

$$s_{\hat{Y}} = s\sqrt{\frac{1}{n} + \frac{(X_p - m_X)^2}{\sum_{i=1}^{n}(X_i - m_X)^2}},$$

where s is the regression standard error (see page 46). Thus, $s_{\hat{Y}}$ tends to be smaller (and our estimates more accurate) when n is large, when X_p is close to its sample mean, m_X, and when the regression standard error, s, is small. Also, of course, a lower level of confidence leads to a narrower confidence interval for E(Y)—for example, a 90% confidence interval will be narrower than a 95% confidence interval (all else being equal).

Returning to the home prices–floor size dataset, $s_{\hat{Y}} = 1.313$ for *Floor*=2 (see statistical software output on the next page), so that the 95% confidence interval for E(*Price*) when *Floor*=2 is

$$\widehat{Price} \pm 97.5\text{th percentile}\,(s_{\hat{Y}}) = 271.9 \pm 3.182 \times 1.313$$
$$= 271.9 \pm 4.2$$
$$= (267.7, 276.1).$$

The relevant statistical software output (see computer help #29 in the software information files available from the book website) is:

Price	Floor	$\widehat{Price}$	$s_{\hat{y}}$	CI-low	CI-up
259.9	1.683	258.985	1.864	253.051	264.918
259.9	1.708	260.005	1.761	254.400	265.610
269.9	1.922	268.736	1.246	264.769	272.703
270.0	2.053	274.081	1.437	269.507	278.655
285.0	2.269	282.894	2.309	275.546	290.242
—	2.000	271.918	1.313	267.739	276.098

The point estimates for the population mean, E(*Price*), at particular *Floor*-values are denoted "$\widehat{Price}$," while the standard errors of estimation are denoted "$s_{\hat{y}}$," and the confidence intervals go from "CI-low" to "CI-up."

Now that we've calculated a confidence interval, what exactly does it tell us? Well, for this home prices–floor size example, *loosely speaking*, we can say that "we're 95% confident that average sale price is between $267,700 and $276,100 for 2,000-square foot homes." To provide a more precise interpretation we would have to say something like: "If we were to take a large number of random samples of size 5 from our population of home prices–floor size numbers and calculate a 95% confidence interval for E(Y) at X = 2 in each, then 95% of those confidence intervals would contain the true (unknown) population mean."

For a lower or higher level of confidence than 95%, the percentile used in the calculation must be changed as appropriate. For example, for a 90% interval (i.e., 5% in each tail), the 95th percentile is needed, whereas for a 99% interval (i.e., 0.5% in each tail), the 99.5th percentile is needed. These percentiles can be obtained from Table B.1. As further practice, calculate a 90% confidence interval for E(Y) when X = 2 for the home prices–floor size example (see Problem 2.6 at the end of this chapter)—you should find that it is (268.8, 275.0).

2.6.2   Prediction interval for an individual Y-value

Now, by contrast, consider predicting an individual Y-value at a particular X-value, X_p, say, based on a linear association between Y and X. To distinguish a prediction from an estimated population mean, E(Y), we call this Y-value that is to be predicted Y^*. Just as with estimating E(Y), our best point estimate for Y^* at X_p is $\hat{Y}^* = \hat{b}_0 + \hat{b}_1 X_p$. For example, in the home prices–floor size dataset, our best point estimate for the actual sale price corresponding to an individual 2,000-square foot home is $\widehat{Price}^* = 190.3 + 40.8 \times 2 = \$271,900$.

How sure are we about this answer? One way to express our uncertainty is with a prediction interval (like a confidence interval, but for a prediction rather than an estimated population mean). For example, a 95% prediction interval for Y^* results from the following:

$$\Pr(-97.5\text{th percentile} < t_{n-2} < 97.5\text{th percentile}) = 0.95$$
$$\Pr\left(-97.5\text{th percentile} < (\hat{Y}^* - Y^*)/s_{\hat{y}*} < 97.5\text{th percentile}\right) = 0.95$$
$$\Pr(\hat{Y}^* - 97.5\text{th percentile}\,(s_{\hat{y}*}) < Y^* < \hat{Y}^* + 97.5\text{th percentile}\,(s_{\hat{y}*})) = 0.95,$$

where $s_{\hat{y}*}$ is the *standard error of prediction* for the simple linear regression response, and the 97.5th percentile comes from the t-distribution with $n-2$ degrees of freedom. In other words, the 95% prediction interval for Y^* can be written $\hat{Y}^* \pm 97.5\text{th percentile}\,(s_{\hat{y}*})$.

While we usually rely on statistical software to calculate $s_{\hat{Y}^*}$ for a particular value of X that we might be interested in, the formula for it is instructive:

$$s_{\hat{Y}^*} = \sqrt{s^2 + s_{\hat{Y}}^2} = s\sqrt{1 + \frac{1}{n} + \frac{(X_p - m_X)^2}{\sum_{i=1}^{n}(X_i - m_X)^2}},$$

where s is the regression standard error (see page 46). Thus, $s_{\hat{Y}^*}$ is always larger than $s_{\hat{Y}}$ (on page 68). This makes sense because it is more difficult to predict an individual Y-value at a particular X-value than it is to estimate the mean of the population distribution of Y for that same X-value. For example, suppose that in the population represented by the home prices–floor size dataset there are a large number of 2,000-square foot homes. Estimating the *average* sale price for these homes is easier than predicting the *individual* sale price for any one of these homes. In other words, uncertainty about an individual prediction is always larger than uncertainty about estimating a population mean, and $s_{\hat{Y}^*} > s_{\hat{Y}}$.

For this dataset, $s_{\hat{Y}^*} = 3.080$ for *Floor* $= 2$, and the 95% prediction interval for *Price** is

$$\widehat{Price}^* \pm 97.5\text{th percentile}\,(s_{\hat{Y}^*}) = 271.9 \pm 3.182 \times 3.080$$
$$= 271.9 \pm 9.8$$
$$= (262.1, 281.7).$$

The relevant statistical software output (see computer help #30 in the software information files available from the book website) is:

Price	Floor	$\widehat{Price}$	PI-low	PI-up
259.9	1.683	258.985	248.315	269.655
259.9	1.708	260.005	249.514	270.496
269.9	1.922	268.736	259.021	278.451
270.0	2.053	274.081	264.103	284.059
285.0	2.269	282.894	271.377	294.410
—	2.000	271.918	262.115	281.722

The point estimates for the predictions, *Price**, at particular *Floor*-values are denoted "$\widehat{Price}$," while the prediction intervals go from "PI-low" to "PI-up."

Now that we've calculated a prediction interval, what does it tell us? For this example, *loosely speaking*, "we're 95% confident that the sale price for an individual 2,000-square foot home is between $262,100 and $281,700." A more precise interpretation would have to say something like: "If we were to take a large number of random samples of size 5 from our population of home price–floor size numbers and calculate a 95% prediction interval for *Price** at *Floor* $= 2$ in each, then 95% of those prediction intervals would contain the true (unknown) sale price for an individual 2,000-square foot home picked at random."

As with the standard error of estimation, $s_{\hat{Y}}$, the standard error of prediction, $s_{\hat{Y}^*}$, tends to be smaller (and our predictions more accurate) when n is large, when X_p is close to its sample mean, m_X, and when the regression standard error, s, is small. Also, of course, a lower level of confidence leads to a narrower prediction interval for Y^*: For example, a 90% prediction interval will be narrower than a 95% prediction interval (all else being equal).

For a lower or higher level of confidence than 95%, the percentile in the calculation must be changed as appropriate. For example, for a 90% interval (i.e., 5% in each tail), the

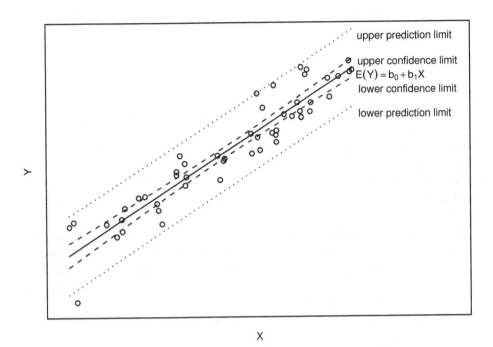

Figure 2.19 Scatterplot illustrating confidence intervals for the mean, E(Y), and prediction intervals for the response, Y.

95th percentile is needed, whereas for a 99% interval (i.e., 0.5% in each tail), the 99.5th percentile is needed. These percentiles can be obtained from Table B.1. As further practice, calculate a 90% prediction interval for an individual value of *Price** when *Floor*$=2$ for the home prices–floor size example (see Problem 2.6 at the end of this chapter)—you should find that it is (264.7, 279.2).

Figure 2.19 illustrates confidence intervals for the population mean, E(Y), and prediction intervals for individual Y-values for a simple linear regression model fit to a simulated dataset. The prediction intervals are clearly wider than the confidence intervals, and both intervals tend to become wider at the extreme values of X rather than in the middle. The regression line passes through the point (m_X, m_Y) and we could think of this as "anchoring" the line at the point of most information/least uncertainty (in the middle of all the data). It would take relatively large changes in the values of Y near here to dramatically change the equation of the line. Contrast this with the most extreme values of X, where the regression line is estimated with the least information/most uncertainty. It wouldn't take much of a change in the values of Y at these extremes to have a dramatic change on the equation of the line. Hence, the standard error of estimation is smallest near m_X and largest at the extreme values of X. It is the same for the standard error of prediction, but less obvious because the standard error of prediction is usually dominated by the random error (rather than the estimation error).

One final note on prediction intervals. Compare Figure 2.19 with Figure 2.7 on page 47. The "$\pm 2s$" interpretation for the regression standard error is actually an approximation of a 95% prediction interval for datasets with a large sample size, n. For large n, the factor

$$\sqrt{1 + \frac{1}{n} + \frac{(X_p - m_X)^2}{\sum_{i=1}^{n}(X_i - m_X)^2}}$$

in $s_{\hat{Y}^*}$ is close to 1, while the 97.5th percentile from the t-distribution with $n-2$ degrees of freedom is close to 2. Thus, the 95% prediction interval for Y^* is approximately $\hat{Y}^* \pm 2s$.

Note that we have now covered four different types of standard error in the context of linear regression. In general terms, a *standard error* in statistics is a sample estimate of a population standard deviation. There are many different types of standard error, because there are many different population standard deviations that we need to estimate for different purposes. In linear regression, the four different types of standard error that we have covered are:

- The regression standard error (Section 2.3.1), which is an estimate of the standard deviation of the error term in the model. Unfortunately, this goes by different names in different software packages. For example, in R it is called the "residual standard error," in SPSS it is the called the "standard error of the estimate," and in SAS it is called the "root mean squared error."

- The standard errors for the regression parameter estimates (Section 2.3.3).

- The standard error of estimation for the multiple linear regression mean (Section 2.6.1).

- The standard error of prediction for the multiple linear regression response (Section 2.6.2).

2.7 CHAPTER SUMMARY

The major concepts that we covered in this chapter relating to simple linear regression are as follows:

Simple linear regression allows us to model a linear association between a response variable, Y, and a predictor variable, X, as $E(Y) = b_0 + b_1 X$.

The method of least squares estimates a regression line, $\hat{Y} = \hat{b}_0 + \hat{b}_1 X$, by minimizing the residual sum of squares (residuals are the differences between the observed Y-values and the fitted $\hat{Y}$-values).

The estimated intercept represents the expected Y-value when $X = 0$ (if zero is a meaningful value and falls close to the range of sample X-values)—denoted $\hat{b}_0$.

The estimated slope represents the expected change in Y for a unit change in X (over the range of sample X-values)—denoted $\hat{b}_1$.

The regression standard error, s, is an estimate of the standard deviation of the random errors. One way to interpret s is to calculate $2s$ and say that when using the simple linear regression model to predict Y from X, we can expect to be accurate to within approximately $\pm 2s$ (at a 95% confidence level).

The coefficient of determination, R^2, represents the proportion of variation in Y (about its sample mean) explained by a linear association between Y and X. In this context, R^2 is also equivalent to the square of the correlation between Y and X.

Hypothesis testing provides a means of making decisions about the likely values of the population slope. The magnitude of the calculated sample test statistic indicates whether we can reject the null hypothesis of a zero slope (no linear association between Y and X) in favor of an alternative hypothesis of a nonzero (or positive, or negative) slope. Roughly speaking, the p-value summarizes the hypothesis test by representing the weight of evidence for the null hypothesis (i.e., small values favor the alternative hypothesis).

Confidence intervals are another method for calculating the sample estimate of the population slope and its associated uncertainty:

$$\hat{b}_1 \pm \text{t-percentile}(s_{\hat{b}_1}).$$

Model evaluation involves assessing how well a model fits sample data. We can do this graphically by informally assessing the strength of the linear association between Y and X in a scatterplot and noting how close the points are to the fitted regression line. We can also do this formally—assuming that a simple linear regression model is appropriate for the data at hand—by considering: (1) the regression standard error, s (smaller values of s generally indicate a better fit), the coefficient of determination; (2) R^2 (larger values of R^2 generally indicate a better fit); (3) the slope test statistic, $t = \hat{b}_1/s_{\hat{b}_1}$ (larger absolute values of t or, equivalently, smaller corresponding p-values generally indicate a better fit).

Model assumptions should be satisfied before we can rely on simple linear regression results. These assumptions relate to the random errors in the model: the probability distribution of the random errors at each value of X should be normal with zero mean and constant variance, and each random error should be independent of every other one. We can check whether these assumptions seem plausible by calculating residuals (estimated errors) and visually assessing residual plots, histograms, and QQ-plots.

Confidence intervals are also used for presenting a sample estimate of the population mean, $E(Y)$, at a particular X-value and its associated uncertainty:

$$\hat{Y} \pm \text{t-percentile}(s_{\hat{Y}}).$$

Prediction intervals, while similar in spirit to confidence intervals, tackle the different problem of predicting an individual Y-value at a particular X-value:

$$\hat{Y}^* \pm \text{t-percentile}(s_{\hat{Y}*}).$$

2.7.1 Review example

We end this chapter by illustrating each of these concepts in a single review example. The National Health and Nutrition Examination Survey is a program of studies designed to assess the health and nutritional status of adults and children in the United States. Consider two body measures: *Height* = standing height (in cm) and *Arm* = upper arm length (in cm). The data for this illustrative example consist of 75 observations randomly selected from 7,217 records from the 2007–2008 survey with *Height* > 140. These data, obtained from the National Center for Health Statistics website at www.cdc.gov/nchs/data_access/ftp_data.htm, are available in the **BODY** data file. To analyze the data, we implement the following steps.

1. **Formulate model.**

   Before looking at the data, we would expect standing height and upper arm length to be positively associated (individuals with relatively long upper arms tend to be taller, and vice versa). If we expect a linear association, a simple linear regression model, $E(Height) = b_0 + b_1 Arm$, might be appropriate.

2. **Construct a scatterplot of *Height* versus *Arm*.**

   Figure 2.20 displays the data for this example.

3. **Estimate model using least squares.**

   Statistical software produces the following output:

   Model Summary

Model	Sample Size	Multiple R Squared	Adjusted R Squared	Regression Std. Error
1 [a]	75	0.6579	0.6532	6.249

   [a] Predictors: (Intercept), *Arm*.

Model		Parameters [a] Estimate	Std. Error	t-stat	Pr(>|t|)	95% Confidence Interval Lower Bound	Upper Bound
1	(Intercept)	53.8404	9.7003	5.55	0.000	34.5078	73.1731
	Arm	3.0428	0.2568	11.85	0.000	2.5310	3.5546

   [a] Response variable: *Height*.

   The estimated regression line is therefore $\widehat{Height} = 53.8 + 3.04\,Arm$.

4. **Evaluate model.**

   (a) **Regression standard error, s**

      Since $2s = 12.5$, *if* we were to use the simple linear regression model to predict *Height* from *Arm*, we can expect to be accurate to within approximately ± 12.5 cm (at a 95% confidence level).

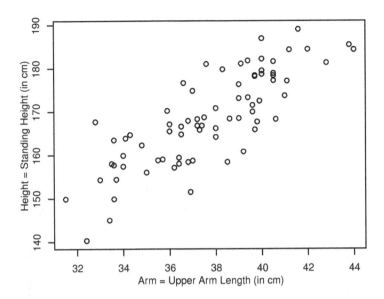

Figure 2.20 Scatterplot of *Height* = standing height (in cm) and *Arm* = upper arm length (in cm) for 75 observations randomly selected from 7,217 records from the 2007–2008 National Health and Nutrition Examination Survey with *Height* > 140.

(b) **Coefficient of determination, R^2**

Since $R^2 = 0.6579$, 65.8% of the variation in *Height* (about its mean) can be explained by a linear association between *Height* and *Arm*.

(c) **Population slope, b_1**

From step 1, we expected the association between *Height* and *Arm* to be positive, so an upper-tail test seems appropriate. Since the slope t-statistic is positive (11.85) and the two-tail p-value for testing a zero population slope is 0.000 (to three decimal places), the upper-tail p-value (half of the two-tail p-value in this case) is significant at the 5% level. We can therefore reject the null hypothesis that there is no linear association between *Height* and *Arm* in favor of the alternative hypothesis that *Height* and *Arm* are positively linearly associated. Alternatively, we can say that we're 95% confident that the population slope, b_1, is between 2.53 and 3.55.

5. **Check model assumptions.**

The residual plot in Figure 2.21 displays a reasonably random pattern, which suggests that the zero mean, constant variance, and independence assumptions seem plausible. The histogram on the left in Figure 2.22 seems bell-shaped and symmetric, while the QQ-plot on the right in Figure 2.22 has points reasonably close to the line—both suggest that the normality assumption seem plausible.

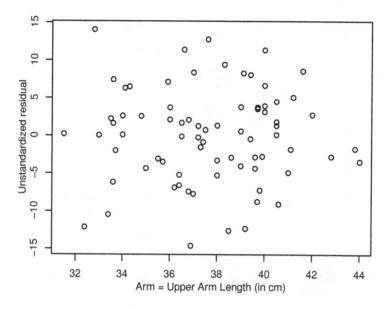

Figure 2.21   Residual plot for the body measurements example.

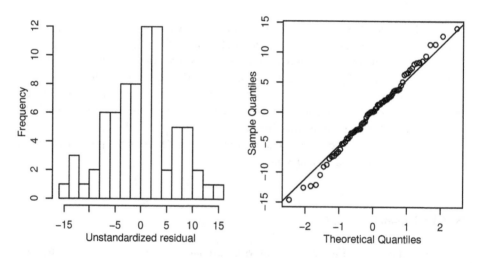

Figure 2.22   Histogram and QQ-plot of residuals for the body measurements example.

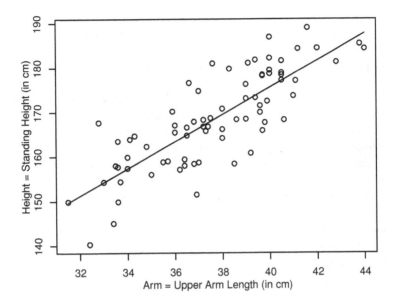

Figure 2.23   Scatterplot of *Height* versus *Arm* for the body measurements example with fitted simple linear regression line.

6. **Interpret model.**

   Figure 2.23 again displays the data for this example, with the fitted simple linear regression line added. In statistical terms, the estimated intercept, $\hat{b}_0 = 53.8$, represents the expected value of *Height* when *Arm* = 0. However, since *Arm* = 0 is not meaningful in this context, $\hat{b}_0 = 53.8$ has no practical interpretation. The estimated slope, $\hat{b}_1 = 3.04$, represents the expected change in *Height* for a unit change in *Arm* (over the range of values of *Arm* in the sample dataset). In other words, for a 1-cm increase in upper arm length (for upper arm lengths between 31.5 cm and 44 cm), the expected increase in standing height is 3.04 cm.

7. **Estimate E(*Height*) and predict *Height*.**

   To illustrate estimation and prediction, consider 95% intervals for *Height* based on the particular value, *Arm* = 38. The relevant statistical software output is:

Arm	$\widehat{Height}$	CI-low	CI-up	PI-low	PI-up
38	169.5	168.0	170.9	156.9	182.0

   Whether we're estimating the population mean, E(*Height*), or predicting an individual value of *Height*, our best point estimate at *Arm* = 38 cm is $\widehat{Height} = 53.8404 + 3.0428 \times 38 = 169.5$ cm. For the population mean, E(*Height*), we're 95% confident that the expected (average) height is between 168.0 and 170.9 cm when the upper arm length is 38 cm. By contrast, we're 95% confident that the height for an individual with an upper arm length of 38 cm is between 156.9 and 182.0 cm.

PROBLEMS

- "Computer help" refers to the numbered items in the software information files available from the book website.
- There are *brief* answers to the even-numbered problems in Appendix E.

2.1 The U.S. Central Intelligence Agency (CIA) 2010 World Factbook (available at the website `https://www.cia.gov/library/publications/the-world-factbook/`) contains information on topics such as geography, people, the economy, communications, and transportation for most countries in the world. For example, the **INTERNET** data file contains data from 2010 relating to gross domestic product (GDP) per capita in US$ thousands (*Gdp*) and the percentage of the population that are Internet users (*Int*) for 212 countries. Here, GDP is based on purchasing power parities to account for between-country differences in price levels. This problem investigates whether there is a linear association between these two variables. In particular, how effective is it to use *Gdp* to predict *Int* using simple linear regression?

   (a) Find the least squares line for the data, that is, use statistical software to find the intercept and slope of the least squares line, and write down the equation of that line [computer help #25].
   Hint: The equation of a least squares line is $\hat{Y} = \hat{b}_0 + \hat{b}_1 X$, with Y replaced by the name of the response variable, X replaced by the name of the predictor variable, and $\hat{b}_0$ and $\hat{b}_1$ replaced by appropriate numbers obtained using statistical software.

   (b) Interpret the estimates of the slope and the intercept in the context of the problem.

   (c) Predict the percentage of Internet users if GDP per capita is US$20,000.

   (d) Draw a scatterplot with *Int* on the vertical axis and *Gdp* on the horizontal axis, and add the least squares line to the plot [computer help #15 and #26].

   (e) Based on the scatterplot, do you think it is appropriate to use this simple linear regression model in this problem, or is the model potentially misleading (and if so, how)? (Problem 4.3 on page 184 investigates some alternative models for this problem.)

2.2 Does a Major League Baseball (MLB) team's batting average relate to the number of games the team wins over a season? The **MLB** data file (downloaded from the ESPN MLB statistics index at `espn.go.com/mlb/statistics`) contains the number of wins (*Win*) and the batting average (*Bat*, calculated as number of hits divided by number of at-bats) for the 2010 regular season. This problem is a highly simplified introduction to *sabermetrics*, which uses baseball statistics to find objective insights into the game. For a fascinating insight into the use of sabermetrics by the Oakland Athletics, see Lewis (2003) and the associated 2011 movie, *Moneyball*.

   (a) Remove the American League teams from the data file to leave just the 16 teams in the National League. Consider using simple linear regression to predict the number of wins from a team's batting average. Would you expect the slope of the resulting least squares line to be positive or negative? Explain.

   (b) Construct an appropriate scatterplot of the data [computer help #15]. Does the pattern of points in the scatterplot agree with your answer from part (a)?

   (c) Find the intercept and slope of the least squares line and write down the equation of that line [computer help #25].

(d) Add the least squares line to the scatterplot from part (b) [computer help #26]. Does the least squares line seem to adequately represent the dominant pattern in the points on the scatterplot?

(e) Does the number of wins appear to be strongly associated with a team's batting average? Explain.

(f) Interpret the estimates of the slope and the intercept in the context of the problem.

(g) Reopen the data file, remove the National League teams from the data file to leave just the 14 teams in the American League, and repeat parts (a) to (f).

2.3 The **ELECTRICITY** data file contains data from the CIA 2010 World Factbook (available at https://www.cia.gov/library/publications/the-world-factbook/) on electricity consumption in billions of kilowatt-hours (*Elec*) and gross domestic product (GDP) in billions of dollars (*Gdp*) for the 30 most populous countries. Here, GDP is based on purchasing power parities to account for between-country differences in price levels. The data file can be used to investigate the claim that there is a linear association between electricity consumption and GDP. For the purposes of this problem, assume that increases in electricity consumption tend to respond to increases in GDP (rather than the other way around).

(a) Say which variable should be the predictor variable (X) and which the response variable (Y) for this problem, and consider the simple linear regression model $E(Y) = b_0 + b_1 X$. *Before looking at the data*, say whether the population value of b_1 would be positive or negative under the claim that electricity consumption tends to increase in response to increases in GDP.

(b) Plot the data in a scatterplot (make sure that you put the appropriate variables on each axis) [computer help #15]. Add the least squares line to the scatterplot [computer help #26]. Briefly describe the dominant pattern in the points on the scatterplot and their position in relation to the least squares line.

(c) Identify the countries with the two highest values of *Gdp* [computer help #18], remove them from the dataset [computer help #19], and redraw the scatterplot and least squares line [computer help #15 and #26]. How does your visual impression of the scatterplot change?
Hint: Focus more on what you see overall, not so much on individual points. In particular, think about whether the overall trend changes and whether the variation of the points about the line changes.

(d) Fit the appropriate simple linear regression model to the dataset that has the countries with the two highest values of *Gdp* removed [computer help #25]. Do the results provide evidence that the claim of a positive association could be true for the remaining 28 countries?
Hint: This problem is asking you to do a hypothesis test. Use a significance level of 5%, make sure you write down the null and alternative hypotheses, and conclude with a statement either supporting or contradicting the claim once you've written up the results of the hypothesis test.

(e) Based on the graph from part (c) and the model from part (d), which country has a particularly *high* electricity consumption and which country has a particularly *low* consumption *relative to the dominant pattern for the remaining countries*? [computer help #18]

2.4   Consider the home prices–floor size example in the **HOMES2** data file.

    (a) As suggested on page 59, calculate a 90% confidence interval for the population slope in a simple linear regression model of $Price =$ sale price in thousands of dollars on $Floor =$ floor size in thousands of square feet. Recall that the estimated slope is $\hat{b}_1 = 40.800$, while the standard error of the slope estimate is $s_{\hat{b}_1} = 5.684$.

    (b) As mentioned on page 58, testing whether a population slope could be zero is usually of more interest than testing some other value. However, sometimes testing some other value is required. For example, suppose that a homeowner is contemplating putting a 500-square foot addition onto her house and wishes to know if doing so could be expected to increase its sale price by $10,000 or more. A $10,000 increase for a 500-square foot addition is equivalent to a $20,000 increase for a 1,000-square foot addition. Thus, this homeowner should conduct an upper-tail hypothesis test for whether the population slope could be greater than 20. What is the result of such a test? Use a 5% significance level.

2.5   The **CARS2** data file contains information for 127 new U.S. front-wheel drive passenger cars for the 2011 model year. These data come from a larger dataset (obtained from www.fueleconomy.gov) which is analyzed more fully in a case study in Section 6.2.

    (a) Transform city miles per gallon into "city gallons per hundred miles." In other words, create a new variable called $Cgphm$ equal to $100/Cmpg$ [computer help #6]. To check that you've done the transformation correctly, confirm that the sample mean of $Cgphm$ is 4.613.

    (b) Consider predicting $Cgphm$ from either engine size in liters (variable Eng) or interior passenger and cargo volume in hundreds of cubic feet (variable Vol). Estimate and compare two alternative simple linear regression models: In the first, Eng is the predictor, while in the second, Vol is the predictor. Write a report about two pages long that compares the two models using *three* numerical methods for evaluating regression models and one graphical method [computer help #25 and #26]. Which of the two models would you recommend for explaining/predicting $Cgphm$?

    (c) Report the regression standard error (s) for the model you recommended in part (b) and use this number to say something about the predictive ability of your model.

2.6   Consider the home prices–floor size example in the **HOMES2** data file.

    (a) As suggested on page 69, calculate a 90% confidence interval for E($Price$) when $Floor = 2$. Recall that the estimated intercept is $\hat{b}_0 = 190.318$, the estimated slope is $\hat{b}_1 = 40.800$, and the standard error of estimation at $Floor = 2$ is $s_{\hat{y}} = 1.313$.

    (b) As suggested on page 71, calculate a 90% prediction interval for $Price^*$ when $Floor = 2$. Recall that the standard error of prediction at $Floor = 2$ is $s_{\hat{y}^*} = 3.080$.

2.7   Consider the **CARS2** data file from Problem 2.5 again.

    (a) You should have found in Problem 2.5 part (b) that of the two simple linear regression models, the model with Eng as the predictor should be used to explain and predict $Cgphm$. You also should have found that the best estimate for the regression slope parameter for Eng is 0.8183. Find a 95% confidence interval for this slope parameter [computer help #27].

(b) Use the model with *Eng* as the predictor to find a 95% confidence interval for the *mean Cgphm* when *Eng* is 3 [computer help #29].

(c) Use the model with *Eng* as the predictor to find a 95% prediction interval for an *individual Cgphm* when *Eng* is 3 [computer help #30].

2.8 Section 2.3 suggests three numerical measures for evaluating a simple linear regression model: the regression standard error, s, the coefficient of determination, R^2, and the t-statistic for the slope parameter, t_{b_1}. The purpose of this problem is to demonstrate that a simple linear regression model with superior measures of fit (i.e., a lower s, a higher R^2, and a higher absolute t-statistic, $|t_{b_1}|$) does not necessarily imply that the model is more appropriate than a model with a higher s, a lower R^2, and a lower absolute t-statistic, $|t_{b_1}|$. These three measures of model fit should always be used in conjunction with a graphical check of the assumptions to make sure that the model is appropriate—see Section 2.4. Download the simulated data from the **COMPARE** data file, which contains a single response variable, Y, and four possible predictor variables, X_1, X_2, X_3, and X_4.

(a) Fit a simple linear regression using X_1 as the predictor [computer help #25]. Make a note of the values of the regression standard error, s, the coefficient of determination, R^2, and the t-statistic for the slope parameter, t_{b_1}. Also construct a scatterplot of Y (vertical) versus X_1 (horizontal) and add the least squares regression line to the plot [computer help #15 and #26].

(b) Repeat part (a), but this time use X_2 as the predictor. Do the values of s, R^2, and $|t_{b_1}|$ suggest a worse or better fit than the model from part (a)? Does the visual appearance of the scatterplot for the X_2 model confirm or contradict the numerical findings?

(c) Repeat part (a), but this time use X_3 as the predictor. Do the values of s, R^2, and $|t_{b_1}|$ suggest a worse or better fit than the model from part (a)? Does the visual appearance of the scatterplot for the X_3 model confirm or contradict the numerical findings?

(d) Repeat part (a), but this time use X_4 as the predictor. Do the values of s, R^2, and $|t_{b_1}|$ suggest a worse or better fit than the model from part (a)? Does the visual appearance of the scatterplot for the X_4 model confirm or contradict the numerical findings?

2.9 The following questions allow you to practice important concepts from Chapter 2 without having to use a computer.

(a) For simple linear regression, state the four assumptions concerning the probability distribution of the random error term.

(b) Is it true that a simple linear regression model allows the expected (or predicted) response values to fall around the regression line while the actual response values must fall on the line? Explain.

(c) Is it true that the coefficient of correlation is a useful measure of the linear association between two variables? Explain.

2.10 A restaurant owner wishes to model the association between mean daily costs, $E(Costs)$, and the number of customers served, *Covers*, using the simple linear regression model, $E(Costs) = b_0 + b_1 \, Covers$. The owner collected a week of data and the values for $n = 7$ days are shown in the accompanying table, followed by computer output for the resulting regression analysis.

Costs = daily operating costs in dollars	1,000	2,180	2,240	2,410	2,590	2,820	3,060
Covers = daily customers served	0	60	120	133	143	175	175

Model Summary

Model	Sample Size	Multiple R Squared	Adjusted R Squared	Regression Std. Error
1[a]	7	0.9052	0.8862	224.0

[a] Predictors: (Intercept), *Covers*.

Parameters[a]

Model		Estimate	Std. Error	t-stat	Pr($>$|t|)
1	(Intercept)	1191.930	185.031	6.442	0.001
	Covers	9.872	1.429	6.909	0.001

[a] Response variable: *Costs*.

(a) State the set of hypotheses that you would test to determine whether costs are positively linearly associated with the number of customers.

(b) Is there sufficient evidence of a positive linear association between costs and number of customers? Use significance level 5%.

(c) Give a practical interpretation of the estimate of the slope of the least squares line.

(d) Give a practical interpretation, if you can, of the estimate of the intercept of the least squares line.

(e) Give a practical interpretation of s, the estimate of the standard deviation of the random error term in the model.

(f) Give a practical interpretation of R^2, the coefficient of determination for the simple linear regression model.

CHAPTER 3

MULTIPLE LINEAR REGRESSION

In the preceding chapter, we considered simple linear regression for analyzing two variables measured on a sample of observations, that is, bivariate data. In this chapter we consider more than two variables measured on a sample of observations, that is, *multivariate data*. In particular, we will learn about *multiple linear regression*, a technique for analyzing certain types of multivariate data. This can help us to understand the association between a response variable and one or more predictor variables, to see how a change in one of the predictor variables is associated with a change in the response variable, and to estimate or predict the value of the response variable knowing the values of the predictor variables.

3.1 PROBABILITY MODEL FOR $(X_1, X_2, \ldots)$ AND Y

The *multiple linear regression model* is represented mathematically as an algebraic relationship between a response variable and one or more predictor variables.

- Y is the *response* variable, which can also go by the name dependent, outcome, or output variable. This variable should be quantitative, having meaningful numerical values. In Chapter 7 we introduce some extensions to qualitative (categorical) response variables.

- $(X_1, X_2, \ldots)$ are the *predictor* variables, which can also go by the name independent or input variables, or covariates. For the purposes of this chapter, these variables

Applied Regression Modeling, Second Edition. By Iain Pardoe
Copyright © 2012 John Wiley & Sons, Inc.

should also be quantitative. In Section 4.3 we will see how to incorporate qualitative information in predictor variables.

We have a sample of n sets of $(X_1, X_2, \ldots, Y)$ values, denoted $(X_{1i}, X_{2i}, \ldots, Y_i)$ for $i = 1, 2, \ldots, n$ (the index i keeps track of the sample observations). The simple linear regression model considered in Chapter 2 represents the special case in which there is just one predictor variable. For any particular problem, it is often clear which variable is the response variable, Y; it often "responds" in some way to a change in the values of the predictor variables, $(X_1, X_2, \ldots)$. Similarly, if our model provides a useful approximation to the association between Y and $(X_1, X_2, \ldots)$, then knowing values for the predictor variables can help us to "predict" a corresponding value for the response variable.

The variables therefore take on very different roles in a multiple linear regression analysis, so it is important, for any particular problem with multivariate data, to identify which is the response variable and which are the predictor variables. For example, consider the problem of quantifying the association between the final exam score of a student taking a course in statistics, and the number of hours spent partying during the last week of the term and the average number of hours per week spent studying for this course. It makes sense to think of the final exam score as responding to time spent partying and time spent studying, so we should set the response variable as exam score (variable *Exam*), and the predictor variables as time spent partying (variable *Party*) and time spent studying (variable *Study*). Furthermore, this will allow us to predict *Exam* for a student with particular values of *Party* and *Study*.

As with simple linear regression models, multiple regression does not require there to be a *causal* link between Y and $(X_1, X_2, \ldots)$. The regression modeling described in this book can really only be used to quantify linear associations and to identify whether a change in one variable is associated with a change in another variable, not to establish whether changing one variable "causes" another to change. For further discussion of causality in the context of regression, see Gelman and Hill (2006).

Having identified the response and predictor variables and defined them carefully, we take a random sample of n observations of $(X_1, X_2, \ldots, Y)$. We then use the observed association between Y and $(X_1, X_2, \ldots)$ in this sample to make statistical inferences about the corresponding population association. [We might think of the population in the final exam score example as a probability distribution of possible $(Party, Study, Exam)$ values for the particular statistics course that we are considering.]

Before we specify the model algebraically, consider the kind of association between Y and $(X_1, X_2, \ldots)$ that we might expect. It is often useful to think about these matters before analyzing the data. Often, expert knowledge can be tapped to find expected associations between variables. For example, there may be theories in the field of application relating to why certain variables tend to have particular associations, or previous research may suggest how certain variables tend to be associated with one another. In the final exam score example, common sense tells us a lot: *Exam* probably decreases as X_1 increases, but increases as X_2 increases (at least we would hope this is the case).

We can express the multiple linear regression model as

$$Y\text{-value} \mid X\text{-values} = \text{deterministic part} + \text{random error}$$
$$Y_i \mid (X_{1i}, X_{2i}, \ldots) = \mathrm{E}(Y \mid (X_{1i}, X_{2i}, \ldots)) + e_i \quad (i = 1, \ldots, n),$$

where the vertical bar, "|," means "given," so that, for example, $E(Y \mid (X_{1i}, X_{2i}, \ldots))$ means "the expected value of Y given that X_1 is equal to X_{1i}, X_2 is equal to X_{2i}, and so on." In other words, each sample Y-value is decomposed into two pieces—a deterministic part depending on the X-values, and a random error part varying from observation to observation.

As an example, suppose that for the final exam score example, the exam score is (on average) 70 minus 1.6 times the number of hours spent partying during the last week of the term plus 2.0 times the average number of hours per week spent studying for this course. In other words, for each additional hour spent partying the final exam score tends to decrease by 1.6, and for each additional hour per week spent studying the score tends to increase by 2.0. The deterministic part of the model for such a situation is thus

$$E(Exam \mid (Party_i, Study_i)) = 70 - 1.6\,Party_i + 2.0\,Study_i \quad (i = 1, \ldots, n).$$

The whole model, including the random error part, is

$$Exam_i \mid (Party_i, Study_i) = 70 - 1.6\,Party_i + 2.0\,Study_i + e_i \quad (i = 1, \ldots, n),$$

although typically only the deterministic part of the model is reported in a regression analysis, usually with the index i suppressed. The random error part of this model is the difference between the value of $Exam$ actually observed with particular observed predictor values and what we expect $Exam$ to be on average for those particular observed predictor values: that is, $e_i = Exam_i - E(Exam \mid (Party_i, Study_i)) = Exam_i - 70 + 1.6\,Party_i - 2.0\,Study_i$. This random error represents variation in $Exam$ due to factors other than $Party$ and $Study$ that we haven't measured. In this example, these might be factors related to quantitative skills, exam-taking ability, and so on.

Figure 3.1 displays a 3D-scatterplot of some hypothetical students, together with a flat surface, called the *regression plane*, going through the data. The values of $Party$ increase from 0 at the "front" of the plot to 10 on the right, while the values of $Study$ increase from 0 at the front of the plot to 10 on the left. The values of $Exam$ increase from 50 at the bottom of the plot to 90 at the top of the plot. The $Exam$-values are also represented by the shading on the regression plane according to the scale on the right.

The regression plane represents $E(Exam \mid (Party, Study)) = 70 - 1.6\,Party + 2.0\,Study$, the deterministic part of the model. For example, a student who parties for 7.5 hours during the last week of the term but studies only 1.3 hours per week for this course has an expected final exam score of $E(Exam \mid (Party, Study)) = 70 - 1.6(7.5) + 2.0(1.3) = 60.6$. If their observed value of $Exam$ is 65, then their random error (shown in the figure) is $e = 65 - 60.6 = 4.4$. Perhaps for that particular student, exam performance was better than expected because of strong analytical skills, for example.

To estimate the expected $Exam$-values and random errors, we need to know the numbers in the deterministic part of the model, that is, "70," "−1.6," and "2.0." Before we see how to obtain these numbers, we need to formalize the representation of a multiple linear regression model in an algebraic expression:

$$Y_i \mid (X_{1i}, X_{2i}, \ldots) = E(Y \mid (X_{1i}, X_{2i}, \ldots)) + e_i \quad (i = 1, \ldots, n),$$
$$\text{where } E(Y \mid (X_{1i}, X_{2i}, \ldots)) = b_0 + b_1 X_{1i} + b_2 X_{2i} + \cdots \quad (i = 1, \ldots, n).$$

We usually write this more compactly without the i indices as

$$Y \mid (X_1, X_2, \ldots) = E(Y \mid (X_1, X_2, \ldots)) + e,$$
$$\text{where } E(Y \mid (X_1, X_2, \ldots)) = b_0 + b_1 X_1 + b_2 X_2 + \cdots.$$

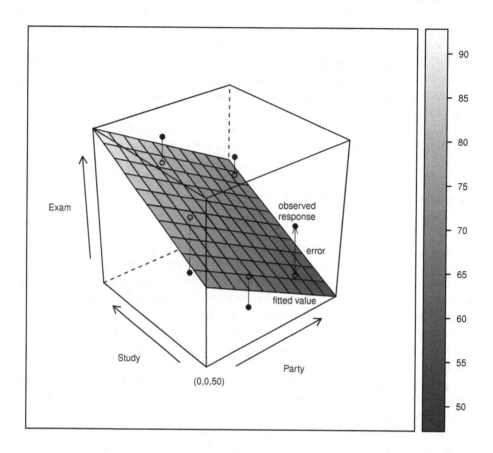

Figure 3.1   Multiple linear regression model with two predictors fitted to a hypothetical population for the final exam score example.

As mentioned earlier, typically only the latter expression—the deterministic part of the model—is reported in a regression analysis. The regression parameter (or regression coefficient) b_0 ("b-zero") is the intercept (the value of Y when $X_1 = 0, X_2 = 0, \ldots$). The regression parameter b_1 ("b-one") is the change in Y for a 1-unit change in X_1 when all the other predictor X-variables are held constant. Similarly, the regression parameter b_2 ("b-two") is the change in Y for a 1-unit change in X_2 when all the other predictor X-variables are held constant, and so on.

For example, in the final exam score model, $b_0 = 70$ represents the expected score for a student who went to no parties during the last week of the term, but who also spent no time studying for this course. If we were to consider two students who spent the same time studying per week, but one spent one more hour than the other partying in the last week of the term, we would expect the former student to score $b_1 = -1.6$ points more on the final exam (in other words, 1.6 points less) than the latter student. On the other hand, if we were to consider two students who spent the same time partying during the last week of the term, but one spent one more hour per week than the other studying, we would expect the former student to score $b_2 = 2.0$ points more on the final exam than the latter student.

Figure 3.1 represents each of these quantities: $b_0 = 70$ is the *Exam*-value for the corner of the regression plane at the front of the graph where *Party* and *Study* are both zero; $b_1 = -1.6$ is the slope of the regression plane in the "*Party*-direction" (i.e., when *Study* is held constant); $b_2 = 2.0$ is the slope of the regression plane in the "*Study*-direction" (i.e., when *Party* is held constant).

The expression $E(Y \mid (X_1, X_2, \ldots)) = b_0 + b_1 X_1 + b_2 X_2 + \cdots$ is called the *regression equation* and is an algebraic representation of the regression plane. When there are more than two predictors, we cannot visualize this plane in three dimensions as we can in Figure 3.1. Nevertheless, the mathematical theory still works in higher dimensions (the regression plane becomes a regression "hyperplane") and the intuitive interpretations we have been using up until now all carry over.

We are now ready to find a systematic, easily calculated way to obtain estimates for b_0, b_1, b_2, ... We explore this in more detail in the next section.

3.2 LEAST SQUARES CRITERION

Figure 3.1 represents a hypothetical population association between Y and $(X_1, X_2, \ldots)$. Usually, we don't get to observe all the values in the population. Rather, we just get to observe the values in the sample, in this case the n observations of $(X_1, X_2, \ldots, Y)$. If we can estimate a "best fit" regression equation to go through our sample $(X_1, X_2, \ldots, Y)$ values, then we can use probability theory results to make inferences about the corresponding regression equation for the population—we will see how to do that in Section 3.3. In the meantime, how can we estimate a "best fit" regression equation for our sample?

Consider tilting the regression plane in Figure 3.1 from side to side and up and down, until the plane is as close to the data points as possible. One way to do this is to make the vertical distances between the data points and the plane as small as possible: These vertical distances are the random errors in the model, that is, $e_i = Y_i - E(Y \mid (X_{1i}, X_{2i}, \ldots)) = Y_i - b_0 - b_1 X_{1i} - b_2 X_{2i} - \cdots$. Since some random errors are positive (corresponding to data points above the plane) and some are negative (data points below the plane), a mathematical way to make the "magnitudes" of the random errors as small as possible is to square them, add them up, and then minimize the resulting *error sum of squares*.

Since we observe the sample only (not the population), we can only find an *estimated* regression equation for the sample. Recall the "hat" notation from Chapter 2, such that estimated quantities in the sample have "hats"; for example, $\hat{e}$ ("*e*-hat") represents an estimated random error (or residual) in the sample. We will again also drop the "$\mid (X_1, X_2, \ldots)$" (given $X_1, X_2, \ldots$) notation, so that from this point on this concept will be implicit in all expressions relating Y and $(X_1, X_2, \ldots)$.

In particular, we write the estimated multiple linear regression model as

$$Y_i = \hat{Y}_i + \hat{e}_i \quad (i = 1, \ldots, n),$$
$$\text{where } \hat{Y}_i = \hat{b}_0 + \hat{b}_1 X_{1i} + \hat{b}_2 X_{2i} + \cdots \quad (i = 1, \ldots, n).$$

Again, we usually write this more compactly without the i indices as

$$Y = \hat{Y} + \hat{e}$$
$$\text{where } \hat{Y} = \hat{b}_0 + \hat{b}_1 X_1 + \hat{b}_2 X_2 + \cdots.$$

$\hat{Y}$ ("*Y*-hat") represents an estimated expected value of *Y*, also known as a *fitted or predicted value* of *Y*. The expression $\hat{Y} = \hat{b}_0 + \hat{b}_1 X_1 + \hat{b}_2 X_2 + \cdots$ is called the estimated (or sample) regression equation and is an algebraic representation of the estimated (or sample) regression hyperplane. To find values for the point estimates, $\hat{b}_0$, $\hat{b}_1$, $\hat{b}_2$, and so on, we minimize the residual sum of squares (RSS):

$$\begin{aligned} \text{RSS} &= \sum_{i=1}^{n} \hat{e}_i^2 \\ &= \sum_{i=1}^{n} (Y_i - \hat{Y}_i)^2 \\ &= \sum_{i=1}^{n} (Y_i - \hat{b}_0 - \hat{b}_1 X_{1i} - \hat{b}_2 X_{2i} - \cdots)^2. \end{aligned}$$

We have already seen this method of *least squares* in Chapter 2. Mathematically, to minimize a function such as this, we can set the partial derivatives with respect to b_0, b_1, b_2, … equal to zero and then solve for b_0, b_1, b_2, …, which leads to relatively simple expressions for the regression parameter estimates, $\hat{b}_0$, $\hat{b}_1$, $\hat{b}_2$, …. Since we will be using statistical software to do the calculations, these expressions are provided for interest only at the end of this section. Again, the intuition behind the expressions is more useful than the formulas themselves.

Recall the home prices–floor size example from Chapter 2, in which we analyzed the association between sale prices and floor sizes for single-family homes in Eugene, Oregon. We continue this example here and extend the analysis to see if an additional predictor, lot size (property land area), helps to further explain variation in sale prices. Again, to keep things simple to aid understanding of the concepts in this chapter, the data for this example consist of a subset of a larger data file containing more extensive information on 76 homes, which is analyzed as a case study in Section 6.1.

The data for the example in this chapter consist of $n = 6$ homes with variables *Price* = sale price in thousands of dollars, *Floor* = floor size in thousands of square feet, and *Lot* = lot size category and are available in the **HOMES3** data file:

Price = Sale price in $ thousands	252.5	259.9	259.9	269.9	270.0	285.0
Floor = Floor size in sq. feet thousands	1.888	1.683	1.708	1.922	2.053	2.269
Lot = Lot size category (see below)	2	5	4	4	3	3

This dataset contains an additional home to the dataset from Chapter 2 (**HOMES2**), which had a strong linear association between sale price and floor size. This additional home has values of *Price* = 252.5 and *Floor* = 1.888, which don't fit the estimated association from Chapter 2 nearly as well as the other five homes (this is evident from Figure 3.2 on the next page). However, we shall see in this chapter that a new association between *Price* and both *Floor* and *Lot* together fits the data for all six homes very well.

The variable *Lot* in this dataset requires some further explanation. It is reasonable to assume that homes built on properties with a large amount of land area command higher sale prices than homes with less land, all else being equal. However, it is also reasonable to suppose that an increase in land area of 2,000 square feet from 4,000 to 6,000 would make a larger difference (to sale price) than going from 24,000 to 26,000. Thus, realtors have constructed lot size "categories," which in their experience correspond to approximately equal-sized increases in sale price. The categories used in this dataset are:

Lot size	0-3k	3-5k	5-7k	7-10k	10-15k	15-20k	20k-1ac	1-3ac	3-5ac	5-10ac	10-20ac
Category	1	2	3	4	5	6	7	8	9	10	11

Lot sizes ending in "k" represent thousands of square feet, while "ac" stands for acres (there are 43,560 square feet in an acre).

It makes sense to look at scatterplots of the data before we start to fit any models. With simple linear regression, a scatterplot with the response, Y, on the vertical axis and the predictor, X, on the horizontal axis provides all the information necessary to identify an association between Y and X. However, with a response variable, Y, but more than one predictor variable, $X_1, X_2, \ldots$, we can use scatterplots only to identify bivariate associations between any two variables (e.g., Y and X_1, Y and X_2, or even X_1 and X_2). We cannot identify a multivariate association between Y and $(X_1, X_2, \ldots)$ just from bivariate scatterplots.

Nevertheless, we can use these bivariate scatterplots to see whether the data have any strange patterns or odd-looking values that might warrant further investigation. For example, data entry errors are often easy to spot in bivariate scatterplots when one data point appears isolated a long way from all the other data points. A useful method for looking at all possible bivariate scatterplots in a multivariate data setting is to construct a *scatterplot matrix*, such as the scatterplot matrix for the home prices dataset in Figure 3.2 (see computer help #16 in the software information files available from the book website). Here, the scatterplot of *Price* versus *Floor* is shown in the top middle part of the matrix, *Price*

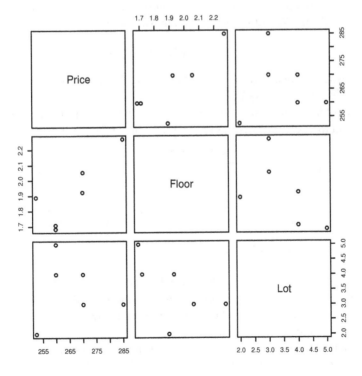

Figure 3.2 Scatterplot matrix for the home prices example.

versus *Lot* is at top right, and *Floor* versus *Lot* is just below. Reflections of these three plots are below the diagonal of this matrix. Scatterplot matrices can be challenging to decipher initially. One key to understanding a particular graph in a scatterplot matrix is to look left or right to see the label for the variable plotted on the vertical axis and to look up or down to see the label for the variable plotted on the horizontal axis. In this respect, scatterplot matrices are similar to scatterplots where the vertical axis label is typically to the left of the graph and the horizontal axis label is typically below the graph.

We can see an increasing pattern between *Price* and *Floor*, an ambiguous pattern between *Price* and *Lot*, and a decreasing pattern between *Floor* and *Lot* in the plots, but such patterns *cannot* tell us whether the multiple linear regression model that we consider below can provide a useful mathematical approximation to these bivariate associations. In Section 3.3.2 we shall see examples of how such thinking can be misleading. The scatterplot matrix is useful primarily for identifying any strange patterns or odd-looking values that might warrant further investigation *before* we start modeling. In this case, there are no data points that appear isolated a long way from all the other data points, so it seems reasonable to proceed.

We propose the following multiple linear regression model:

$$Price = \mathrm{E}(Price) + e$$
$$= b_0 + b_1\,Floor + b_2\,Lot + e,$$

with *Price*, *Floor*, and *Lot* defined as above. The random errors, e, represent variation in *Price* due to factors other than *Floor* and *Lot* that we haven't measured. In this example, these might be factors related to numbers of bedrooms/bathrooms, property age, garage size, or nearby schools. We estimate the deterministic part of the model, $\mathrm{E}(Price)$, as

$$\widehat{Price} = \hat{b}_0 + \hat{b}_1\,Floor + \hat{b}_2\,Lot,$$

by using statistical software to find the values of b_0, b_1, and b_2 that minimize RSS $= \sum_{i=1}^{n}(Price_i - \hat{b}_0 - \hat{b}_1 X_{1i} - \hat{b}_2 X_{2i})^2$.

Here is part of the output produced by statistical software when a multiple linear regression model is fit to this home prices example (see computer help #31):

Parameters[a]

Model		Estimate	Std. Error	t-stat	$\Pr(> |t|)$
1	(Intercept)	122.357	14.786	8.275	0.004
	Floor	61.976	6.113	10.139	0.002
	Lot	7.091	1.281	5.535	0.012

[a] Response variable: *Price*.

The regression parameter estimates $\hat{b}_0$, $\hat{b}_1$, and $\hat{b}_2$ are in the column headed "Estimate," with $\hat{b}_0$ in the row labeled "(Intercept)," $\hat{b}_1$ in the row labeled with the name of the corresponding predictor, "*Floor*" in this case, and $\hat{b}_2$ in the row labeled "*Lot*." We discuss the other numbers in the output later in the book.

Having obtained these estimates, how can we best interpret these numbers? Overall, we have found that *if* we were to model *Price*, *Floor*, and *Lot* for a housing market population represented by this sample with the multiple linear regression model, $\mathrm{E}(Price) = b_0 + b_1\,Floor + b_2\,Lot$, then the best-fitting model is $\widehat{Price} = 122.36 + 61.98\,Floor + 7.09\,Lot$.

This association holds only over the range of the sample predictor values, that is, *Floor* from 1,683 to 2,269 square feet and *Lot* from lot size category 2 to 5. In the next section we discuss whether this multiple linear regression model is appropriate for this example.

What of the particular values for $\hat{b}_0$, $\hat{b}_1$, and $\hat{b}_2$? Since $\hat{b}_0$ is the estimated *Price*-value when *Floor* $= 0$ and *Lot* $= 0$, it makes practical sense to interpret this estimate only if X-values of zero make sense for the particular situation being considered, *and* if we have some data *close* to *Floor* $= 0$ and *Lot* $= 0$. In this example, it does not make practical sense to estimate sale price when floor size and lot size category are both zero. Also, we don't have any sample data particularly close to *Floor* $= 0$ and *Lot* $= 0$. So, in this case, it is *not* appropriate to interpret $\hat{b}_0 = 122.36$ in practical terms.

The estimate $\hat{b}_1 = 61.98$ represents the change in *Price* for a 1-unit increase in *Floor* when all the other predictor variables are held constant. In particular, we can say that we expect sale price to increase by \$62,000 for each 1,000-square foot increase in floor size when lot size is held constant. A more meaningful interpretation in this example is that we expect sale price to increase by \$6,200 for each 100-square foot increase in floor size when lot size is held constant. Similarly, the estimate $\hat{b}_2 = 7.09$ represents the change in *Price* for a 1-unit increase in *Lot*, when all the other predictor variables are held constant. In particular, we can say that we expect sale price to increase by \$7,090 for each 1-category increase in lot size when floor size is held constant. It is important to state the units of measurement for *Price*, *Floor*, and *Lot* when making these interpretations. Again, these interpretations are valid only over the range of the sample predictor values, that is, *Floor* from 1,683 to 2,269 square feet and *Lot* from lot size category 2 to 5.

The estimates $\hat{b}_1 = 61.98$ and $\hat{b}_2 = 7.09$ can be combined to find changes in sale price for different changes in floor size and lot size together. For example, since *Price* is measured in thousands of dollars, we would expect that a 200-square foot increase in floor size coupled with an increase of one lot size category would lead to an increase in sale price of $\$1,000 \times (0.2 \times 61.98 + 1 \times 7.09) = \$19,500$.

Remember that $\hat{b}_1 = 61.98$ and $\hat{b}_2 = 7.09$ cannot be given *causal* interpretations. The regression modeling described in this book can really only be used to quantify linear associations and to identify whether a change in one variable is associated with a change in another variable, not to establish whether changing one variable "causes" another to change.

Some statistical software will calculate *standardized* regression parameter (or coefficient) estimates in addition to the (unstandardized) estimates considered here. These are the parameter estimates that would result if the response variable and predictor variables were first standardized to have mean 0 and standard deviation 1. Their interpretations are then in terms of "standard deviations." For example, a standardized regression parameter estimate for a particular predictor variable would represent the standard deviation change in the response variable for a 1-standard deviation increase in that predictor variable when all the other predictor variables are held constant. In some circumstances, use of standardized regression parameter estimates can enable comparisons of the relative contributions of each predictor in explaining the overall variation in the response variable. However, this use of standardized estimates does not carry over to models with transformations, interactions, and indicator variables, which we consider in Chapter 4. In Section 5.5 we introduce a graphical method that can be used to compare the relative contributions of predictors for such models.

In the next section we cover methods for evaluating whether a multiple linear regression model is appropriate for a particular multivariate dataset.

Optional—formula for regression parameter estimates. Suppose that our multiple linear regression model has k predictor variables. Consider all the observations for each predictor variable as column vectors (each length n, the sample size). Then put all these column vectors side by side into a matrix, typically called $\mathbf{X}$. Note that this matrix has an additional column of 1's at the far left (representing the intercept term in the model). Thus, $\mathbf{X}$ has n rows and $k+1$ columns. Also let $\mathbf{Y}$ be a column vector representing all the observations of the response variable. Then calculate the following vector, $(\mathbf{X}^{\mathrm{T}}\mathbf{X})^{-1}\mathbf{X}^{\mathrm{T}}\mathbf{Y}$, where $\mathbf{X}^{\mathrm{T}}$ is the transpose of $\mathbf{X}$ and $(\mathbf{X}^{\mathrm{T}}\mathbf{X})^{-1}$ is the inverse of $\mathbf{X}^{\mathrm{T}}\mathbf{X}$. The $k+1$ entries of this vector are the regression parameter estimates, $\hat{b}_0, \hat{b}_1, \hat{b}_2, \ldots, \hat{b}_k$. In practice, statistical software uses a variation of this method, which centers the variables about their means first.

3.3 MODEL EVALUATION

Before making the kinds of interpretations discussed at the end of the preceding section, we need to be reasonably confident that our multiple linear regression model provides a useful approximation to the actual association between Y and $(X_1, X_2, \ldots)$. All we have to base that decision on are the results of the fit of the model to the sample data. It is important to be able to present unambiguous numerical justification for whether or not the model provides a good fit.

We used three standard methods to evaluate numerically how well a simple linear regression model fits some sample data. Two of those methods—the regression standard error, s, and the coefficient of determination, R^2—carry over essentially unchanged. The last method which focused on the slope parameter, b_1, becomes a little more complicated since we now have a series of regression parameters, $b_1, b_2, \ldots$. It turns out that we can tackle this issue globally (looking at all the regression parameters, $b_1, b_2, \ldots$, simultaneously), in subsets (looking at two or more of the regression parameters at a time), or individually (considering just one of the regression parameters at a time). We consider each of these methods—s, R^2, regression parameters globally, regression parameters in subsets, and regression parameters individually—in turn.

3.3.1 Regression standard error

Suppose that our multiple linear regression model has k predictor X-variables. For example, $k=2$ for the home prices dataset above with predictors *Floor* = floor size and *Lot* = lot size. Recall the least squares method used for estimating the regression parameters, $b_0, b_1, b_2, \ldots, b_k$. The estimates $\hat{b}_0, \hat{b}_1, \hat{b}_2, \ldots, \hat{b}_k$ are the values that minimize the residual sum of squares,

$$\mathrm{RSS} = \sum_{i=1}^{n} \hat{e}_i^2 = \sum_{i=1}^{n} (Y_i - \hat{Y}_i)^2 = \sum_{i=1}^{n} (Y_i - \hat{b}_0 - \hat{b}_1 X_{1i} - \hat{b}_2 X_{2i} - \cdots - \hat{b}_k X_{ki})^2.$$

We can use this minimum value of RSS to say how far (on average) the actual observed values, Y_i, are from the model-based fitted values, $\hat{Y}_i$, by calculating the regression standard error, s:

$$s = \sqrt{\frac{RSS}{n-k-1}},$$

which is an estimate of the standard deviation of the random errors in the multiple linear regression model. The $n-k-1$ denominator in this expression generalizes from the simple linear regression case when it was $n-2$ (and k was 1). The unit of measurement for s is the same as the unit of measurement for Y. For example, the value of the regression standard error for the home prices dataset is $s=2.48$ (see the statistical software output below). In other words, loosely speaking, the actual observed $Price$-values are, on average, a distance of \$2,480 from the model-based fitted values, $\widehat{Price}$.

Here is the output produced by statistical software that displays the value of s for the multiple linear regression model fit to the home prices example (see computer help #31 in the software information files available from the book website):

Model Summary

Model	Sample Size	Multiple R Squared	Adjusted R Squared	Regression Std. Error
1 [a]	6	0.9717	0.9528	2.475

[a] Predictors: (Intercept), *Floor*, *Lot*.

The value of the regression standard error, s, is in the column headed "Regression Std. Error." This can go by a different name depending on the statistical software used. For example, in SPSS it is called the "standard error of the estimate," while in SAS it is called the "root mean squared error," and in R it is called the "residual standard error." We discuss the other numbers in the output later in the book.

A multiple linear regression model is more effective the closer the observed Y-values are to the fitted $\hat{Y}$-values. Thus, for a particular dataset, we would prefer a *small* value of s to a large one. How small is small depends on the scale of measurement of Y (since Y and s have the same unit of measurement). Thus, s is most useful for comparing one model to another *for the same response variable Y.* For example, suppose that we have alternative possible predictors to use instead of floor size and lot size, say, numbers of bedrooms and property age. We might fit a multiple linear regression model with the response variable, $Price =$ sale price in thousands of dollars, and predictor variables, number of bedrooms and property age, and find that the value of the regression standard error for this model is $s=9.33$. Thus, the observed $Price$-values are farther away (on average) from the fitted $\widehat{Price}$-values for this model than they were for the model that used floor size and lot size (which had $s=2.48$). In other words, the random errors tend to be larger, and consequently, the deterministic part of the model that uses numbers of bedrooms and property age must be less accurate (on average).

Just as with simple linear regression, another way of interpreting s is to multiply its value by 2 to provide an approximate level of "prediction uncertainty." In particular, approximately 95% of the observed Y-values lie within plus or minus $2s$ of their fitted $\hat{Y}$-values (recall from Section 2.3.1 that this is an approximation derived from the central limit theorem). In other words, if we use a multiple linear regression model to predict an unobserved Y-value from potential X-values, we can expect to be accurate to within approximately $\pm 2s$ (at a 95% confidence level). Returning to the home prices example, $2s=4.95$, so if we use a multiple linear regression model to predict an unobserved sale price for an individual home with particular floor size and lot size values, we can expect to be accurate to within approximately ±$4,950 (at a 95% confidence level).

3.3.2 Coefficient of determination—R^2

Another way to evaluate the fit of a multiple linear regression model is to contrast the model with a situation in which the predictor X-variables are not available. If there are no predictors, then all we would have is a list of Y-values; that is, we would be in the situation that we found ourselves in for Chapter 1. Then, when predicting an individual Y-value, we found that the sample mean, m_Y, is the best point estimate in terms of having no bias and relatively small sampling variability. One way to summarize how well this univariate model fits the sample data is to compute the sum of squares of the differences between the Y_i-values and this point estimate m_Y; this is known as the *total sum of squares* (TSS):

$$\text{TSS} = \sum_{i=1}^{n}(Y_i - m_Y)^2.$$

This is similar to the residual sum of squares (RSS) on page 88, but it measures how far off our observed Y-values are from predictions, m_Y, which *ignore* the predictor X-variables.

For the multiple linear regression model, the predictor X-variables should allow us to predict an individual Y-value more accurately. To see how much more accurately the predictors help us to predict an individual Y-value, we can see how much we can reduce the random errors between the observed Y-values and our new predictions, the fitted $\hat{Y}$-values. Recall from page 88 that RSS for the multiple linear regression model is

$$\text{RSS} = \sum_{i=1}^{n}(Y_i - \hat{Y}_i)^2.$$

Just as in simple linear regression, to quantify how much smaller RSS is than TSS, we can calculate the proportional reduction from TSS to RSS, known as the coefficient of determination or R^2 ("R-squared"):

$$\text{R}^2 = \frac{\text{TSS} - \text{RSS}}{\text{TSS}} = 1 - \frac{\text{RSS}}{\text{TSS}} = 1 - \frac{\sum_{i=1}^{n}(Y_i - \hat{Y}_i)^2}{\sum_{i=1}^{n}(Y_i - m_Y)^2}.$$

To fully understand what R^2 measures, think of multiple linear regression as a method for using predictor X-variables to help explain the variation in a response variable, Y. The "total variation" in Y is measured by TSS (which ignores the X-variables) and considers how far the Y-values are from their sample mean, m_Y. The multiple linear regression model predicts Y through the estimated regression equation, $\hat{Y} = \hat{b}_0 + \hat{b}_1 X_1 + \hat{b}_2 X_2 + \cdots$. Any differences between observed Y-values and fitted $\hat{Y}$-values remains "unexplained" and is measured by RSS. The quantity TSS − RSS therefore represents the variation in Y-values (about their sample mean) that has been "explained" by the multiple linear regression model.

In other words, R^2 is the proportion of variation in Y (about its mean) explained by a multiple linear regression association between Y and $(X_1, X_2, \ldots)$. In practice, we can obtain the value for R^2 directly from statistical software for any particular multiple linear regression model. For example, here is the output that displays the value of R^2 for the home prices dataset (see computer help #31 in the software information files available from the book website):

Model Summary

Model	Sample Size	Multiple R Squared	Adjusted R Squared	Regression Std. Error
1 [a]	6	0.9717	0.9528	2.475

[a] Predictors: (Intercept), *Floor, Lot.*

The value of the coefficient of determination, $R^2 = 0.9717$, is in the column headed "Multiple R Squared." (We have already discussed the "Regression Std. Error" in Section 3.3.1 and discuss the other numbers in the output later in the book.) To interpret this number, it is standard practice to report the value as a percentage. In this case we would conclude that 97.2% of the variation in sale price (about its mean) can be explained by a multiple linear regression association between sale price and (floor size, lot size).

Since TSS and RSS are both nonnegative (i.e., greater than or equal to 0), and TSS is always greater than or equal to RSS, the value of R^2 must always be between 0 and 1. Higher values of R^2 correspond to better-fitting multiple linear regression models. However, there is no "reference value" such that R^2 greater than this value suggests a "good model" and R^2 less than this suggests a "poor model." This type of judgment is very context dependent, and an apparently "low" value of $R^2 = 0.3$ may actually correspond to a useful model in a setting where the response variable is particularly hard to predict with the available predictors. Thus, R^2 by itself cannot tell us whether a particular model is good, but it can tell us something useful (namely, how much variation in Y can be explained).

Adjusted R^2 This chapter is concerned with the mechanics of the multiple linear regression model and how the various concepts from simple linear regression carry over to this new multivariate setting. For the simple linear regression model of Chapter 2, we discussed just one possible way to model the association between a response variable, Y, and a single predictor, X: using a straight line. By contrast, when there are two or more potential predictors we have more possibilities for how we model the association between Y and $(X_1, X_2, \ldots)$. We could include just one of the potential predictors (i.e., use a simple linear regression model), or all of the potential predictors, or something in between (i.e., a subset of the potential predictors). In addition, we could "transform" some or all of the predictors, or create "interactions" between predictors. We discuss these topics, part of *model building*, in detail in Chapters 4 and 5.

However, some model building topics arise naturally as we explore how to adapt concepts from simple linear regression to the multiple linear regression setting—R^2 is one such concept. We have seen that R^2 tells us the proportion of variation in Y (about its mean) explained by a multiple linear regression association between Y and $(X_1, X_2, \ldots)$. Since higher values of R^2 are better than lower values of R^2, all other things being equal, perhaps R^2 can be used as a criterion for guiding model building (i.e., out of a collection of possible models, the one with the highest value of R^2 is the "best"). Unfortunately, R^2 *cannot* be used in this way to guide model building because of a particular property that it has. This property dictates that if one model—model A, say—has a value of R^2 equal to R_A^2, then R^2 for a second model with the same predictors as model A *plus* one or more additional predictors will be greater than (or equal to) R_A^2. In other words, as we add predictors to a model, R^2 either increases or stays the same.

While we can justify this result mathematically, a geometrical argument is perhaps more intuitive. Consider a bivariate scatterplot of a response variable, Y, versus a predictor variable, X_1, with a regression line going through the points. Call the model represented by this line "model A," so that the line minimizes the residual sum of squares, RSS_A. If we add a second predictor variable, X_2, to this model, we can think of adding a third axis to this scatterplot (much like in Figure 3.1) and moving the data points out along this axis according to their values for X_2. The regression model, "model B" say, is now represented by a plane rather than a line, with the plane minimizing the residual sum of squares, RSS_B. Whereas for model A we can only tilt the regression line in two dimensions (represented by the Y-axis and the X_1-axis), for model B we can tilt the regression plane in three dimensions (represented by the Y-axis, the X_1-axis, and the X_2-axis). So, we can always make RSS_B less than (or at least equal to) RSS_A. This in turn makes R_B^2 always greater than (or equal to) R_A^2. This result holds in higher dimensions also, for any model B with the same predictors as model A plus one or more additional predictors.

Consider a collection of *nested models*, that is, a sequence of models with the next model in the sequence containing the same predictor variables as the preceding model in the sequence plus one or more additional predictor variables. Then *if* we were to use R^2 as a criterion for assessing the "best" model, R^2 would pick the last model in the sequence, that is, the one with the most predictor variables. This model certainly does the best job of getting closest to the *sample* data points (since it has the smallest RSS of all the models), but that does not necessarily mean that it does the best job of modeling the *population*. Often, the model with all the potential predictor variables will "overfit" the sample data so that it reacts to every slight twist and turn in the sample associations between the variables. A simpler model with fewer predictor variables will then be preferable *if* it can capture the major, important population associations between the variables without getting distracted by minor, unimportant sample associations.

Since R^2 is inappropriate for finding such a model (one that captures the major, important population associations), we need an alternative criterion, which penalizes models that contain too many unimportant predictor variables. The *adjusted R^2* measure does just this:

$$\text{adjusted } R^2 = 1 - \left(\frac{n-1}{n-k-1}\right)(1-R^2).$$

As the number of predictors (k) in the model increases, R^2 increases (which also causes adjusted R^2 to increase), but the factor "$-(n-1)/(n-k-1)$" causes adjusted R^2 to decrease. This trade-off penalizes models that contain too many unimportant predictor variables, and allows us to use adjusted R^2 to help find models that do a reasonable job of finding the population association between Y and $(X_1, X_2, \dots)$ without overcomplicating things.

In practice, we can obtain the value for adjusted R^2 directly from statistical software for any particular multiple linear regression model. For example, here is the output that displays adjusted R^2 for the home prices model with predictors, *Floor* and *Lot* (see computer help #31 in the software information files available from the book website):

Model Summary

Model	Sample Size	Multiple R Squared	Adjusted R Squared	Regression Std. Error
1 [a]	6	0.9717	0.9528	2.475

[a] Predictors: (Intercept), *Floor*, *Lot*.

The value of adjusted $R^2 = 0.9528$ is in the column headed "Adjusted R Squared." Contrast the output if we just use *Floor* as a single predictor:

Model Summary

Model	Sample Size	Multiple R Squared	Adjusted R Squared	Regression Std. Error
2 [a]	6	0.6823	0.6029	7.178

[a] Predictors: (Intercept), *Floor*.

In this case, since adjusted R^2 for this single-predictor model is 0.6029 and adjusted R^2 for the two-predictor model is 0.9528, this suggests that the two-predictor model is better than the single-predictor model (at least according to this criterion). In other words, there is no indication that adding the variable *Lot* to the model causes overfitting.

As a further example of adjusted R^2 for a multiple linear regression analysis, consider the following example, adapted from McClave et al. (2005) and based on accounting methods discussed in Horngren et al. (1994). The **SHIPDEPT** data file contains 20 weeks of a firm's accounting and production records on cost information about the firm's shipping department—see Table 3.1.

Suppose that we propose the following multiple linear regression model:

$$E(Lab) = b_0 + b_1\,Tws + b_2\,Pst + b_3\,Asw + b_4\,Num.$$

Table 3.1   Shipping data with response variable *Lab* = weekly labor hours and four potential predictor variables: *Tws* = total weight shipped in thousands of pounds, *Pst* = proportion shipped by truck, *Asw* = average shipment weight in pounds, and *Num* = week number.

Lab	Tws	Pst	Asw	Num
100	5.1	0.90	20	1
85	3.8	0.99	22	2
108	5.3	0.58	19	3
116	7.5	0.16	15	4
92	4.5	0.54	20	5
63	3.3	0.42	26	6
79	5.3	0.12	25	7
101	5.9	0.32	21	8
88	4.0	0.56	24	9
71	4.2	0.64	29	10
122	6.8	0.78	10	11
85	3.9	0.90	30	12
50	2.8	0.74	28	13
114	7.5	0.89	14	14
104	4.5	0.90	21	15
111	6.0	0.40	20	16
115	8.1	0.55	16	17
100	7.0	0.64	19	18
82	4.0	0.35	23	19
85	4.8	0.58	25	20

Here is the output produced by statistical software that displays the relevant results for this model (see computer help #31):

Model Summary

Model	Sample Size	Multiple R Squared	Adjusted R Squared	Regression Std. Error
1 [a]	20	0.8196	0.7715	9.103

[a] Predictors: (Intercept), *Tws*, *Pst*, *Asw*, *Num*.

Contrast these results with those for the following two-predictor model:

$$E(Lab) = b_0 + b_1\,Tws + b_3\,Asw.$$

Model Summary

Model	Sample Size	Multiple R Squared	Adjusted R Squared	Regression Std. Error
2 [a]	20	0.8082	0.7857	8.815

[a] Predictors: (Intercept), *Tws*, *Asw*.

Whereas R^2 decreases from 0.8196 to 0.8082 (from the four-predictor model to the two-predictor model), adjusted R^2 increases from 0.7715 to 0.7857. In other words, although the four-dimensional regression hyperplane can get a little closer to the sample data points, it appears to do so at the expense of overfitting by including apparently redundant predictor variables. The adjusted R^2 criterion suggests that the simpler two-predictor model does a better job than the four-predictor model of finding the population association between *Lab* and *(Tws,Pst,Asw,Num)*.

Since we've considered all four predictor variables in the analysis, we should mention them all in reporting the results. We found that the population association between *Lab* and *(Tws,Pst,Asw,Num)* is summarized well by a multiple linear regression model with just *Tws* and *Asw*. But, because we started by including *Pst* and *Num* in our analysis, we have effectively averaged over the sample values of *Pst* and *Num* to reach this conclusion. So, it is the population association between *Lab* and *(Tws,Pst,Asw,Num)* that we have modeled, even though *Pst* and *Num* do not appear in our two-predictor model regression equation. If we had ignored *Pst* and *Num* all along, we could say that we are modeling the population association between *Lab* and *(Tws,Asw)*, but that is not what we have done here.

Although R^2 and adjusted R^2 are related, they measure different things and both have their uses when fitting multiple linear regression models.

- R^2 has a clear interpretation since it represents the proportion of variation in Y (about its mean) explained by a multiple linear regression association between Y and $(X_1, X_2, \dots)$.

- Adjusted R^2 is useful for identifying which models in a sequence of nested models provide a good fit to sample data without overfitting. We can use adjusted R^2 to guide model building since it tends to decrease in value when extra, unimportant predictors have been added to the model. It is not a foolproof measure, however, and should be used with caution, preferably in conjunction with other model building criteria.

Another such criterion is the regression standard error, s, which in the shipping example is 9.103 for the four-predictor model but only 8.815 for the two-predictor model. This finding reinforces the conclusion that the two-predictor model may be preferable to the four-predictor model for this dataset. Since $s = \sqrt{\mathrm{RSS}/(n-k-1)}$ increases as k increases (all else being equal), we can see that this criterion also penalizes models that contain too many unimportant predictor variables.

Multiple correlation In simple linear regression, the concepts of R^2 and correlation are distinct but related. However, whereas the concept of R^2 carries over directly from simple linear regression to multiple linear regression, the concept of correlation does not. In fact, intuition about correlation can be seriously misleading when it comes to multiple linear regression.

Consider the simulated data represented by the scatterplot of *Sales* versus *Advert* in Figure 3.3, where *Sales* represents annual sales in millions of dollars for a small retail business and X_1 represents total annual spending on advertising in millions of dollars (**SALES2** data file). The correlation between *Sales* and *Advert* here is very low (in fact, it is 0.165). This means that *Advert* is unlikely to be a useful predictor of *Sales* in a *simple* linear regression model. Nevertheless, it is possible for *Advert* to be a useful predictor of *Sales* in a *multiple* linear regression model (if there are other predictors that have a particular association with *Sales* and *Advert*). For example, there is a second predictor for the dataset represented in Figure 3.3 that produces just such an outcome— in Section 3.3.5 we see exactly how this happens.

This simulated example demonstrates that low correlation between a response variable and a predictor variable does *not* imply that this predictor cannot be useful in a multiple linear regression model. Unfortunately, intuition about correlation can break down in the other direction also: High correlation between a response variable and a predictor variable

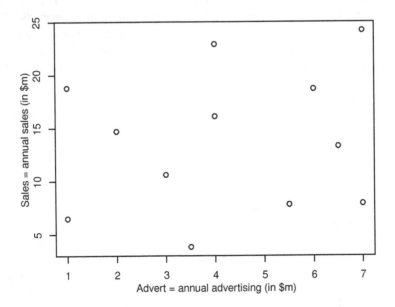

Figure 3.3 Scatterplot of simulated data with low correlation between *Sales* and *Advert*.

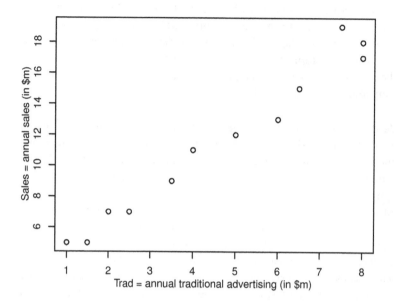

Figure 3.4   Scatterplot of simulated data with high correlation between *Sales* and *Trad*.

does *not* imply that this predictor will be useful in a multiple linear regression model. For example, consider a second simulated dataset represented by the scatterplot of *Sales* versus *Trad* in Figure 3.4, where *Sales* represents annual sales in millions of dollars for a small high-tech business and *Trad* represents annual spending on traditional advertising (TV, print media, etc.) in millions of dollars (**SALES3** data file).

The correlation between *Sales* and *Trad* here is very high (in fact, it is 0.986). This means that *Trad* is likely to be a useful predictor of *Sales* in a *simple* linear regression model. Nevertheless, it is possible for *Trad* to apparently be a poor predictor of *Sales* in a *multiple* linear regression model (if there are other predictors that have a particular association with *Sales* and *Trad*). For example, there is a second predictor for the dataset represented in Figure 3.4 that produces just such an outcome—in Section 3.3.5 we see exactly how this happens.

The only correlation coefficient that should not cause confusion when it comes to multiple linear regression is *multiple R*, or the multiple correlation coefficient. It is defined as the correlation between the observed Y-values and the fitted $\hat{Y}$-values from the model. It is related to R^2 in the following way:

$$\text{multiple R} = +\sqrt{R^2}.$$

If R^2 is high (close to 1), multiple R is also high (close to 1), and there is a strong positive linear association between the observed Y-values and the fitted $\hat{Y}$-values.

In practice, we can find the value for multiple R for any particular multiple linear regression model by calculating the positive square root of R^2. For example, recall that $R^2 = 0.9717$ for the home prices model with predictors, *Floor* and *Lot*. Since the positive square root of 0.9717 is 0.986, multiple R—or the correlation between the observed values of *Price* and the fitted values of $\widehat{Price}$ from the model—is 0.986. Since there is a direct relationship between multiple R and R^2, and, as we have seen in the two examples above, the concept of correlation can cause problems in multiple linear regression, in this book we tend to prefer the use of R^2 rather than multiple R.

3.3.3   Regression parameters—global usefulness test

Suppose that our population multiple linear regression model has k predictor X-variables:

$$E(Y) = b_0 + b_1 X_1 + b_2 X_2 + \cdots + b_k X_k,$$

which we estimate from our sample by

$$\hat{Y} = \hat{b}_0 + \hat{b}_1 X_1 + \hat{b}_2 X_2 + \cdots + \hat{b}_k X_k.$$

Before interpreting the values $\hat{b}_1$, $\hat{b}_2$, ..., $\hat{b}_k$, we would like to quantify our uncertainty about the corresponding regression parameters in the population, b_1, b_2, ..., b_k. For example, is it possible that all k regression parameters in the population could be zero? If this were the case, it would suggest that our multiple linear regression model contains very little useful information about the population association between Y and $(X_1, X_2, \ldots, X_k)$. So to potentially save a lot of wasted effort (interpreting a model that contains little useful information), we should test this assertion before we do anything else. The *global usefulness test* is a hypothesis test of this assertion.

   To see how to apply this test, we need to introduce a new probability distribution, the *F-distribution*. This probability distribution is always positive and is skewed to the right (i.e., has a peak closer to the left side than the right with a long right-hand tail that never quite reaches the horizontal axis at zero). Figure 3.5 shows a typical density curve for an F-distribution. The relative position of the peak and the thickness of the tail are controlled by two degrees of freedom values—the *numerator* degrees of freedom and the *denominator* degrees of freedom. These degrees of freedom values dictate which specific F-distribution gets used for a particular hypothesis test. Since the F-distribution is always positive, hypothesis tests that use the F-distribution are always upper-tail tests.

   The critical value, significance level, test statistic, and p-value shown in the figure feature in the following global usefulness hypothesis test.

- State null hypothesis: NH: $b_1 = b_2 = \cdots = b_k = 0$.

- State alternative hypothesis: AH: at least one of $b_1, b_2, \ldots, b_k$ is not equal to zero.

- Calculate test statistic: global F-statistic $= \dfrac{(\text{TSS} - \text{RSS})/k}{\text{RSS}/(n-k-1)} = \dfrac{R^2/k}{(1-R^2)/(n-k-1)}$.

  The first formula provides some insight into how the hypothesis test works. If the difference between TSS and RSS is small, then the predictors $(X_1, X_2, \ldots, X_k)$ are unable to reduce the random errors between the Y-values and the fitted $\hat{Y}$-values very much, and we may as well use the sample mean, m_Y, for our model. In such a case, the F-statistic will be small and will probably not be in the rejection region (so the null hypothesis is more plausible). On the other hand, if the difference between TSS and RSS is large, then the predictors $(X_1, X_2, \ldots, X_k)$ are able to reduce the random errors between the Y-values and the fitted $\hat{Y}$-values sufficiently that we should use at least one of the predictors in our model. In such a case, the F-statistic will be large and will probably be in the rejection region (so the alternative hypothesis is more plausible).

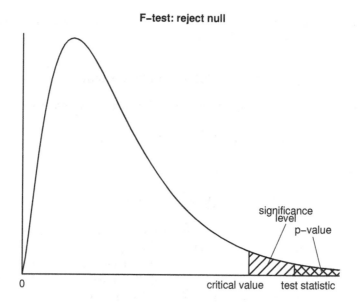

Figure 3.5   Relationships between critical values, significance levels, test statistics, and p-values for hypothesis tests based on the F-distribution. The relative positions of these quantities in this figure would lead to rejecting the null hypothesis. If the positions of the quantities were reversed, then the null hypothesis would not be rejected.

The second formula allows calculation of the global F-statistic using the value of R^2 and shows how a large value for R^2 tends to produce a high value for the global F-statistic (and vice versa).

Statistical software can provide the value of the global F-statistic, as well as the values for R^2, TSS, and RSS to allow calculation by hand.

- Set significance level (e.g., 5%).
- Look up a critical value or a p-value using an F-distribution (use computer help #8 or #9 in the software information files available from the book website):
  - critical value: A particular percentile of the F-distribution with k numerator degrees of freedom and $n-k-1$ denominator degrees of freedom; for example, the rejection region for a significance level of 5% is any global F-statistic greater than the 95th percentile.
  - p-value: The area to the right of the global F-statistic for the F-distribution with k numerator degrees of freedom and $n-k-1$ denominator degrees of freedom.
- Make decision:
  - If the global F-statistic falls in the rejection region, or the p-value is less than the significance level, then we reject the null hypothesis in favor of the alternative (Figure 3.5 provides an illustration of this situation).
  - If the global F-statistic does not fall in the rejection region, or the p-value is more than the significance level, then we cannot reject the null hypothesis in favor of the alternative (it should be clear how Figure 3.5 would need to be redrawn to correspond to this situation).

- Interpret in the context of the situation: Rejecting the null hypothesis in favor of the alternative means that at least one of $b_1, b_2, \ldots, b_k$ is not equal to zero (i.e., at least one of the predictors, $X_1, X_2, \ldots, X_k$, is linearly associated with Y); failing to reject the null hypothesis in favor of the alternative means that we cannot rule out the possibility that $b_1 = b_2 = \cdots = b_k = 0$ (i.e., it is plausible that none of the predictors, $X_1, X_2, \ldots, X_k$, are linearly associated with Y).

To conduct a global usefulness test we need to be able to look up a critical value or a p-value using the F-distribution with k numerator degrees of freedom and $n-k-1$ denominator degrees of freedom. While it is possible to consult tables similar to Table B.1 on page 291, we will find it easier from now on to use computer software to find the necessary information (see computer help #8 or #9 in the software information files available from the book website). Alternatively, statistical software provides the p-value for this test directly. For example, here is software output that displays the results of the global usefulness test for the home prices dataset (see computer help #31):

ANOVA[a]

Model		Sum of Squares	df	Mean Square	Global F-stat	Pr(>F)
1	Regression	630.259	2	315.130	51.434	0.005[b]
	Residual	18.381	3	6.127		
	Total	648.640	5			

[a] Response variable: *Price*.
[b] Predictors: (Intercept), *Floor, Lot*.

The heading "ANOVA" for this output stands for *analysis of variance* and relates to the comparison of RSS and TSS in the global F-statistic formula. Thus the global usefulness test is an example of an ANOVA test. RSS and TSS are in the column headed "Sum of Squares," with RSS in the row labeled "Residual" and TSS in the row labeled "Total." The degrees of freedom values are in the column headed "df," with numerator degrees of freedom, k, in the row labeled "Regression" and denominator degrees of freedom, $n-k-1$, in the row labeled "Residual." The global F-statistic itself is in the column labeled "Global F-stat," while the p-value for the global usefulness test is in the column labeled "Pr(>F)."

By hand, we can calculate the global F-statistic as follows:

$$\text{global F-statistic} = \frac{(\text{TSS}-\text{RSS})/k}{\text{RSS}/(n-k-1)} = \frac{(648.640-18.381)/2}{18.381/(6-2-1)}$$

$$= \frac{R^2/k}{(1-R^2)/(n-k-1)} = \frac{0.97166/2}{(1-0.97166)/(6-2-1)}$$

$$= 51.4.$$

The value of R^2 for the second formula was obtained from $R^2 = (\text{TSS}-\text{RSS})/\text{TSS}$.

Suppose that we choose a significance level of 5% for this test. Using computer help #8, the 95th percentile of the F-distribution with $k=2$ numerator degrees of freedom and $n-k-1=3$ denominator degrees of freedom is 9.55. Since the global F-statistic of 51.4 is larger than this critical value, it is in the rejection region and we reject the null hypothesis in favor of the alternative. Alternatively, using the quicker p-value method, since the p-value of 0.005 (from the statistical software output above) is less than our significance level

(0.05), we reject the null hypothesis in favor of the alternative. Thus, at least one of b_1 or b_2 is not equal to zero: that is, at least one of the predictors, (X_1, X_2), is linearly associated with Y.

As in this example, the global usefulness test *usually* (but not always) results in concluding that at least one of the predictors, $(X_1, X_2, \ldots, X_k)$, is linearly associated with the response, Y. This is reassuring since it means that we can go on to analyze and interpret the multiple linear regression model confident that we've found something of interest in the association between Y and $(X_1, X_2, \ldots, X_k)$.

We'll put the home prices dataset to one side for now, but we shall return to it in some examples and problems in Chapter 4 and again in a case study in Section 6.1. Instead, consider the shipping department example again (data file **SHIPDEPT**), and fit the following multiple linear regression model:

$$\text{E}(Lab) = b_0 + b_1\,Tws + b_2\,Pst + b_3\,Asw + b_4\,Num.$$

Here is the output produced by statistical software that displays the results of the global usefulness test for this model (see computer help #31):

ANOVA [a]

Model		Sum of Squares	df	Mean Square	Global F-stat	Pr(>F)
1	Regression	5646.052	4	1411.513	17.035	0.000[b]
	Residual	1242.898	15	82.860		
	Total	6888.950	19			

[a] Response variable: *Lab*.
[b] Predictors: (Intercept), *Tws, Pst, Asw, Num*.

Suppose that we choose significance level 5% for this test. Since the p-value of 0.000 (from the statistical software output above) is less than our significance level (0.05), we reject the null hypothesis (NH: $b_1 = b_2 = b_3 = b_4 = 0$) in favor of the alternative hypothesis (AH: at least one of b_1, b_2, b_3, or b_4 is not equal to zero). Thus, at least one of the predictors, (Tws, Pst, Asw, Num), is linearly associated with *Lab*. We could also manually check that the global F-statistic of 17.035 is larger than the 95th percentile of the F-distribution with $k = 4$ numerator degrees of freedom and $n - k - 1 = 15$ denominator degrees of freedom, but there is no need to do so (remember, there are always two ways to do a hypothesis test, and they will always give the same result if done correctly).

3.3.4 Regression parameters—nested model test

Suppose that we have fit a multiple linear regression model, and a global usefulness test has suggested that at least one of the predictors, $(X_1, X_2, \ldots, X_k)$, is linearly associated with the response, Y. From the application of adjusted R^2 to the shipping department example on page 98, we saw that it is possible that a simpler model with fewer than k predictor variables may be preferable to the full k-predictor model. This can occur when, for example, a subset of the predictor variables provides very little information about the response, Y, *beyond* the information provided by the other predictor variables.

A *nested model test* formally investigates such a possibility. Suppose that the full k-predictor model, also known as the *complete model*, has an RSS value equal to RSS_C.

Consider removing a subset of the predictor variables that we suspect provides little information about the response, Y, beyond the information provided by the other predictors. Removing this subset leads to a *reduced* model with r predictors (i.e., $k-r$ predictors are removed). Since the reduced model is nested in the complete model (i.e., it contains a subset of the complete model predictors), it will have an RSS value, say, RSS_R, that is greater than or equal to RSS_C; see page 95 for a geometrical argument for why this is so.

Intuitively, if the difference between RSS_R and RSS_C is small, then the explanatory power of the two models is similar, and we would prefer the simpler reduced model since the complete model seems to be overfitting the sample data. On the other hand, if the difference between RSS_R and RSS_C is large, we would prefer the complete model since the $k-r$ extra predictors in the complete model do appear to provide useful information about the response, Y, beyond the information provided by the r reduced model predictors.

To turn this intuition into a formal hypothesis test, we need to find a test statistic proportional to $RSS_R - RSS_C$, whose sampling distribution we know under a null hypothesis that states the reduced and complete models are equivalent in the population. The F-distribution we introduced in Section 3.3.3 serves this purpose in the following nested model test.

- Write the reduced model as $E(Y) = b_0 + b_1 X_1 + \cdots + b_r X_r$.

- Write the complete model as $E(Y) = b_0 + b_1 X_1 + \cdots + b_r X_r + b_{r+1} X_{r+1} + \cdots + b_k X_k$.

- State null hypothesis: NH: $b_{r+1} = \cdots = b_k = 0$.

- State alternative hypothesis: AH: at least one of $b_{r+1}, \ldots, b_k$ is not equal to zero.

- Calculate test statistic: nested F-statistic $= \dfrac{(RSS_R - RSS_C)/(k-r)}{RSS_C/(n-k-1)}$.

  Statistical software can provide the value of the nested F-statistic, as well as the values for RSS_R and RSS_C to allow calculation by hand.

- Set significance level (e.g., 5%).

- Look up a critical value or a p-value using an F-distribution (use computer help #8 or #9 in the software information files available from the book website):

  - critical value: A particular percentile of the F-distribution with $k-r$ numerator degrees of freedom and $n-k-1$ denominator degrees of freedom; for example, the rejection region for a significance level of 5% is any nested F-statistic greater than the 95th percentile.

  - p-value: The area to the right of the nested F-statistic for the F-distribution with $k-r$ numerator degrees of freedom and $n-k-1$ denominator degrees of freedom.

- Make decision:

  - If the nested F-statistic falls in the rejection region, or the p-value is less than the significance level, then we reject the null hypothesis in favor of the alternative (Figure 3.5 on page 102 provides an illustration of this situation).

  - If the nested F-statistic does not fall in the rejection region, or the p-value is more than the significance level, then we cannot reject the null hypothesis in favor of the alternative (it should be clear how Figure 3.5 would need to be redrawn to correspond to this situation).

- Interpret in the context of the situation: Rejecting the null hypothesis in favor of the alternative means that at least one of $b_{r+1}, \ldots, b_k$ is not equal to zero (i.e., at least one of the extra predictors in the complete model, $X_{r+1}, \ldots, X_k$, appears to provide useful information about the response, Y, beyond the information provided by the r predictor variables in the reduced model); failing to reject the null hypothesis in favor of the alternative means that we cannot rule out the possibility that $b_{r+1} = \cdots = b_k = 0$ (i.e., none of the extra predictors in the complete model, $X_{r+1}, \ldots, X_k$, appear to provide useful information about the response, Y, beyond the information provided by the r predictor variables in the reduced model).

To do a nested model test we need to look up a critical value or a p-value for the F-distribution with $k-r$ numerator degrees of freedom and $n-k-1$ denominator degrees of freedom. As with the global usefulness test, we will find it easier to use computer software to find the necessary information (see computer help #8 or #9).

Recall the shipping department example from page 97. Suppose that we propose the following (complete) multiple linear regression model:

$$E(Lab) = b_0 + b_1\,Tws + b_2\,Pst + b_3\,Asw + b_4\,Num.$$

Here is part of the output produced by statistical software that displays some results for this model (see computer help #31):

Parameters[a]

Model		Estimate	Std. Error	t-stat	Pr($>$ |t|)
C	(Intercept)	95.415	30.036	3.177	0.006
	Tws	6.074	2.662	2.281	0.038
	Pst	8.435	8.870	0.951	0.357
	Asw	−1.746	0.760	−2.297	0.036
	Num	−0.124	0.380	−0.328	0.748

[a] Response variable: Lab.

We will see in Section 3.3.5 that these results suggest that perhaps neither *Pst* nor *Num* provides useful information about the response, *Lab*, beyond the information provided by *Tws* and *Asw*. To test this formally, we do a nested model test of the following hypotheses:

- NH: $b_2 = b_4 = 0$.

- AH: at least one of b_2 or b_4 is not equal to zero.

Statistical software output for the complete model (computer help #31) is:

ANOVA[a]

Model		Sum of Squares	df	Mean Square	Global F-stat	Pr($>$F)
C	Regression	5646.052	4	1411.513	17.035	0.000[b]
	Residual	1242.898	15	82.860		
	Total	6888.950	19			

[a] Response variable: Lab.
[b] Predictors: (Intercept), Tws, Pst, Asw, Num.

RSS_C is in the "Sum of Squares" column, while $n-k-1$ is in the "df" column, both in the row labeled "Residual," while k is in the "df" column in the row labeled "Regression." Contrast these results with those for the reduced two-predictor model:

$$E(Lab) = b_0 + b_1\,Tws + b_3\,Asw.$$

ANOVA[a]

Model		Sum of Squares	df	Mean Square	Global F-stat	Pr(>F)
R	Regression	5567.889	2	2783.945	35.825	0.000[b]
	Residual	1321.061	17	77.709		
	Total	6888.950	19			

[a] Response variable: *Lab*.

[b] Predictors: (Intercept), *Tws*, *Asw*.

RSS_R is in the "Sum of Squares" column in the row labeled "Residual," while r is in the "df" column in the row labeled "Regression." By hand, we can calculate the nested F-statistic as follows:

$$\text{nested F-statistic} = \frac{(RSS_R - RSS_C)/(k-r)}{RSS_C/(n-k-1)} = \frac{(1321.061 - 1242.898)/(4-2)}{1242.898/(20-4-1)}$$
$$= 0.472.$$

Suppose that we choose a significance level of 5% for this test. Using computer help #8, the 95th percentile of the F-distribution with $k-r=2$ numerator degrees of freedom and $n-k-1=15$ denominator degrees of freedom is 3.68. Since the nested F-statistic of 0.472 is smaller than this critical value (so it is not in the rejection region), we cannot reject the null hypothesis in favor of the alternative hypothesis. Thus, it is plausible that both b_2 and b_4 are equal to zero in the population; that is, neither *Pst* nor *Num* appears to provide useful information about the response, *Lab*, beyond the information provided by *Tws* and *Asw*.

Intuitively, whereas the four-dimensional regression hyperplane can get a little closer to the sample data points, it appears to do so at the expense of overfitting by including apparently redundant predictor variables. The nested model test suggests that the reduced two-predictor model does a better job of finding the population association between *Lab* and (Tws, Pst, Asw, Asw) than the complete four-predictor model.

Alternatively, statistical software provides the p-value for this test directly. For example, here is the output that displays the results of the nested model test above for the shipping department dataset (see computer help #34):

Model Summary

Model	R Squared	Adjusted R Squared	Regression Std. Error	Change Statistics			
				F-stat	df1	df2	Pr(>F)
R[a]	0.8082	0.7857	8.815				
C[b]	0.8196	0.7715	9.103	0.472	2	15	0.633

[a] Predictors: (Intercept), *Tws*, *Asw*.

[b] Predictors: (Intercept), *Tws*, *Pst*, *Asw*, *Num*.

The nested F-statistic is in the second row of the column headed "F-stat." The associated p-value is in the second row of the column headed "Pr(>F)." (Ignore any numbers in the first rows of these columns.) Since the p-value of 0.633 is more than our significance level (0.05), we cannot reject the null hypothesis in favor of the alternative hypothesis—the same conclusion (necessarily) that we made with the rejection region method above.

The complete and reduced models considered in this section are "nested" in the sense that the complete model includes all of the predictors in the reduced model as well as some additional predictors unique to the complete model. Equivalently, the reduced model is similar to the complete model except that the regression parameters for these additional predictors are all zero in the reduced model. More generally, one model is nested in another if they each contain the same predictors but the first model constrains some of its regression parameters to be fixed numbers (zero in the examples above). The nested model test then determines whether the second model provides a significant improvement over the first (small p-value); if not, then the constrained values of the regression parameters are plausible (large p-value).

With nested models like this, the residual sum of squares for the complete model is always lower than (or the same as) the residual sum of squares for the reduced model (see page 95). Thus, R^2 for the complete model is always higher than (or the same as) R^2 for the reduced model. However, we can think of the nested model test as telling us whether the complete model R^2 is *significantly* higher—if not, we prefer the reduced model. Another way to see which model is favored is to consider the regression standard error, s (we would generally prefer the model with the smaller value of s), and adjusted R^2 (we would generally prefer the model with the larger value of adjusted R^2). To summarize:

- R^2 is always higher for the complete model than for the reduced model (so this tells us nothing about which model we prefer).

- The regression standard error, s, may be higher or lower in the reduced model than the complete model, but if it is lower, then we would prefer the reduced model according to this criterion.

- Adjusted R^2 may be higher or lower in the reduced model than in the complete model, but if it is higher, then we would prefer the reduced model according to this criterion.

In many cases, these three methods for comparing nested models—nested model test, comparing regression standard errors, and comparing values of adjusted R^2—will agree, but it is possible for them to conflict. In conflicting cases, a reasonable conclusion might be that the two models are essentially equivalent. Alternatively, other model comparison tools are available—see Section 5.4.

It is easy to confuse the global usefulness and nested model tests since both involve F-statistics and both are examples of ANOVA tests. In summary, the *global usefulness test* considers whether all the regression parameters are equal to zero. In other words, if the test ends up rejecting the null hypothesis, then we conclude that at least one of the predictor variables has a linear association with the response variable. This test is typically used just once at the beginning of a regression analysis to make sure that there will be something worth modeling (if none of the predictor variables has a linear association with the response variable, then a multiple linear regression model using those predictors is probably not appropriate). By contrast, the *nested model test* considers whether a subset of the regression parameters are equal to zero. In other words, if the test ends up failing to reject the null hypothesis, then we conclude that the corresponding subset of predictor variables has no linear association with the response variable once the association with the remaining predictors left in the model has been accounted for. This generally means that we can then drop the subset of predictor variables being tested. This test is typically used one or more times during a regression analysis to discover which, if any, predictor variables are redundant given the presence of the other predictor variables and so are better left out of the model.

However, the global usefulness and nested model tests are related to one another. The formulas for the two F-statistics (on pages 101 and 105) show that the global usefulness F-statistic is a special case of the nested model F-statistic, in which the subset of predictors being tested consists of *all* the predictors and the reduced model has no predictor variables at all (i.e., RSS for the reduced model is the same as TSS and $r = 0$).

It is also possible to use the nested model F-test to test whether a subset of regression parameters in a multiple linear regression model could be equal to one another—see the optional section at the end of Section 4.3.2 on page 182 and also the case study in Section 6.2.

3.3.5 Regression parameters—individual tests

Suppose that we have fit a multiple linear regression model, and a global usefulness test has suggested that at least one of the predictors, $(X_1, X_2, \ldots, X_k)$, has a linear association with the response, Y. We have seen in Sections 3.3.2 and 3.3.4 that it is possible that a reduced model with fewer than k predictor variables may be preferable to the complete k-predictor model. This can occur when, for example, a subset of the predictor variables provides very little information about the response, Y, *beyond* the information provided by the other predictor variables. We have seen how to use a nested model test to remove a subset of predictors from the complete model, but how do we identify which predictors should be in this subset? One possible approach is to consider the regression parameters individually. In particular, what do the estimated sample estimates, $\hat{b}_1, \hat{b}_2, \ldots, \hat{b}_k$, tell us about likely values for the population parameters, $b_1, b_2, \ldots, b_k$?

Since we assume that the sample has been randomly selected from the population, under repeated sampling we would expect the sample estimates and population parameters to match *on average*, but for any particular sample they will probably differ. We cannot be sure how much they will differ, but we can quantify this uncertainty using the sampling distribution of the estimated regression parameters, $\hat{b}_1, \hat{b}_2, \ldots, \hat{b}_k$. Recall that when analyzing simple linear regression models we calculated a test statistic:

$$\text{slope t-statistic} = \frac{\hat{b}_1 - b_1}{s_{\hat{b}_1}}.$$

Under very general conditions, this slope t-statistic has an approximate t-distribution with $n - 2$ degrees of freedom. We used this result to conduct hypothesis tests and construct confidence intervals for the population slope, b_1.

We can use a similar result to conduct hypothesis tests and construct confidence intervals for each of the population regression parameters, $b_1, b_2, \ldots, b_k$, in multiple linear regression. In particular, we can calculate the following test statistic for the pth regression parameter, b_p:

$$\text{regression parameter t-statistic} = \frac{\hat{b}_p - b_p}{s_{\hat{b}_p}},$$

where $s_{\hat{b}_p}$ is the standard error of the regression parameter estimate. The formula for this standard error is provided for interest at the end of this section. Under very general conditions, this regression parameter t-statistic has an approximate t-distribution with $n - k - 1$ degrees of freedom.

Regression parameter hypothesis tests Suppose that we are interested in a particular value of the pth regression parameter for a multiple linear regression model. Usually, "zero" is an interesting value for the regression parameter since this would be equivalent to there being no linear association between Y and X_p once the linear association between Y and the other $k-1$ predictors has been accounted for. Another way of saying this is that there is no linear association between Y and X_p when we hold the other $k-1$ predictors fixed at constant values. One way to test this could be to literally keep the other $k-1$ predictors fixed at constant values and vary X_p to see if Y changes. This is not usually possible with observational data (see page 36) and even with experimental data would be very time consuming and expensive to do for each predictor in turn. Alternatively, we can easily do hypothesis tests to see if the information in our sample supports population regression parameters of zero or whether it favors some alternative values.

For the shipping data example, *before* looking at the sample data we might have reason to believe that there is a linear association between weekly labor hours, *Lab*, and the total weight shipped in thousands of pounds, *Tws*, once the linear association between *Lab* and *Pst* (proportion shipped by truck), *Asw* (average shipment weight), and *Num* (week number) has been accounted for (or holding *Pst*, *Asw*, and *Num* constant). To see whether the sample data provide compelling evidence that this is the case, we should conduct a two-tail hypothesis test for the population regression parameter b_1 in the model $E(Lab) = b_0 + b_1 Tws + b_2 Pst + b_3 Asw + b_4 Asw$.

- State null hypothesis: NH: $b_1 = 0$.
- State alternative hypothesis: AH: $b_1 \neq 0$.
- Calculate test statistic: t-statistic $= \frac{\hat{b}_1 - b_1}{s_{\hat{b}_1}} = \frac{6.074 - 0}{2.662} = 2.28$ ($\hat{b}_1$ and $s_{\hat{b}_1}$ can be obtained using statistical software—see output below—while b_1 is the value in NH).
- Set significance level: 5%.
- Look up t-table:
    - critical value: The 97.5th percentile of the t-distribution with $20-4-1=15$ degrees of freedom is 2.13 (see computer help #8 in the software information files available from the book website); the rejection region is therefore any t-statistic greater than 2.13 or less than -2.13 (we need the 97.5th percentile in this case because this is a two-tail test, so we need half the significance level in each tail).
    - p-value: The sum of the areas to the right of the t-statistic (2.28) and to the left of the negative of the t-statistic (-2.28) for the t-distribution with 15 degrees of freedom is 0.038 (use computer help #9).
- Make decision:
    - Since the t-statistic of 2.28 falls in the rejection region, we reject the null hypothesis in favor of the alternative.
    - Since the p-value of 0.038 is less than the significance level of 0.05, we reject the null hypothesis in favor of the alternative.
- Interpret in the context of the situation: The 20 sample observations suggest that a population regression parameter, b_1, of zero seems implausible and the sample data favor a nonzero value (at a significance level of 5%); in other words, there does appear to be a linear association between *Lab* and *Tws* once *Pst*, *Asw*, and *Num* have been accounted for (or holding *Pst*, *Asw*, and *Num* constant).

Hypothesis tests for the other regression parameters, b_2, b_3, and b_4, are similar. Sometimes, we may have a particular interest in doing an upper- or lower-tail test rather than a two-tail test. The test statistic is the same value for all three flavors of test, but the significance level represents an area in just one tail rather than getting split evenly between both tails (this affects where the critical values for the rejection regions are), and the p-value also represents an area in just one tail rather than getting split evenly between both tails.

In practice, we can do population regression parameter hypothesis tests in multiple linear regression directly using statistical software. For example, here is the relevant output for the shipping dataset (see computer help #31):

Parameters[a]

Model		Estimate	Std. Error	t-stat	$\Pr(>|t|)$
1	(Intercept)	95.415	30.036	3.177	0.006
	Tws	6.074	2.662	2.281	0.038
	Pst	8.435	8.870	0.951	0.357
	Asw	−1.746	0.760	−2.297	0.036
	Num	−0.124	0.380	−0.328	0.748

[a] Response variable: *Lab*.

The regression parameter estimates $\hat{b}_p$ are in the column headed "Estimate" and the row labeled with the name of the predictor. The standard errors of the estimates, $s_{\hat{b}_p}$, are in the column headed "Std. Error," while the t-statistics are in the column headed "t-stat," and the two-tail p-values are in the column headed "$\Pr(>|t|)$" (meaning "the probability that a t random variable with $n-k-1$ degrees of freedom could be larger than the absolute value of the t-statistic or smaller than the negative of the absolute value of the t-statistic"). In general:

- To carry out a two-tail hypothesis test for a zero value for the pth population regression parameter in multiple linear regression, decide on the significance level (e.g., 5%), and check to see whether the two-tail p-value ("$\Pr(>|t|)$" in the statistical software output) is smaller than this significance level. If it is, reject NH: $b_p = 0$ in favor of AH: $b_p \neq 0$ and conclude that the sample data favor a nonzero regression parameter (at the chosen significance level). Otherwise, there is insufficient evidence to reject NH: $b_p = 0$ in favor of AH: $b_p \neq 0$, and we conclude that a zero population parameter cannot be ruled out (at the chosen significance level).

- For an upper-tail hypothesis test, set the significance level (e.g., 5%) and calculate the upper-tail p-value. The upper-tail p-value is the area to the right of the t-statistic under the appropriate t-distribution density curve. For a positive t-statistic, this area is equal to the two-tail p-value divided by 2. Then, if the upper-tail p-value is smaller than the chosen significance level, reject NH: $b_p = 0$ in favor of AH: $b_p > 0$. Otherwise, there is insufficient evidence to reject NH: $b_p = 0$ in favor of AH: $b_p > 0$.

- For a lower-tail hypothesis test, set the significance level (e.g., 5%) and calculate the lower-tail p-value. The lower-tail p-value is the area to the left of the t-statistic under the appropriate t-distribution density curve. For a negative t-statistic, this area is equal to the two-tail p-value divided by 2. Then, if the lower-tail p-value is smaller than the chosen significance level, reject NH: $b_p = 0$ in favor of AH: $b_p < 0$. Otherwise, there is insufficient evidence to reject NH: $b_p = 0$ in favor of AH: $b_p < 0$.

However, be careful when an upper-tail hypothesis test has a negative t-statistic or a lower-tail test has a positive t-statistic. In such situations the p-value must be at least 0.5 (draw a picture to convince yourself of this), so it is also going to be larger than any reasonable significance level that we might have picked. Thus, we won't be able to reject NH: $b_p = 0$ in favor of AH: $b_1 > 0$ (for an upper-tail test) or AH: $b_1 < 0$ (for a lower-tail test).

For the shipping example, since the two-tail p-value for *Tws* is 0.038, we reject NH: $b_1 = 0$ in favor of AH: $b_1 \neq 0$ and conclude that the sample data favor a nonzero regression parameter (at a significance level of 5%). For an upper-tail test, since the t-statistic is positive, the upper-tail p-value is 0.019 and we reject NH: $b_1 = 0$ in favor of AH: $b_1 > 0$ and conclude that the sample data favor a positive regression parameter (at a significance level of 5%). For a lower-tail test, the positive t-statistic means that the lower-tail p-value is at least 0.5, so we cannot reject NH: $b_1 = 0$ in favor of AH: $b_1 < 0$.

We present all three flavors of hypothesis test here (two-tail, upper-tail, and lower-tail), but in real-life applications we would usually conduct only one—selected before looking at the data. Remember also that each time we do a hypothesis test, we have a chance of making a mistake (either rejecting NH when we should not have, or failing to reject NH when we should have)—see page 24. Thus, when trying to decide which predictors should remain in a particular multiple linear regression model, we should use as few hypothesis tests as possible. One potential strategy is to identify a subset of predictors with relatively high two-tail p-values, and then use the nested model test of Section 3.3.4 to formally decide whether this subset of predictors provides information about the response, *Y*, beyond the information provided by the other predictor variables. For example, for the shipping data, the relatively high p-values for *Pst* (0.357) and *Num* (0.748) on page 111 suggest that we should do the nested model test we conducted on page 107.

Also keep in mind that the p-value of 0.357 for *Pst* in this example suggests that there is no linear association between *Lab* and *Pst* once the linear association between *Lab* and *Tws*, *Asw*, and *Num* has been accounted for (or holding *Tws*, *Asw*, and *Num* constant). In other words, *Pst* may be redundant in the model as long as *Tws*, *Asw*, and *Num* remain in the model. Similarly, the p-value of 0.748 for *Num* suggests that there is no linear association between *Lab* and *Num* once *Tws*, *Pst*, and *Asw* have been accounted for (or holding *Tws*, *Pst*, and *Asw* constant). In other words, *Num* may be redundant in the model as long as *Tws*, *Pst*, and *Asw* remain in the model.

This is *not* quite the same as the conclusion for the nested model test, which was that there does not appear to be a linear association between *Lab* and (*Pst*, *Num*) once *Tws* and *Asw* have been accounted for (or holding *Tws* and *Asw* constant). In other words, *Pst* and *Num* may be redundant in the model as long as *Tws* and *Asw* remain in the model.

Thus, we can do individual regression parameter t-tests to remove just one redundant predictor at a time or to identify which predictors to investigate with a nested model F-test. However, we need the nested model test to actually remove more than one redundant predictor at a time. Using nested model tests allows us to use fewer hypothesis tests overall to help identify redundant predictors (so that the remaining predictors appear to explain the response variable adequately); this also lessens the chance of making any hypothesis test errors.

It is also possible to conduct a hypothesis test for the intercept parameter, b_0; the procedure is exactly the same as for the regression parameters, b_1, b_2, ..., b_k. For example, if the intercept p-value is greater than a significance level of 5%, then we cannot reject the null hypothesis that the intercept (b_0) is zero (at that significance level). In other words, a zero intercept is quite plausible. Whether this makes sense depends on the practical context. For example, in some physical systems we might know that when the predictors are zero, the response must necessarily be zero. In contexts such as this it might make sense to drop the intercept from the model and fit what is known as *regression through the origin*. In most practical applications, however, testing a zero intercept (and dropping the intercept from the model if the p-value is quite high) is relatively rare (it generally occurs only in contexts where an intercept exactly equal to zero is expected). Thus we won't dwell any further on this.

As with simple linear regression, don't confuse being unable to meaningfully interpret the estimated intercept parameter, $\hat{b}_0$, with dropping the intercept from the model—see page 58.

In principle, we can also test values other than zero for the population regression parameters, b_1, b_2, ..., b_k—just plug the appropriate value for b_p into the t-statistic formula on page 109 and proceed as usual. This is quite rare in practice since testing whether the population regression parameters could be zero is usually of most interest. We can also test values other than zero for b_0—again, this is quite rare in practice.

It turns out that the individual regression parameter t-test considered in this section is related to the nested model F-test considered in the preceding section. It can be shown that if we square a *t*-distributed random variable with d degrees of freedom, then we obtain an *F*-distributed random variable with 1 numerator degree of freedom and d denominator degrees of freedom. This relationship leads to the following result. Although the nested model F-test is usually used to test more than one regression parameter, *if* we use it to test just a single parameter, then the F-statistic is equal to the square of the corresponding individual regression parameter t-statistic. Also, the F-test p-value and two-tail t-test p-value are identical. Try this for the shipping example [you should find that the nested model F-statistic for b_1 in the model $E(Lab) = b_0 + b_1\,Tws + b_2\,Pst + b_3\,Asw + b_4\,Asw$ is 5.204, which is the square of the individual regression parameter t-statistic, 2.281; the p-values for both tests are 0.038].

Regression parameter confidence intervals   Another way to express our level of uncertainty about the population regression parameters, b_1, b_2, ..., b_k, is with confidence intervals. For example, a 95% confidence interval for b_p results from the following:

$$\Pr\left(-97.5\text{th percentile} < t_{n-k-1} < 97.5\text{th percentile}\right) = 0.95$$

$$\Pr\left(-97.5\text{th percentile} < (\hat{b}_p - b_p)/s_{\hat{b}_p} < 97.5\text{th percentile}\right) = 0.95$$

$$\Pr\left(\hat{b}_p - 97.5\text{th percentile}\,(s_{\hat{b}_p}) < b_p < \hat{b}_p + 97.5\text{th percentile}\,(s_{\hat{b}_p})\right) = 0.95,$$

where the 97.5th percentile is from the t-distribution with $n-k-1$ degrees of freedom. In other words, the 95% confidence interval is $\hat{b}_p \pm 97.5$th percentile $(s_{\hat{b}_p})$. For example, the 95% confidence interval for b_1 in the four-predictor shipping model is

$$\hat{b}_1 \pm 97.5\text{th percentile}\,(s_{\hat{b}_1}) = 6.074 \pm 2.131 \times 2.662$$

$$= 6.074 \pm 5.673$$

$$= (0.40, 11.75).$$

The values for $\hat{b}_1 = 6.074$ and $s_{\hat{b}_1} = 2.662$ come from the statistical software output on page 111, while the 97.5th percentile from the t-distribution with $n-k-1 = 15$ degrees of freedom is obtained using computer help #8 in the software information files available from the book website. We can calculate the confidence intervals for b_2, b_3, and b_4 similarly. Here is statistical software output displaying 95% confidence intervals for the regression parameters, $b_1, \ldots, b_4$, in the four-predictor shipping model (see computer help #27):

Model		Parameters[a]				95% Confidence Interval	
		Estimate	Std. Error	t-stat	$Pr(>|t|)$	Lower Bound	Upper Bound
1	(Intercept)	95.415	30.036	3.177	0.006		
	Tws	6.074	2.662	2.281	0.038	0.399	11.749
	Pst	8.435	8.870	0.951	0.357	−10.472	27.341
	Asw	−1.746	0.760	−2.297	0.036	−3.366	−0.126
	Num	−0.124	0.380	−0.328	0.748	−0.934	0.685

[a] Response variable: Lab.

The confidence intervals are in the columns headed "95% Confidence Interval" and the row labeled with the name of the predictor.

Now that we've calculated some confidence intervals, what exactly do they tell us? Well, for this example, *loosely speaking*, we can say that "we're 95% confident that the population regression parameter, b_1, is between 0.40 and 11.75 in the model $E(Lab) = b_0 + b_1 Tws + b_2 Pst + b_3 Asw + b_4 Asw$." In other words, "we're 95% confident that labor hours increases by between 0.40 and 11.75 for each 1,000-pound increase in *Tws* (weight shipped) once *Pst* (proportion shipped by truck), *Asw* (average shipment weight), and *Num* (week number) have been accounted for (or holding *Pst*, *Asw*, and *Num* constant)." A more precise interpretation would be something like "if we were to take a large number of random samples of size 20 from our population of shipping numbers and calculate a 95% confidence interval for b_1 in each, then 95% of those confidence intervals would contain the true (unknown) population regression parameter." We can provide similar interpretations for the confidence intervals for b_2, b_3, and b_4. As further practice, find 95% intervals for the regression parameters, b_1 and b_3, in the two-predictor shipping model—you should find that they are narrower (more precise) than in the four-predictor model (which contains two unimportant predictors).

For confidence levels other than 95%, percentiles in the calculations must be changed as appropriate. For example, 90% intervals (5% in each tail) need the 95th percentile, whereas 99% intervals (0.5% in each tail) need the 99.5th percentile. These percentiles can be obtained using computer help #8. As further practice, calculate 90% intervals for b_1 for the four- and two-predictor shipping models (see Problem 3.2 at the end of this chapter)—you should find that they are (1.41, 10.74) and (1.07, 8.94), respectively. Some statistical software automatically calculates 95% intervals only for this type of confidence interval.

Correlation revisited   Recall from page 99 the warning about using the concept of correlation in multiple linear regression. We return to the two earlier simulated examples to see how we can be led astray if we're not careful. First, consider the simulated data represented by the scatterplot in Figure 3.3 on page 99, where *Sales* is annual sales in $ millions for a small retail business, and *Advert* is total annual spending on advertising in $ millions (**SALES2** data file). The correlation between *Sales* and *Advert* here is very low (in fact, it is 0.165), but there is a second predictor, *Stores* (the number of retail stores

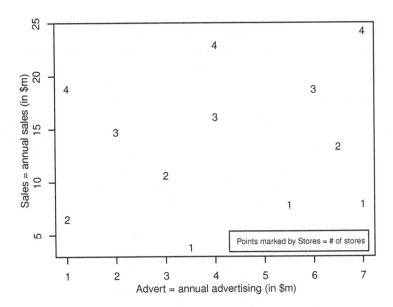

Figure 3.6 Scatterplot of simulated data with low correlation between *Sales* and *Advert*, but a strong positive linear association between *Sales* and *Advert* when *Stores* is fixed at a constant value.

operated by the company), which enables *Advert* to be a useful predictor in the multiple linear regression model:

$$E(Sales) = b_0 + b_1 Advert + b_2 Stores.$$

Statistical software output for this model (see computer help #31 in the software information files available from the book website) is:

Parameters[a]

Model		Estimate	Std. Error	t-stat	Pr(> |t|)
1	(Intercept)	−4.769	0.820	−5.818	0.000
	Advert	1.053	0.114	9.221	0.000
	Stores	5.645	0.215	26.242	0.000

[a] Response variable: *Sales*.

Since the two-tail p-value for *Advert* is 0.000, we know that there is a strong linear association between *Sales* and *Advert*, holding *Stores* constant. Similarly, since the two-tail p-value for *Stores* is 0.000, we know that there is a strong linear association between *Sales* and *Stores*, holding *Advert* constant. Thus, the low correlation between *Sales* and *Advert* is irrelevant to the outcome of the multiple linear regression model. Figure 3.6 shows why this is the case (see computer help #17). When *Stores* is held constant at 1, the points represented by 1's have an increasing trend. When *Stores* is held constant at 2, the points represented by 2's have a similarly increasing trend. It is a similar story when *Stores* is held constant at 3 or 4. The estimated regression parameter, $\hat{b}_1 = 1.0530$, represents the common slope of all these associations.

This simulated example demonstrates that low correlation between a response variable and a predictor variable does *not* imply that this predictor variable cannot be useful in a multiple linear regression model. Unfortunately, intuition about correlation can break down in the other direction also: High correlation between a response variable and a predictor variable does *not* imply that this predictor variable will necessarily be useful in a multiple linear regression model. For example, consider the simulated dataset represented by the scatterplot of *Sales* versus *Trad* in Figure 3.4 on page 100, where *Sales* represents annual sales in millions of dollars for a small high-tech business and *Trad* represents annual spending on traditional advertising (TV, print media, etc.) in $ millions (**SALES3** data file).

The correlation between *Sales* and *Trad* here is very high (in fact, it is 0.986), but there is a second predictor, *Int* (annual spending on Internet advertising in $ millions), which results in *Trad* apparently being a poor predictor of *Sales* in the multiple linear regression model:

$$E(Sales) = b_0 + b_1\,Trad + b_2\,Int.$$

Statistical software output for this model (see computer help #31) is:

Parameters[a]

Model		Estimate	Std. Error	t-stat	$\Pr(>|t|)$
1	(Intercept)	1.992	0.902	2.210	0.054
	Trad	1.275	0.737	1.730	0.118
	Int	0.767	0.868	0.884	0.400

[a] Response variable: *Sales*.

The relatively large two-tail p-value for *Trad* (0.118) means that any linear association between *Sales* and *Trad*, holding *Int* constant, is weak. Similarly, the relatively large two-tail p-value for *Int* (0.400) means that any linear association between *Sales* and *Int*, holding *Trad* constant, is weak. Thus, the high correlation between *Sales* and *Trad* is irrelevant to the outcome of the multiple linear regression model. Figure 3.7 shows why this is the case (see computer help #16). Since *Trad* and *Int* are highly correlated, when *Int* is held constant there is little variation in *Trad* and hence relatively little variation possible in *Sales*. Thus, there is only a weak linear association between *Sales* and *Trad*, holding *Int* constant. Similarly, with *Trad* constant, there is little variation in *Int* or *Sales*. Thus, there is only a weak linear association between *Sales* and *Int*, holding *Trad* constant. This problem is known as *multicollinearity*, which we shall return to in Section 5.2.3. One way to address this problem is to drop one of the highly correlated predictors, for example, *Int* (with the larger p-value) in this case. Now, the significant linear association between *Sales* and *Trad* reveals itself:

Parameters[a]

Model		Estimate	Std. Error	t-stat	$\Pr(>|t|)$
2	(Intercept)	2.624	0.542	4.841	0.001
	Trad	1.919	0.104	18.540	0.000

[a] Response variable: *Sales*.

Another way to consider the role of an individual predictor, X_1 say, in a multiple linear regression model with response variable Y is to adjust for all the other predictor variables

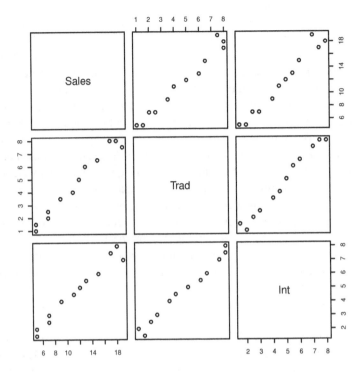

Figure 3.7   Scatterplot matrix for simulated data with high correlation between *Sales* and *Trad*, but also high correlation between *Sales* and *Int* and between *Trad* and *Int*.

as follows. First regress Y on all the other predictor variables except X_1 and calculate the residuals from that model. Then regress X_1 on all the other predictor variables except X_1 and calculate the residuals from that model. The correlation between these two sets of residuals is called the *partial correlation* between Y and X_1. For example, the partial correlation between *Sales* and *Advert* for the **SALES2** data is 0.951 (recall that the ordinary correlation was just 0.165). Conversely, the partial correlation between *Sales* and *Trad* for the **SALES3** data is just 0.500 (recall that the ordinary correlation was 0.986). To illustrate partial correlation, we could construct a scatterplot of the two sets of residuals, known as an *added variable plot* or *partial regression plot*. For more details, see Weisberg (2005) and Cook (1996).

Predictor selection   We can use a global usefulness test to determine whether any of the potential predictors in a dataset are useful for modeling the response variable. Assuming that this is the case, we can then use nested model F-tests and individual regression parameter t-tests to identify the most important predictors. We should employ these tests judiciously to avoid conducting too many tests and reduce our chance of making a mistake (by excluding important predictors or failing to exclude unimportant ones). If possible, identification of the important predictors should be guided not just by the results of statistical tests, but also by practical considerations and background knowledge about the application.

For some applications, the number of predictors, k, is very large, so determining which are the most important can be a challenging problem. Statistical software provides some automated methods for predictor selection in such cases. Examples include *forward selection* (predictors are added sequentially to an initial zero-predictor model in order of their individual significance), *backward elimination* (predictors are excluded sequentially from the full k-predictor model in order of their individual significance), and a combined *stepwise* method (which can proceed forwards or backwards at each stage). The "final" model selected depends on the particular method used and the model evaluation criterion used at each step. There are also alternative computer-intensive approaches to predictor selection that have been developed in the *machine learning* and *data mining* fields.

While automated predictor selection methods can be quick and easy to use, in applications with a manageable number of potential predictors (say, less than 10), manual selection of the important ones through practical considerations, background knowledge, and judiciously chosen hypothesis tests should usually lead to good results. In Section 5.3 we provide practical guidelines for implementing this approach. In larger applications with tens (or even hundreds) or potential predictors, automated methods can be useful for making an initial pass through the data to identify a smaller, more manageable set of potentially useful predictors. This smaller set can then be evaluated more carefully in the usual way.

We consider some of these ideas about predictor selection further in Section 5.4.

Optional—formula for regression parameter standard errors. Define the predictor matrix $\mathbf{X}$ as on page 92. Then calculate the matrix $s^2(\mathbf{X}^T\mathbf{X})^{-1}$, where s is the regression standard error. The square roots of the $k+1$ diagonal entries of this matrix are the regression parameter standard errors, $s_{\hat{b}_0}, s_{\hat{b}_1}, s_{\hat{b}_2}, \ldots, s_{\hat{b}_k}$.

3.4 MODEL ASSUMPTIONS

The multiple linear regression model relies on a number of assumptions being satisfied in order for it to provide a reliable approximation to the true association between a response variable, Y, and predictor variables, $(X_1, X_2, \ldots, X_k)$. These assumptions describe the probability distributions of the random errors in the model:

$$\text{random error} = e = Y - \mathrm{E}(Y) = Y - b_0 - b_1 X_1 - \cdots - b_k X_k.$$

In particular, there are four assumptions about these random errors, e:

- The probability distribution of e at each set of values $(X_1, X_2, \ldots, X_k)$ has a **mean of zero** (in other words, the data points are balanced on both sides of the regression "hyperplane" so that the random errors average out to zero at each set of X-values).

- The probability distribution of e at each set of values $(X_1, X_2, \ldots, X_k)$ has **constant variance**, sometimes called *homoscedasticity* (in other words, the data points spread out evenly around the regression hyperplane so that the (vertical) variation of the random errors remains similar at each set of X-values).

- The probability distribution of e at each set of values $(X_1, X_2, \ldots, X_k)$ is **normal** (in other words, the data points are more likely to be closer to the regression hyperplane than farther away and have a gradually decreasing chance of being farther away).

- The value of e for one observation is **independent** of the value of e for any other observation (in other words, knowing the value of one random error gives us no information about the value of another).

Figure 2.13 on page 60 illustrates these assumptions for simple linear regression. It is difficult to illustrate them for multiple regression, but we can check them in a similar way.

3.4.1 Checking the model assumptions

The model assumptions relate to the random errors in the population, so we are left with the usual statistical problem. Is there information in the sample data that we can use to ascertain what is likely to be going on in the population? One way to address this is to consider the estimated random errors from the multiple linear regression model fit to the sample data. We can calculate these estimated errors or *residuals* as:

$$\text{residual} = \hat{e} = Y - \hat{Y} = Y - \hat{b}_0 - \hat{b}_1 X_1 - \cdots - \hat{b}_k X_k.$$

These numbers represent the distances between sample Y-values and fitted $\hat{Y}$-values perpendicular to the corresponding regression hyperplane. We can construct *residual plots*, which are scatterplots with $\hat{e}$ along the vertical axis and a function of $(X_1, X_2, \ldots, X_k)$ along the horizontal axis. Examples of functions of $(X_1, X_2, \ldots, X_k)$ to put on the horizontal axis include:

- the fitted $\hat{Y}$-values, that is, $\hat{b}_0 + \hat{b}_1 X_1 + \cdots + \hat{b}_k X_k$;
- each predictor variable in the model;
- potential predictor variables that have not been included in the model;
- a variable representing the order in which data values were observed if the data were collected over time (see also Section 5.2.2 on "autocorrelation").

We can construct residual plots for each of these horizontal axis quantities—the more horizontal axis quantities we can assess, the more confidence we can have about whether the model assumptions have been satisfied. For example, suppose that we fit a two-predictor model in a dataset with three predictors. Then, we should construct four residual plots with different quantities on the horizontal axis: one with the fitted $\hat{Y}$-values, two with the predictors in the model, and one with the predictor that isn't in the model. The reason for constructing a residual plot for the predictor that isn't in the model is that such a plot can sometimes help determine whether that predictor really ought to be included in the model. We illustrate this in the **MLRA** example on page 120. We can then assess each plot by eye to see whether it is plausible that the four model assumptions described on page 118 could hold *in the population*. Since this can be somewhat subjective, to help build intuition refer back to Figure 2.14 on page 62, which displays residual plots generated from simulated populations in which the four model assumptions hold. By contrast, Figure 2.15 on page 63 displays residual plots in which the four model assumptions fail. We can use each residual plot to assess the assumptions as follows:

- To assess the **zero mean** assumption, visually divide each residual plot into five or six vertical slices and consider the approximate average value of the residuals in each slice. The five or six within-slice averages should each be "close" to zero (the horizontal lines in Figures 2.14 and 2.15). We should seriously question the zero mean assumption only if some of the within-slice averages are clearly different from zero.

- To assess the **constant variance** assumption, again visually divide each residual plot into five or six vertical slices, but this time consider the spread of the residuals in each slice. The variation should be approximately the same within each of the five or six slices. We should seriously question the constant variance assumption only if there are clear changes in variation between some of the slices.

- The **normality** assumption is quite difficult to check with the "slicing" technique since there are usually too few residuals within each slice to assess normality for each. Instead, we can use histograms and QQ-plots as described below.

- To assess the **independence** assumption, take one final quick look at each residual plot. If any nonrandom patterns jump out at you, then the independence assumption may be in doubt. Otherwise, the independence assumption is probably satisfied.

While residual plots work well for assessing the zero mean, constant variance, and independence assumptions, histograms and QQ-plots are more useful for assessing normality. Refer back to Figure 2.16 on page 65, which displays residual histograms that look sufficiently normal in the upper row of plots but that suggest violation of the normality assumption in the lower row. Similarly, Figure 2.17 on page 65 displays reasonably normal residual QQ-plots in the upper row of plots but nonnormal QQ-plots in the lower row.

Multiple linear regression is reasonably robust to mild violations of the four assumptions. Although we should rely on model results only when we can be reasonably confident that the assumptions check out, we really need to worry only when there is a clear violation of an assumption. Recognizing clear violations can be challenging—one approach is to be concerned only if a pattern we see in a residual plot "jumps off the screen and slaps us in the face." If we find a clear violation, we can fit an alternative model (e.g., with a different subset of available predictors) and recheck the assumptions for this model. In Chapter 4 we introduce some additional strategies for dealing with such a situation (see also the remedies suggested on page 190).

A graphical tool in statistical software can make the process of checking the zero mean assumption a little easier. Consider the simulated **MLRA** data file, in which Y depends potentially on three predictor variables: X_1, X_2, and X_3. Consider the following model first:

$$\text{Model 1}: \quad E(Y) = b_0 + b_1 X_1 + b_2 X_2.$$

We can assess the four multiple linear regression model assumptions with the following:

- a residual plot with the fitted $\hat{Y}$-values on the horizontal axis (check zero mean, constant variance, and independence assumptions);

- residual plots with each predictor in turn (X_1, X_2, and X_3) on the horizontal axis (check zero mean, constant variance, and independence assumptions);

- a histogram and QQ-plot of the residuals (check normality assumption).

Most of these graphs (not shown here) lend support to the four assumptions, but Figure 3.8 shows the residual plot with X_3 on the horizontal axis, which indicates violation of the zero mean assumption. To help make this call, the left-hand graph in Figure 3.8 adds a *loess fitted line* to this residual plot (see computer help #36 in the software information files available from the book website). This line, essentially a computational method for applying the "slicing and averaging" smoothing technique described on page 120, is sufficiently

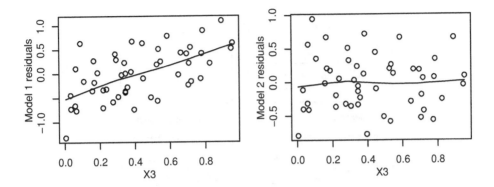

Figure 3.8 Residual plots for the **MLRA** example, with model 1 on the left and model 2 on the right. X_3 is on the horizontal axes, a predictor that is not in model 1 but that is in model 2. The loess fitted line on the left-hand plot is sufficiently different from a horizontal line at zero to suggest that the zero mean assumption is violated for model 1. By contrast, the loess fitted line on the right-hand plot is sufficiently close to a horizontal line at zero to suggest that the zero mean assumption seems reasonable for model 2.

different from a horizontal line at zero to violate the zero mean assumption for this model. Some statistical software uses a *lowess fitted line* instead of (or as an alternative to) a loess fitted line. Both types of line generally produce similar results, although technically they are based on slightly different smoothing methods.

Since X_3 is not in model 1, one possible remedy to try is to include X_3 in a new model:

$$\text{Model 2}: \quad \text{E}(Y) = b_0 + b_1 X_1 + b_2 X_2 + b_3 X_3.$$

We can then assess the four assumptions for this model using the same set of graphs as for model 1 (but using the residuals for model 2 instead). The right-hand graph in Figure 3.8 shows the new residual plot with X_3 on the horizontal axis for model 2. In contrast with the model 1 residuals, there is no clear upward trend in the model 2 residuals.

While the visual impression of a single graph can raise doubt about a model assumption, to have confidence in all four assumptions we need to consider all the suggested residual plots, histograms, and QQ-plots—see Figures 3.9 and 3.10. The four left-hand residual plots in Figure 3.9 include loess fitted lines, which should be reasonably flat to satisfy the zero mean assumption. For the four right-hand plots (similar to the left-hand plots except without loess fitted lines), the average vertical variation of the points should be reasonably constant across each plot to satisfy the constant variance assumption. The four right-hand plots should also have no clear nonrandom patterns to satisfy the independence assumption. In Figure 3.10 the histogram should be reasonably bell-shaped and symmetric, and the QQ-plot points should lie reasonably close to the line to satisfy the normality assumption.

Note that residual plots with a predictor on the horizontal axis are generally most useful for *quantitative* predictor variables. We'll see in Section 4.3 how to incorporate *qualitative* predictor variables into a multiple linear regression model. Such variables have a small number (often two, three, or four) of categories or levels. We can use such variables in residual plots but must adapt our approach slightly. For example, a residual plot with the categories of a qualitative predictor variable on the horizontal axis can be useful if we "jitter" the points horizontally. This simply adds a small amount of random noise to the

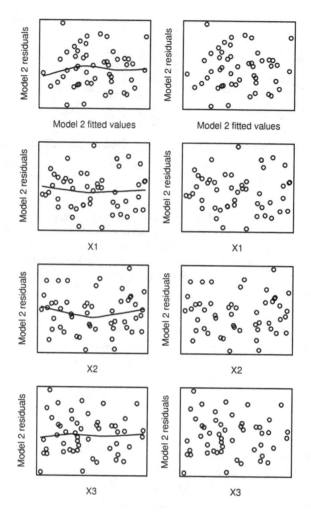

Figure 3.9   Model 2 residual plots for the **MLRA** example. Moving across each plot from left to right, the residuals appear to average close to zero and remain equally variable, providing support for the zero mean and constant variance assumptions. The lack of clear nonrandom patterns supports the independence assumption.

points to spread them out so that we can see them more easily. Otherwise, the points are likely to overlap one another, which makes it harder to see how many of them there are. Once we've done this we can check the zero mean assumption by assessing whether the means of the residuals within each category are approximately zero. We can check the constant variance assumption by assessing whether the residuals are approximately equally variable within each category. To check the normality assumption, we can see whether the residuals are approximately normal within each category using histograms or QQ-plots (although this can be difficult to assess if there are small numbers of observations within the categories). Finally, it is difficult to check the independence assumption in residual plots with a qualitative predictor variable on the horizontal axis since the pattern of the discrete categories in these plots is so dominant.

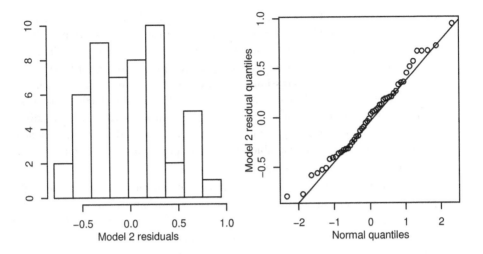

Figure 3.10 Histogram and QQ-plot of the model 2 residuals for the **MLRA** example. The approximately bell-shaped and symmetric histogram and QQ-plot points lying close to the line support the normality assumption.

For checking the zero mean and constant variance assumptions, an alternative to the jittered scatterplot with a qualitative predictor variable on the horizontal axis is to use boxplots of the residuals for the different categories of the qualitative predictor variable. Keep in mind, however, that boxplots work with medians and interquartile ranges rather than means and variances.

As discussed on page 66 in relation to simple linear regression, a note of caution regarding residual plots relates to the sample size, n. If n is either very small (say, less than 20) or very large (say, in the thousands or more), then residual plots can become very difficult to use. One possible approach for the latter case is to take an appropriate random sample of, say, 100 residuals and then to use the methods of this section as normal. More comprehensive still would be to take a few such random samples and assess the assumptions for each sample.

In Section 5.2.1 we discuss the constant variance assumption in more detail and outline some remedies for situations where this assumption is in doubt. In Section 5.2.2 we cover the most common way in which the independence assumption can be violated, which is when there is autocorrelation (or serial correlation).

3.4.2   Testing the model assumptions

In many cases, visually assessing residual plots, histograms, and QQ-plots to check the four multiple linear regression assumptions is sufficient. However, in ambiguous cases or when we wish to complement our visual conclusions with quantitative evidence, there are a number of tests available that address each of the assumptions. A full discussion of these tests lies beyond the scope of this book, but brief details and references follow.

- One simple way to test the **zero mean** assumption is to see whether there is a significant linear trend in a residual plot in which the horizontal axis quantity is a potential predictor variable that has not been included in the model. In particular, we can conduct a t-test for that predictor variable in a simple linear regression model applied to the residual plot. For example, for the **MLRA** dataset, if we fit a simple linear regression model with the residuals from model 1 as the response variable and X_3 as the predictor variable, we obtain a significant p-value for X_3 of 0.000. This result confirms the visual impression of the left-hand plot in Figure 3.8 on page 121. If we try to apply this test to a residual plot in which the horizontal axis quantity is a predictor in the model, we'll always obtain a nonsignificant p-value of essentially 1. We can illustrate this by fitting a simple linear regression model with the residuals from model 2 as the response variable and X_3 as the predictor variable, and observing the p-value for X_3 to be essentially 1.

- For residual plots in which the horizontal axis quantity is one of the predictors in the model, we can extend the previous test by fitting a *quadratic model* (see page 145) to the residual plot and applying a t-test to the squared term. For example, if we fit a multiple linear regression model with the residuals from model 2 as the response variable and two predictor variables, X_3 and X_3^2, we obtain a nonsignificant p-value for X_3^2 of 0.984. In other words, there is no evidence from this test that there is a quadratic trend in this residual plot. In practice, we probably would not have applied this test since that there is no suggestion of a quadratic trend in the plot. The test is most useful in cases where we suspect a significant nonlinear trend in a residual plot, such as those in the upper row of residual plots in Figure 2.15 on page 63. When applying the test for a residual plot in which the horizontal axis quantity is the fitted values from the model, we use the standard normal distribution to conduct the test rather than a t-distribution and the test is called "Tukey's test for nonadditivity."

- The most common test for the **constant variance** assumption is the test independently developed by Breusch and Pagan (1979) and Cook and Weisberg (1983)—see Section 5.2.1 for further details.

- There are a variety of statistics available to test the **normality** assumption, including the Shapiro and Wilk (1965) W statistic (see also Royston, 1982a, 1982b, and 1995); the Anderson-Darling test (see Thode, 2002, Sec. 5.1.4, and Stephens, 1986); the Cramer-von Mises test (see Thode, 2002, Sec. 5.1.3, and Stephens, 1986); the Lilliefors (Kolmogorov-Smirnov) test (see Thode, 2002, Sec. 5.1.1; Stephens, 1974; and Dallal and Wilkinson, 1986); the Pearson chi-square test (see Thode, 2002, Sec. 5.2, and Moore, 1986); and the Shapiro-Francia test (see Thode, 2002, Sec. 2.3.2, and Royston, 1993).

- A common way to test the **independence** assumption is to test for autocorrelation (or serial correlation) using such tests as the Wald and Wolfowitz (1940) runs test, the Durbin and Watson (1950, 1951, 1971) test, and the Breusch (1978) and Godfrey (1978) test—see Section 5.2.2 for further details.

3.5 MODEL INTERPRETATION

Once we are satisfied that the four multiple linear regression assumptions seem plausible, we can interpret the model results. This requires relating the numerical information from the statistical software output back to the subject matter. For example, in the shipping dataset, we saw in Section 3.3.4 that the two-predictor model, $\mathrm{E}(Lab) = b_0 + b_1\,Tws + b_3\,Asw$, was preferable to the four-predictor model, $E(Lab) = b_0 + b_1 Tws + b_2 Pst + b_3 Asw + b_4 Num$ (see computer help #34 in the software information files available from the book website):

Model Summary

Model	R Squared	Adjusted R Squared	Regression Std. Error	F-stat	Change Statistics df1	df2	Pr($>$F)
2^a	0.8082	0.7857	8.815				
1^b	0.8196	0.7715	9.103	0.472	2	15	0.633

[a] Predictors: (Intercept), *Tws*, *Asw*.

[b] Predictors: (Intercept), *Tws*, *Pst*, *Asw*, *Num*.

Checking regression assumptions with a sample size of 20 is challenging, but there are no clear violations in any residual plots (not shown here) for the two-predictor model. Statistical software output for this model (see computer help #31) is:

Model Summary

Model	Sample Size	Multiple R Squared	Adjusted R Squared	Regression Std. Error
2^a	20	0.8082	0.7857	8.815

[a] Predictors: (Intercept), *Tws*, *Asw*.

Parameters[a]					**95% Confidence Interval**	
Model	Estimate	Std. Error	t-stat	Pr($>$|t|)	Lower Bound	Upper Bound
2 (Intercept)	110.431	24.856	4.443	0.000		
Tws	5.001	2.261	2.212	0.041	0.231	9.770
Asw	−2.012	0.668	−3.014	0.008	−3.420	−0.604

[a] Response variable: *Lab*.

The corresponding practical interpretations of the results are as follows:

- There is no evidence at the 5% significance level that *Pst* (proportion shipped by truck) or *Num* (week number) provide useful information about the response, *Lab* (weekly labor hours), beyond the information provided by *Tws* (total weight shipped in thousands of pounds) and *Asw* (average shipment weight in pounds). (*Nested model test for the regression parameters, b_2 and b_4.*)

- There is a linear association between *Lab* and *Tws*, holding *Asw* constant, that is statistically significant at the 5% significance level. (*Hypothesis test for the regression parameter, b_1.*)

- There is a linear association between *Lab* and *Asw*, holding *Tws* constant, that is statistically significant at the 5% significance level. (*Hypothesis test for the regression parameter, b_3.*)

- We expect weekly labor hours to increase by 5.00 for each 1,000-pound increase in total weight shipped when average shipment weight remains constant (for total shipment weights of 2,000–10,000 pounds and average shipment weights of 10–30 pounds). To express our uncertainty due to sampling variation, we could say that we're 95% confident that labor hours increase by between 0.23 and 9.77 for each

1,000-pound increase in total weight shipped when average shipment weight remains constant. (*Point estimate and confidence interval for the regression parameter,* b_1.)

- We expect weekly labor hours to decrease by 2.01 for each 1-pound increase in average shipment weight when total weight shipped remains constant (for total shipment weights of 2,000–10,000 pounds and average shipment weights of 10–30 pounds). To express our uncertainty due to sampling variation, we could say that we're 95% confident that labor hours decrease by between 0.60 and 3.42 for each 1-pound increase in average shipment weight when total weight shipped remains constant. (*Point estimate and confidence interval for the regression parameter,* b_3.)

- If we use a multiple linear regression model to predict weekly labor hours from potential total weight shipped and average shipment weight values, we can expect to be accurate to within approximately ±17.6 (at a 95% confidence level). (*Regression standard error, s.*)

- 80.8% of the variation in weekly labor hours (about its mean) can be explained by a multiple linear regression association between labor hours and (total weight shipped, average shipment weight). (*Coefficient of determination,* R^2.)

3.6   ESTIMATION AND PREDICTION

As with simple linear regression, there is a distinction between a confidence interval for the population mean, $E(Y)$, at particular values of the predictor variables, $(X_1, X_2, \ldots, X_k)$, and a prediction interval for an individual Y-value at those same values of the predictor variables, $(X_1, X_2, \ldots, X_k)$.

3.6.1   Confidence interval for the population mean, E(Y)

Consider estimating the mean (or expected) value of Y at particular values of the predictor variables, $(X_1, X_2, \ldots, X_k)$, based on a multiple linear regression association between Y and $(X_1, X_2, \ldots, X_k)$. Since we have estimated the association to be $\hat{Y} = \hat{b}_0 + \hat{b}_1 X_1 + \cdots + \hat{b}_k X_k$, our best point estimate for $E(Y)$ is $\hat{Y}$. For example, suppose that for the two-predictor model for the shipping dataset we would like to estimate the average level of weekly labor hours corresponding to total weight shipped of 6,000 pounds and average shipment weight of 20 pounds. Our best point estimate for $E(Lab)$ at $Tws = 6$ and $Asw = 20$ is $\widehat{Lab} = 110.431 + 5.001 \times 6 - 2.012 \times 20 = 100.2$.

How sure are we about this answer? One way to express our uncertainty is with a confidence interval. For example, a 95% confidence interval for $E(Y)$ results from the following:

$$\text{Pr}\left(-97.5\text{th percentile} < t_{n-k-1} < 97.5\text{th percentile}\right) = 0.95$$

$$\text{Pr}\left(-97.5\text{th percentile} < \frac{\hat{Y} - E(Y)}{s_{\hat{Y}}} < 97.5\text{th percentile}\right) = 0.95$$

$$\text{Pr}\left(\hat{Y} - 97.5\text{th percentile}\,(s_{\hat{Y}}) < E(Y) < \hat{Y} + 97.5\text{th percentile}\,(s_{\hat{Y}})\right) = 0.95,$$

where $s_{\hat{Y}}$ is the *standard error of estimation* for the multiple linear regression mean, and the 97.5th percentile comes from the t-distribution with $n - k - 1$ degrees of freedom. In other words, the 95% confidence interval for $E(Y)$ can be written as $\hat{Y} \pm 97.5\text{th percentile}\,(s_{\hat{Y}})$.

We can use statistical software to calculate $s_{\hat{Y}}$ for particular X-values that we might be interested in, or just use the software to calculate the confidence interval for $E(Y)$ directly. For interested readers, a formula for $s_{\hat{Y}}$ is provided at the end of this section. The formula shows that $s_{\hat{Y}}$ tends to be smaller (and our estimates more accurate) when n is large, when the particular X-values we are interested in are close to their sample means, and when the regression standard error, s, is small. Also, of course, a lower level of confidence leads to a narrower confidence interval for $E(Y)$—for example, a 90% confidence interval will be narrower than a 95% confidence interval (all else being equal).

Returning to the shipping dataset, $s_{\hat{Y}} = 2.293$ for $Tws = 6$ and $Asw = 20$ (see statistical software output below), so that the 95% confidence interval for $E(Lab)$ at $Tws = 6$ and $Asw = 20$ is

$$\widehat{Lab} \pm 97.5\text{th percentile}\,(s_{\hat{Y}}) = 100.2 \pm 2.110 \times 2.293$$
$$= 100.2 \pm 4.838$$
$$= (95.4, 105.0).$$

The 97.5th percentile, 2.110, comes from the t-distribution with $n-k-1 = 17$ degrees of freedom (see computer help #8 in the software information files available from the book website). The relevant statistical software output (see computer help #29) is:

Lab	Tws	Asw	$\widehat{Lab}$	$s_{\hat{Y}}$	CI-low	CI-up
—	6	20	100.192	2.293	95.353	105.031

The point estimate for the population mean, $E(Lab)$, at particular predictor values is denoted "$\widehat{Lab}$," while the standard error of estimation is denoted "$s_{\hat{Y}}$," and the confidence interval goes from "CI-low" to "CI-up."

Now that we've calculated a confidence interval, what exactly does it tell us? Well, for this shipping example, *loosely speaking*, we can say "we're 95% confident that expected weekly labor hours is between 95.4 and 105.0 when total weight shipped is 6,000 pounds and average shipment weight is 20 pounds." To provide a more precise interpretation we would have to say something like "if we were to take a large number of random samples of size 20 from our population of shipping numbers and calculate a 95% confidence interval for $E(Lab)$ at $Tws = 6$ and $Asw = 20$ in each, then 95% of those confidence intervals would contain the true (unknown) population mean."

For a lower or higher level of confidence than 95%, the percentile used in the calculation must be changed as appropriate. For example, for a 90% interval (i.e., with 5% in each tail), the 95th percentile would be needed, whereas for a 99% interval (i.e., with 0.5% in each tail), the 99.5th percentile would be needed. These percentiles can be obtained using computer help #8. As further practice, calculate a 90% confidence interval for $E(Lab)$ at $Tws = 6$ and $Asw = 20$ for the shipping example (see Problem 3.4 at the end of this chapter)—you should find that it is (96.2, 104.2).

3.6.2 Prediction interval for an individual Y-value

Now, by contrast, consider predicting an individual Y-value at particular values of the predictor variables, $(X_1, X_2, \ldots, X_k)$, based on a multiple linear regression association between Y

and $(X_1, X_2, \ldots, X_k)$. To distinguish a prediction from an estimated population mean, $E(Y)$, we will call this Y-value to be predicted Y^*. Just as with estimating $E(Y)$, our best point estimate for Y^* is $\hat{Y} = \hat{b}_0 + \hat{b}_1 X_1 + \cdots + \hat{b}_k X_k$. For example, suppose that for the shipping dataset we would like to predict the actual level of weekly labor hours corresponding to total weight shipped of 6,000 pounds and average shipment weight of 20 pounds. Our best point estimate for Lab^* at $Tws = 6$ and $Asw = 20$ is $\widehat{Lab} = 110.431 + 5.001 \times 6 - 2.012 \times 20 = 100.2$.

How sure are we about this answer? One way to express our uncertainty is with a prediction interval (like a confidence interval, but for a prediction rather than an estimated population mean). For example, a 95% prediction interval for Y^* results from the following:

$$\Pr\left(-97.5\text{th percentile} < t_{n-k-1} < 97.5\text{th percentile}\right) = 0.95$$

$$\Pr\left(-97.5\text{th percentile} < \frac{\hat{Y}^* - Y^*}{s_{\hat{Y}^*}} < 97.5\text{th percentile}\right) = 0.95$$

$$\Pr\left(\hat{Y}^* - 97.5\text{th percentile}\,(s_{\hat{Y}^*}) < Y^* < \hat{Y}^* + 97.5\text{th percentile}\,(s_{\hat{Y}^*})\right) = 0.95,$$

where $s_{\hat{Y}^*}$ is the *standard error of prediction* for the multiple linear regression response, and the 97.5th percentile comes from the t-distribution with $n-k-1$ degrees of freedom. In other words, the 95% prediction interval for Y^* can be written $\hat{Y}^* \pm 97.5\text{th percentile}\,(s_{\hat{Y}^*})$.

We can use statistical software to calculate $s_{\hat{Y}^*}$ for particular X-values that we might be interested in, or just use the software to calculate the prediction interval for Y^* directly. For interested readers, a formula for $s_{\hat{Y}^*}$ is provided at the end of this section. The formula shows that $s_{\hat{Y}^*}$ is always larger than $s_{\hat{Y}}$ (on page 126) for any particular set of X-values. This makes sense because it is more difficult to predict an individual Y-value at a particular set of X-values than to estimate the mean of the population distribution of Y at those same X-values. Consider the following illustrative example. Suppose that the business for the shipping dataset plans to ship 6,000 pounds with an average shipment weight of 20 pounds each week over the next quarter. Estimating the average weekly labor hours over the quarter is easier than predicting the actual weekly labor hours in any individual week. In other words, our uncertainty about an individual prediction is always larger than our uncertainty about estimating a population mean, and $s_{\hat{Y}^*} > s_{\hat{Y}}$.

For the shipping dataset, $s_{\hat{Y}^*} = 9.109$ for $Tws = 6$ and $Asw = 20$, and the 95% prediction interval for Lab^* is

$$\widehat{Lab}^* \pm 97.5\text{th percentile}\,(s_{\hat{Y}^*}) = 100.2 \pm 2.110 \times 9.109$$
$$= 100.2 \pm 19.220$$
$$= (81.0, 119.4).$$

The 97.5th percentile, 2.110, comes from the t-distribution with $n-k-1 = 17$ degrees of freedom (see computer help #8 in the software information files available from the book website). The relevant statistical software output (see computer help #30) is:

Lab	Tws	Asw	$\widehat{Lab}$	PI-low	PI-up
—	6	20	100.192	80.974	119.410

The point estimate for the prediction, Lab^*, at particular predictor values is denoted "$\widehat{Lab}$," while the prediction interval goes from "PI-low" to "PI-up."

Now that we've calculated a prediction interval, what does it tell us? For this shipping example, *loosely speaking*, "we're 95% confident that actual labor hours in a week is between 81.0 and 119.4 when total weight shipped is 6,000 pounds and average shipment weight is 20 pounds." A more precise interpretation would have to say something like "if we were to take a large number of random samples of size 20 from our population of shipping numbers and calculate a 95% prediction interval for Lab^* at $Tws=6$ and $Asw=20$ in each, then 95% of those prediction intervals would contain the true (unknown) labor hours for an individual week picked at random when $Tws=6$ and $Asw=20$."

As with the standard error of estimation, $s_{\hat{Y}}$, the standard error of prediction, $s_{\hat{Y}*}$, tends to be smaller (and our predictions more accurate) when n is large, when the particular X-values we are interested in are close to their sample means and when the regression standard error, s, is small. Also, of course, a lower level of confidence leads to a narrower prediction interval for Y^*—for example, a 90% prediction interval will be narrower than a 95% prediction interval (all else being equal).

For a lower or higher level of confidence than 95%, the percentile used in the calculation must be changed as appropriate. For example, for a 90% interval (i.e., with 5% in each tail), the 95th percentile would be needed, whereas for a 99% interval (i.e., with 0.5% in each tail), the 99.5th percentile would be needed. These percentiles can be obtained using computer help #8. As further practice, calculate a 90% prediction interval for Lab^* at $Tws=6$ and $Asw=20$ for the shipping example (see Problem 3.4 at the end of this chapter); you should find that it is (84.4, 116.1).

One final note on prediction intervals. The "$\pm 2s$" interpretation we discussed for the regression standard error in Section 3.3.1 is based on an approximation of a 95% prediction interval for datasets with a large sample size, n. For sufficiently large n, $s_{\hat{Y}*}$ is approximately equal to s, while the 97.5th percentile from the t-distribution with $n-k-1$ degrees of freedom is close to 2. Thus, the 95% prediction interval for Y^* can be written approximately as $\hat{Y}^* \pm 2s$.

As for simple linear regression, we have now covered four different types of standard error in the context of multiple linear regression:

- the regression standard error, s, which is an estimate of the standard deviation of the error term in the model (Section 3.3.1)

- the standard errors for the regression parameter estimates, $s_{\hat{b}_0}$, $s_{\hat{b}_1}$, $s_{\hat{b}_2}$, ..., $s_{\hat{b}_k}$ (Section 3.3.5)

- the standard error of estimation for the multiple linear regression mean, $s_{\hat{Y}}$ (Section 3.6.1)

- the standard error of prediction for the multiple linear regression response, $s_{\hat{Y}*}$ (Section 3.6.2)

Optional—formulas for standard errors of estimation and prediction. Define the predictor matrix $\mathbf{X}$ as on page 92 and write the particular X-values we are interested in as a vector, $\mathbf{x}$ (including a "1" as the first entry to represent the intercept term). Then the standard error of estimation for the multiple linear regression mean at $\mathbf{x}$ is $s_{\hat{Y}} = s\sqrt{\mathbf{x}^T(\mathbf{X}^T\mathbf{X})^{-1}\mathbf{x}}$, where s is the regression standard error. The standard error of prediction for the multiple linear regression response at $\mathbf{x}$ is $s_{\hat{Y}*} = s\sqrt{1+\mathbf{x}^T(\mathbf{X}^T\mathbf{X})^{-1}\mathbf{x}}$. The two standard errors are linked by the formula, $s_{\hat{Y}*} = \sqrt{s^2 + s_{\hat{Y}}^2}$.

3.7 CHAPTER SUMMARY

The major concepts that we covered in this chapter relating to multiple linear regression are as follows:

Multiple linear regression allows us to model the association between a response variable, Y, and predictor variables, $(X_1, X_2, \ldots, X_k)$, as $E(Y) = b_0 + b_1 X_1 + \cdots + b_k X_k$.

The method of least squares provides an estimated regression equation, $\hat{Y} = \hat{b}_0 + \hat{b}_1 X_1 + \cdots + \hat{b}_k X_k$, by minimizing the residual sum of squares (residuals are the differences between the observed Y-values and the fitted $\hat{Y}$-values).

The estimated intercept represents the expected Y-value when $X_1 = X_2 = \cdots = X_k = 0$ (if such values are meaningful and fall within the range of X-values in the sample dataset)—denoted $\hat{b}_0$.

The estimated regression parameters represent expected changes in Y for a unit change in X_p, holding the other X's constant (over the range of X-values in the sample dataset)—denoted $\hat{b}_p$ $(p = 1, \ldots, k)$.

The regression standard error, s, is an estimate of the standard deviation of the random errors. One way to interpret s is to calculate $2s$ and say that when using the multiple linear regression model to predict Y from $(X_1, X_2, \ldots, X_k)$, we can expect to be accurate to within approximately $\pm 2s$ (at a 95% confidence level).

The coefficient of determination, R^2, represents the proportion of variation in Y (about its sample mean) explained by a multiple linear regression association between Y and $(X_1, X_2, \ldots, X_k)$. This is also equivalent to the square of multiple R, the correlation between the observed Y-values and the fitted $\hat{Y}$-values. Adjusted R^2 is a variant on R^2, which takes into account the number of predictor variables in a model to facilitate model comparisons.

Hypothesis testing provides a means of making decisions about the likely values of the regression parameters, $b_1, b_2, \ldots, b_k$. The magnitude of the calculated sample test statistic indicates whether we can reject a null hypothesis in favor of an alternative hypothesis. Roughly speaking, the p-value summarizes the hypothesis test by representing the weight of evidence for the null hypothesis (i.e., small values favor the alternative hypothesis):

Global usefulness tests have null hypothesis $b_1 = b_2 = \cdots = b_k = 0$ and test whether *any* of the predictors have a linear association with Y.

Nested model tests have null hypothesis $b_{r+1} = b_{r+2} = \cdots = b_k = 0$ (i.e., a subset of the parameters set equal to zero) and test whether the corresponding *subset* of predictors have a linear association with Y, once the linear association with the other r predictors has been accounted for.

Individual tests have null hypothesis $b_p = 0$ and test whether an *individual* predictor has a linear association with Y, once the other $k - 1$ predictors have been accounted for.

Confidence intervals are another method for calculating the sample estimate of a population regression parameter and its associated uncertainty:

$$\hat{b}_p \pm \text{t-percentile}(s_{\hat{b}_p}).$$

Model assumptions should be satisfied before we can rely on the multiple linear regression results. These assumptions relate to the random errors in the model: The probability distributions of the random errors at each set of values of $(X_1, X_2, \ldots, X_k)$ should be normal with zero means and constant variances, and each random error should be independent of every other. We can check whether these assumptions seem plausible by calculating residuals (estimated errors) and visually assessing residual plots, histograms, and QQ-plots.

Confidence intervals are also used for presenting sample estimates of the population mean, $\text{E}(Y)$, at particular X-values and their associated uncertainty:

$$\hat{Y} \pm \text{t-percentile}(s_{\hat{Y}}).$$

Prediction intervals, while similar in spirit to confidence intervals tackle the different problem of predicting individual Y-values at particular X-values:

$$\hat{Y}^* \pm \text{t-percentile}(s_{\hat{Y}^*}).$$

PROBLEMS

- "Computer help" refers to the numbered items in the software information files available from the book website.

- There are *brief* answers to the even-numbered problems in Appendix E.

3.1 The **MOVIES** data file contains data on 25 movies from "The Internet Movie Database" (www.imdb.com). Based on this dataset, we wish to investigate whether all-time U.S. box office receipts (*Box*, in millions of U.S. dollars unadjusted for inflation) are associated with any of the following variables:

> *Rate* = Internet Movie Database user rating (out of 10)
> *User* = Internet Movie Database users rating the movie (in thousands)
> *Meta* = "metascore" based on 35 critic reviews (out of 100)
> *Len* = runtime (in minutes)
> *Win* = award wins
> *Nom* = award nominations

Theatrical box office receipts (movie ticket sales) may include theatrical re-release receipts, but exclude video rentals, television rights, and other revenues.

(a) Write out an equation (like the equation at the bottom of page 85) for a multiple linear regression model for predicting response *Box* from just three predictors: *Rate*, *User*, and *Meta*.
Hint: This question is asking you to write an equation for $E(Box)$.

(b) Use statistical software to fit this model [computer help #31] and write out the estimated multiple linear regression equation [i.e., replace the b's in part (a) with numbers].
Hint: This question is asking you to write an equation for $\widehat{Box}$.

(c) Interpret the estimated regression parameter for *Rate* in the context of the problem [i.e., put the appropriate number from part (b) into a meaningful sentence, remembering to include the correct units for any variables that you use in your answer].
Hint: See the bottom of page 125 for an example of the type of sentence expected.

3.2 Consider the shipping example in the **SHIPDEPT** data file, introduced in Section 3.3.2.

(a) As suggested on page 114, calculate a 90% confidence interval for the population regression parameter b_1 in the four-predictor multiple linear regression model of *Lab* (weekly labor hours) on *Tws* (total weight shipped in thousands of pounds), *Pst* (proportion shipped by truck), *Asw* (average shipment weight in pounds), and *Num* (week number). Recall that the estimated regression parameter for this model is $\hat{b}_1 = 6.074$, while the standard error of this estimate is $s_{\hat{b}_1} = 2.662$.

(b) Also calculate a 90% confidence interval for the population regression parameter b_1 in the *two*-predictor multiple linear regression model of *Lab* on *Tws* and *Asw*. Recall that the estimated regression parameter for this model is $\hat{b}_1 = 5.001$, while the standard error of the estimate is $s_{\hat{b}_1} = 2.261$. Compare the width of the resulting interval with the width of the interval in part (a). Which is narrower, and why?

3.3 Consider the **MOVIES** data file from Problem 3.1 again.

    (a) Use statistical software to fit the following (complete) model for *Box* as a function of all six predictor variables [computer help #31]:

$$E(Box) = b_0 + b_1 \, Rate + b_2 \, User + b_3 \, Meta + b_4 \, Len + b_5 \, Win + b_6 \, Nom.$$

    Write down the residual sum of squares for this model.

    (b) Use statistical software to fit the following (reduced) model [computer help #31]:

$$E(Box) = b_0 + b_1 \, Rate + b_2 \, User + b_3 \, Meta.$$

    [This is the model from Problem 3.1 part (b).] Write down the residual sum of squares for this model.

    (c) Using the results from parts (a) and (b) together with the nested model test F-statistic formula on page 105, test the null hypothesis NH: $b_4 = b_5 = b_6 = 0$ in the complete model, using significance level 5%. Write out all the hypothesis test steps and interpret the result in the context of the problem.
    Hint: To solve this part you may find the following information useful. The 95th percentile of the F-distribution with 3 numerator degrees of freedom and 18 denominator degrees of freedom is 3.160.

    (d) Check your answer for part (c) by using statistical software to do the nested model test directly [computer help #34]. State the values of the F-statistic and the p-value, and draw an appropriate conclusion.

    (e) Another way to see whether we should prefer the reduced model for this example is to see whether the regression standard error (s) is smaller for the reduced model than for the complete model and whether adjusted R^2 is higher for the reduced model than for the complete model. Confirm whether these relationships hold in this example (i.e., compare the values of s and adjusted R^2 in the reduced and complete models).

3.4 Consider the shipping example in the **SHIPDEPT** data file from Problem 3.2 again.

    (a) As suggested on page 127, calculate a 90% confidence interval for $E(Lab)$ at $Tws = 6$ and $Asw = 20$ in the two-predictor multiple linear regression model of Lab on Tws and Asw. Recall that the estimated regression parameters for this model are $\hat{b}_0 = 110.431$, $\hat{b}_1 = 5.001$, and $\hat{b}_3 = -2.012$, and the standard error of estimation at $Tws = 6$ and $Asw = 20$ is $s_{\hat{Y}} = 2.293$.

    (b) As suggested on page 129, calculate a 90% prediction interval for Lab^* at $Tws = 6$ and $Asw = 20$ in the two-predictor multiple linear regression model. Use the estimated regression parameters from part (a) and recall that the standard error of prediction at $Tws = 6$ and $Asw = 20$ is $s_{\hat{Y}*} = 9.109$.

3.5 De Rose and Galarza (2000) used multiple linear regression to study Att = average attendance in thousands, from the first few years of Major League Soccer (MLS, the professional soccer league in the United States). The 12 MLS teams at the time ranged in average attendance from 10,000 to 22,000 per game. De Rose and Galarza used the following predictor variables:

        Pop = total population of metropolitan area within 40 miles (millions)
        $Teams$ = number of (male) professional sports teams in the four major sports
        $Temp$ = average temperature (April–September, °F)

The regression results reported in the study were:

Predictor variable	Parameter estimate	Two tail p-value
Intercept	28.721	0.001
Pop	1.350	0.001
Teams	−0.972	0.037
Temp	−0.238	0.012

(a) Write out the estimated least squares (regression) equation for predicting *Att* from *Pop*, *Teams*, and *Temp*.

(b) R^2 was 91.4%, suggesting that this model may be useful for predicting average attendance (for expansion teams, say). Test the global usefulness of the model using a significance level of 5%.

Hint: You will need to use the second formula for the global F-statistic on page 101 to solve this part. Also, you may find the following information useful: the 95th percentile of the F-distribution with 3 numerator degrees of freedom and 8 denominator degrees of freedom is 4.07.

(c) Test, at a 5% significance level, whether the regression parameter estimate for *Teams* suggests that increasing the number of (male) professional sports teams in the four major sports (football, baseball, basketball, hockey) in a city is associated with a decrease in average MLS attendance in that city (all else being equal).

Hint: You'll need to do a lower-tail hypothesis test using the p-value method, but be careful because the p-values given in the table are two-tailed.

(d) According to the model results, how much does average attendance differ for two cities with the same population and average temperature when one city has one fewer (male) professional sports teams in the four major sports?

Hint: Write out the equation from part (a) for predicted average attendance in thousands for one city (plug in Teams = 1, say) and then do the same for the other city (plug in Teams = 2). The difference between the two equations gives you the answer to the problem. You should find that as long as you plug in values for Teams that differ by 1, you'll always get the same answer.

(e) One purpose for the study was to predict attendance for future expansion teams. Since the study was published, some of the included cities are no longer represented in MLS and have been replaced by others. In one case, beginning with the 2006 season, the San Jose Earthquakes MLS franchise relocated to Houston, Texas, which was one of the potential cities considered in the study. A 95% prediction interval for average attendance for a potential Houston MLS team based on the model came to (10,980, 15,340). Briefly discuss how studies like this can help to inform decisions about future expansion teams for professional leagues like MLS.

3.6 Researchers at General Motors analyzed data on 56 U.S. Standard Metropolitan Statistical Areas (SMSAs) to study whether air pollution contributes to mortality. These data are available in the **SMSA** data file and were obtained from the "Data and Story Library" at lib.stat.cmu.edu/DASL/ (the original data source is the U.S. Department of Labor Statistics). The response variable for analysis is *Mort* = age adjusted mortality per 100,000 population (a mortality rate statistically modified to eliminate the effect of different age distributions in different population groups). The dataset includes predictor variables measuring demographic characteristics of the cities, climate characteristics, and concentrations of the air pollutant nitrous oxide (NO_x).

    (a) Fit the (complete) model $E(Mort) = b_0 + b_1 Edu + b_2 Nwt + b_3 Jant + b_4 Rain + b_5 Nox + b_6 Hum + b_7 Inc$, where *Edu* is median years of education, *Nwt* is percentage nonwhite, *Jant* is mean January temperature in degrees Fahrenheit, *Rain* is annual rainfall in inches, *Nox* is the natural logarithm of nitrous oxide concentration in parts per billion, *Hum* is relative humidity, and *Inc* is median income in thousands of dollars [computer help #31]. Report the least squares (regression) equation.

    (b) Do a nested model F-test (using a significance level of 5%) to see whether *Hum* and *Inc* provide significant information about the response, *Mort*, beyond the information provided by the other predictor variables [computer help #34]. Use the fact that the 95th percentile of the F-distribution with 2 numerator degrees of freedom and 48 denominator degrees of freedom is 3.19 [computer help #8].

    (c) Do individual t-tests (using a significance level of 5%) for each predictor in the (reduced) model $E(Mort) = b_0 + b_1 Edu + b_2 Nwt + b_3 Jant + b_4 Rain + b_5 Nox$ [computer help #31]. Use the fact that the 97.5th percentile of the t-distribution with 50 degrees of freedom is 2.01 [computer help #8].

    (d) Check the four model assumptions for the model from part (c) [computer help #35, #28, #15, #36, #14, and #22].
Hint: Ideally, you should look at six residual plots (five with each of the five predictors on the horizontal axis in turn and one with the predicted values on the horizontal axis) to check the zero mean, constant variance, and independence assumptions. You should also use a histogram and/or QQ-plot to check the normality assumption. The more comprehensive you can be in checking the assumptions, the more confident you can be about the validity of your model.

    (e) Write out the least squares equation for the model from part (c). Do the signs of the estimated regression parameters make sense in this context?

    (f) Based on the model from part (c), calculate a 95% confidence interval for $E(Mort)$ for cities with the following characteristics: $Edu = 10$, $Nwt = 15$, $Jant = 35$, $Rain = 40$, and $Nox = 2$ [computer help #29].

    (g) Based on the model from part (c), calculate a 95% prediction interval for $Mort^*$ for a city with the following characteristics: $Edu = 10$, $Nwt = 15$, $Jant = 35$, $Rain = 40$, and $Nox = 2$ [computer help #30].

CHAPTER 4

REGRESSION MODEL BUILDING I

In Chapter 3 we learned about multiple linear regression, a technique for analyzing certain types of multivariate data. This can help us to understand the linear association between a response variable, Y, and k predictor variables, $(X_1, X_2, \ldots, X_k)$, to see how a change in one of the predictors is associated with a change in the response, and to estimate or predict the value of the response knowing the values of the predictors. We estimated the model as

$$\hat{Y} = \hat{b}_0 + \hat{b}_1 X_1 + \hat{b}_2 X_2 + \cdots + \hat{b}_k X_k.$$

We began thinking about model building by using hypothesis tests for the regression parameters, $(b_1, b_2, \ldots, b_k)$. These tests can identify which predictors are most useful for modeling Y and should be included in the model. They can also identify which predictors seem redundant—provide little information about Y beyond the information provided by the other predictors—and should be removed from the model. We should strive to *include* all the relevant and important predictors because omitting important predictors can cause the associations between Y and the included predictors to be estimated incorrectly (i.e., estimated effects of included predictors can be biased). We should also *exclude* unimportant predictors whose presence in the model can overcomplicate things unnecessarily and can increase our uncertainty about the magnitudes of the effects for the important predictors.

While deciding which predictors to include and exclude offers a variety of ways to model the association between Y and $(X_1, X_2, \ldots, X_k)$, we sometimes need more flexible methods for modeling that association. In particular, the model above presupposes a "linear" association between the response variable and each predictor variable (holding

Applied Regression Modeling, Second Edition. By Iain Pardoe
Copyright © 2012 John Wiley & Sons, Inc.

other predictor variables constant). For example, the expected change in Y for a unit change in X_1 (holding $X_2, \ldots, X_k$ constant) is b_1, a constant number that does not depend on the values of $X_1, X_2, \ldots, X_k$. But what if a unit change in X_1 is associated with a small increase in Y at low values of X_1, but a large increase in Y at high values of X_1? To see how to model such "nonlinear" predictor effects, Section 4.1 considers transformations of the variables in a multiple linear regression model.

As a further example, suppose that a unit change in X_1 is associated with an increase in Y at low values of another predictor, X_2, but a decrease in Y at high values of X_2. This is an example of interaction between X_1 and X_2 in the way they are associated with Y together—Section 4.2 covers this topic.

Since the multiple linear regression model is a mathematical relationship between a response variable, Y, and predictor variables, $(X_1, X_2, \ldots, X_k)$, each of the variables must be measured using a "quantitative" scale. But what if one of our predictor variables has a "qualitative" or "categorical" scale, for example, gender with values "male" and "female" or sport with values "basketball," "football," and "baseball"? Section 4.3 introduces a method for incorporating qualitative predictor variables into a multiple linear regression model.

4.1 TRANSFORMATIONS

4.1.1 Natural logarithm transformation for predictors

Consider the **TVADS** data file in Table 4.1, obtained from the "Data and Story Library" at lib.stat.cmu.edu/DASL/. These data appeared in the *Wall Street Journal* on March 1, 1984. Twenty-one TV commercials were selected by Video Board Tests, Inc., a New York ad-testing company, based on interviews with 20,000 adults. *Impress* measures millions of retained impressions of those commercials per week, based on a survey of 4,000 adults. *Spend* measures the corresponding 1983 TV advertising budget in millions of dollars.

A *transformation* is a mathematical function applied to a variable in our dataset. For example, $\log_e(Spend)$ measures the natural logarithms of the advertising budgets. Mathematically, it is possible that there is a stronger linear association between *Impress* and

Table 4.1 TV commercial data: *Spend* = spending in $m, *Impress* = millions of retained impressions.

Firm	Spend	Impress	Firm	Spend	Impress
Miller Lite	50.1	32.1	Bud Lite	45.6	10.4
Pepsi	74.1	99.6	AT&T/Bell	154.9	88.9
Stroh's	19.3	11.7	Calvin Klein	5.0	12.0
Federal Express	22.9	21.9	Wendy's	49.7	29.2
Burger King	82.4	60.8	Polaroid	26.9	38.0
Coca-Cola	40.1	78.6	Shasta	5.7	10.0
McDonald's	185.9	92.4	Meow Mix	7.6	12.3
MCI	26.9	50.7	Oscar Meyer	9.2	23.4
Diet-Cola	20.4	21.4	Ford	166.2	40.1
Crest	32.4	71.1	Kibbles 'n Bits	6.1	4.4
Levi's	27.0	40.8			

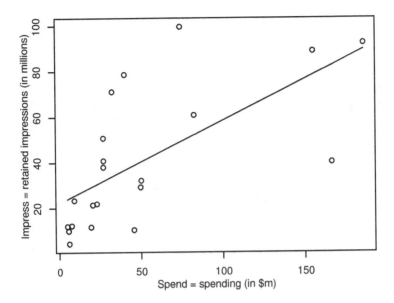

Figure 4.1   Scatterplot of *Impress* versus *Spend* for the TV commercial example with fitted line from $E(Impress) = b_0 + b_1\,Spend$.

$\log_e(Spend)$ than between *Impress* and *Spend*. If so, we should make use of this when trying to apply a linear regression model to explain any association between the success of particular TV commercials and their cost, or when trying to predict the future success of a commercial with a particular budget.

Figure 4.1 displays the data together with a fitted line from a model that uses the untransformed predictor variable, *Spend* (see computer help #15 and #26 in the software information files available from the book website):

$$\text{Model 1}: \quad E(Impress) = b_0 + b_1\,Spend.$$

Statistical software output for this model (see computer help #25) is:

Model Summary

Model	Sample Size	Multiple R Squared	Adjusted R Squared	Regression Std. Error
1[a]	21	0.4240	0.3936	23.50

[a] Predictors: (Intercept), *Spend*.

Parameters[a]

Model		Estimate	Std. Error	t-stat	$\Pr(>|t|)$
1	(Intercept)	22.163	7.089	3.126	0.006
	Spend	0.363	0.097	3.739	0.001

[a] Response variable: *Impress*.

Contrast this with Figure 4.2, which displays *Impress* versus the natural logarithm of *Spend*,

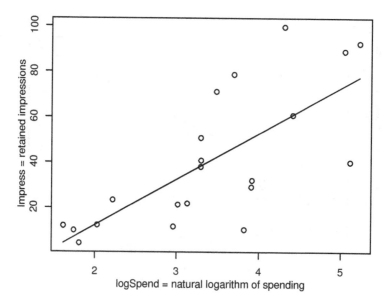

Figure 4.2   Scatterplot of *Impress* versus $\log_e(Spend)$ for the TV commercial example with fitted line from $E(Impress) = b_0 + b_1 \log_e(Spend)$.

$\log_e(Spend)$, together with a fitted line from this model (see computer help #15 and #26):

$$\text{Model 2}: \quad E(Impress) = b_0 + b_1 \log_e(Spend).$$

Statistical software output for this model (see computer help #25) is:

Model Summary

Model	Sample Size	Multiple R Squared	Adjusted R Squared	Regression Std. Error
2[a]	21	0.5323	0.5077	21.18

[a] Predictors: (Intercept), $\log_e(Spend)$.

Parameters[a]

Model		Estimate	Std. Error	t-stat	Pr(> |t|)
2	(Intercept)	−28.050	15.441	−1.817	0.085
	$\log_e(Spend)$	20.180	4.339	4.650	0.000

[a] Response variable: *Impress*.

We know from Section 3.3 that model 2 [which uses $\log_e(Spend)$] provides a more useful description of the association between the success of particular TV commercials and their cost than that of model 1 (which uses untransformed *Spend*). Why? Well, the regression standard error, s, is lower in model 2 (21.2) than in model 1 (23.5), so we expect model 2 to be more accurate when used to predict the future success of a commercial with a particular budget. Also, R^2 is higher in model 2 (53.2%) than in model 1 (42.4%), so we can explain more of the variation in *Impress* using $\log_e(Spend)$ in a regression model than

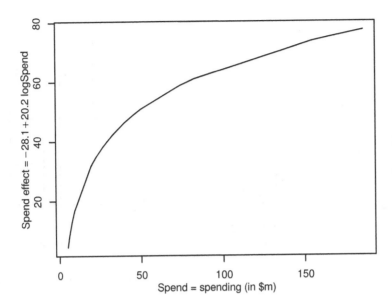

Figure 4.3 Predictor effect plot of *Spend* effect $= -28.1 + 20.2\log_e(Spend)$ versus *Spend* for the TV commercial example.

using *Spend*. Finally, the (two-tail) p-value for testing whether the regression parameter b_1 could be zero in the population is lower in model 2 (0.000) than in model 1 (0.001), so there is more evidence of a nonzero value for b_1 when using $\log_e(Spend)$ in a regression model than when using *Spend*.

The one downside to all this is that the interpretation of b_1 in model 2 is not as straightforward as it is in model 1. For model 1, b_1 represents the expected change in *Impress* for a unit change in *Spend*. For model 2, b_1 represents the expected change in *Impress* for a unit change in $\log_e(Spend)$. But a unit change in $\log_e(Spend)$ does not really correspond to anything sensible in practical terms. There is an algebraic method to derive a more meaningful interpretation for b_1 in model 1 (see the optional part at the end of this section on page 158), but a more intuitive approach might be to graph the "effect of changes in *Spend* on *Impress*." Figure 4.3 shows such a *predictor effect plot* for this example (see computer help #42).

We discuss plots like this more fully in Section 5.5, but this example simply displays the association between $\widehat{Impress} = -28.1 + 20.2\log_e(Spend)$ and *Spend*. We can use the graph axes to find approximate answers to questions like "What happens to *Impress* when *Spend* increases from A to B?" This can be sufficient in circumstances where all we need is a quick visual impression of the modeled association between *Impress* and *Spend*. More precise answers just require some simple arithmetic. For example, as *Spend* increases from 10 to 20, we expect *Impress* to increase from $-28.1 + 20.2\log_e(10) = 18.4$ to $-28.1 + 20.2\log_e(20) = 32.4$: this corresponds to the steeply rising line on the left of Figure 4.3. By contrast, as *Spend* increases from 150 to 160, we expect *Impress* to increase from $-28.1 + 20.2\log_e(150) = 73.1$ to $-28.1 + 20.2\log_e(160) = 74.4$; this corresponds to the more gradual rise on the right of the plot.

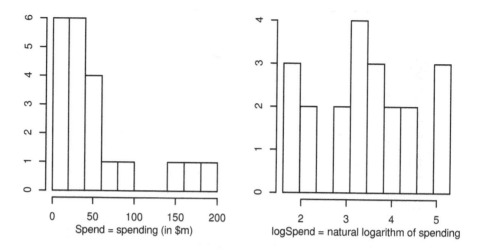

Figure 4.4   Histograms of *Spend* (left) and $\log_e(Spend)$ (right) for the TV commercial dataset.

This example illustrates how we can use variable transformations to improve our ability to explain an association between a response variable and one or more predictor variables, or to predict a value of the response variable at particular values of the predictor variables. One potential disadvantage of using transformations is that the resulting models tend to be more difficult to interpret, but graphical approaches such as Figure 4.3 can remedy this.

When it comes to selecting specific transformations to use, there are many possibilities: We can apply transformations to any of the predictor variables and/or to the response variable, and the transformations can be any mathematical function we can think of. We will focus on the more common transformations found in regression applications. These are often suggested by particular theories in the field of application. At other times, we might observe specific empirical associations in sample datasets, and we can try out various variable transformations to see how to best model the data. Again, there are particular transformations that seem to work well in practice and that we will focus on.

We have just seen the *natural logarithm transformation* applied to a single predictor variable. This transformation works well for positively skewed variables with a few values much higher than the majority, since it tends to spread out the more common lower values and "pull in" the less common higher values. To illustrate, compare the histograms of *Spend* and $\log_e(Spend)$ for the TV commercial dataset in Figure 4.4 (see computer help #14).

Multiple linear regression models often work better when the predictor variables are closer to a normal distribution (approximately symmetric and bell-shaped) than highly skewed. Although the histogram for $\log_e(Spend)$ in Figure 4.4 does not appear to be particularly normal, it is certainly more symmetric than the histogram for *Spend*, which is very skewed. Many regression applications involve skewed variables (e.g., personal income where there is a handful of individuals who earn much more than the majority), which can benefit from being transformed logarithmically before analysis.

More generally, we can apply the natural logarithm transformation to any (or all) of the positive-valued predictor variables in a multiple linear regression model. For example, in a dataset with three potential predictors, X_1, X_2, and X_3, we might find that the following model

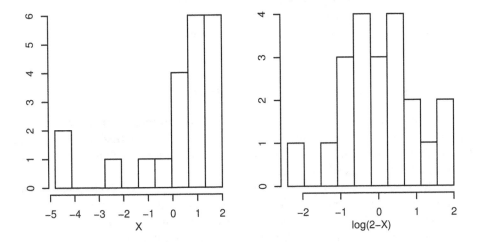

Figure 4.5 Histograms of X (left) and $\log_e(2-X)$ (right) for a simulated dataset.

best represents the association between a response variable, Y, and the three predictors:

$$\mathrm{E}(Y) = b_0 + b_1 \log_e(X_1) + b_2 X_2 + b_3 \log_e(X_3).$$

In other words, there is a curved association between Y and X_1 (holding X_2 and X_3 constant), but a linear association between Y and X_2 (holding X_1 and X_3 constant). There is also a curved association between Y and X_3 (holding X_1 and X_2 constant). This model may have been suggested by an economic theory that hypothesized these associations. Or, when analyzing the data, we may have noticed that X_1 and X_3 seem positively skewed, so that we might obtain a better model by using $\log_e(X_1)$ and $\log_e(X_3)$.

This model is still a multiple *linear* regression model—the word "linear" is a mathematical quality which signifies that the model is represented by a "linear equation." The only requirement for a linear equation is that the regression parameters—b_1, b_2, and b_3 in this case— multiply functions of the predictor variables, and then these "parameter × predictor function" terms are all added together (as above). The models that we considered in Chapter 3 represented the special case in which the predictor functions were each equal to the predictors themselves (in mathematics, these functions are known as "identity functions"). Thus, we can incorporate curved associations within the multiple linear regression model framework that we developed in Chapter 3.

Although the natural logarithm transformation works only on positive-valued variables, we can also apply it to negative-valued or negatively skewed variables by first adding a constant and/or changing the sign. To illustrate, compare the histograms of X and $\log_e(2-X)$ for a simulated dataset in Figure 4.5 (see computer help #14).

The natural logarithm is not the only kind of logarithmic transformation. Mathematically, logarithms have what is known as a "base," and it just so happens that the base for natural logarithms is the constant known as "e" (numerically, $e \approx 2.718$). For the purposes of this book, the base of the particular logarithmic transformations that we use is not important. In many regression applications, natural logarithms (with base e) are common. However, in some fields of application, other bases are more common. For example, in biological applications base-2 logarithms are used (since these have a nice "doubling" interpretation), and in some physical science or engineering applications base-10 logarithms are used (since these have a nice "order of magnitude" interpretation).

It is also possible to apply a logarithmic transformation to the response variable. We cover this in Section 4.1.4.

4.1.2 Polynomial transformation for predictors

The *polynomial transformation* for predictor variables is also common in many regression applications. For example, consider the data available in the **HOMES4** data file. These data are for 76 single-family homes in south Eugene, Oregon in 2005 and were provided by Victoria Whitman, a realtor in Eugene.

In earlier chapters we used information on some of these homes to explore associations between *Price* = the sale price of a home (in thousands of dollars) and the floor and lot sizes. Here we consider the age of a home and investigate whether this factors into the sale price in this particular housing market. For the sake of illustration, we ignore the other predictors (floor size, lot size, etc.) for now and focus solely on *Age* = age, defined as 2005 minus the year the home was built. (We'll revisit this application in Section 6.1 when we consider a complete analysis of the whole dataset.)

Figure 4.6 displays the sale price and age data together with a straight fitted line from a model that uses the untransformed *Age* variable (see computer help #15 and #26 in the software information files available from the book website):

$$\text{Model 1}: \quad \text{E}(Price) = b_0 + b_1 Age.$$

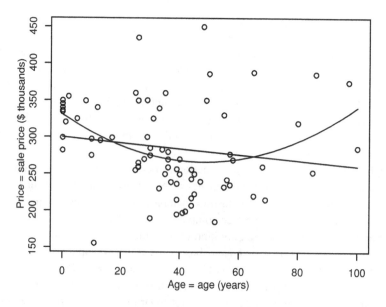

Figure 4.6   Scatterplot of *Price* versus *Age* for the home prices–age example with fitted lines from $\text{E}(Price) = b_0 + b_1 Age$ (straight) and from $\text{E}(Price) = b_0 + b_1 Age + b_2 Age^2$ (curved).

Statistical software output for this model (see computer help #25) is:

Parameters[a]

Model		Estimate	Std. Error	t-stat	$\Pr(>|t|)$
1	(Intercept)	299.886	12.555	23.887	0.000
	Age	−0.396	0.295	−1.342	0.184

[a] Response variable: *Price*.

Contrast this with the curved fitted line from this *quadratic* model (see computer help #32):

$$\text{Model 2}: \quad E(Price) = b_0 + b_1 Age + b_2 Age^2.$$

Statistical software output for this model (see computer help #31) is:

Model Summary

Model	Sample Size	Multiple R Squared	Adjusted R Squared	Regression Std. Error
2[a]	76	0.1468	0.1235	56.49

[a] Predictors: (Intercept), *Age*, Age^2.

Parameters[a]

Model		Estimate	Std. Error	t-stat	$\Pr(>|t|)$
2	(Intercept)	330.410	15.103	21.877	0.000
	Age	−2.653	0.749	−3.542	0.001
	Age^2	0.027	0.008	3.245	0.002

[a] Response variable: *Price*.

Model 1 seems inappropriate here since the straight fitted line misses an apparent curved association in the points, whereas the curved (quadratic) fitted line of model 2 seems to capture that association quite effectively. Furthermore, we know from Section 3.3 that model 2 appears to usefully describe the association between sale price and home age. The regression standard error, $s = 56.49$, suggests that we can use this model to predict sale price to within approximately ±$113,000 at a 95% confidence level (over the range of sample home ages). $R^2 = 0.1468$ suggests that this model explains 14.7% of the variation in sample sale prices. The (two-tail) p-value for testing whether the regression parameter b_2 could be zero in the population is 0.002, so there is strong evidence that it is not zero. Thus, we strongly favor model 2 over model 1, since if we were to set b_2 equal to zero in model 2, then it simplifies to model 1—the hypothesis test results do not support this simplification.

We haven't mentioned the hypothesis test for whether the regression parameter b_1 could be zero in the population. Usually, if we decide that we would like to retain a squared predictor term, X^2, in a model (i.e., when there is a low p-value for the corresponding regression parameter), then we would retain X in the model also, *regardless of the p-value for its corresponding regression parameter*. In this case, this means that there is no need to look at the p-value for b_1—we retain *Age* in the model regardless of this p-value since we have already decided to retain Age^2. This is known as preserving *hierarchy* and is the preferred approach in many regression applications. The only time we can include a squared predictor term, X^2, in a model without also including X is when we happen to know that the fitted regression line (or regression hyperplane) should level off to be completely flat (in the X-direction) when $X = 0$. It is probably safest to assume that we don't know this for sure in most applications.

Note that this notion of hierarchy does not apply to the natural logarithm transformation that we considered in Section 4.1.1. In other words, there is no particular reason why we would retain X in a model that includes $\log_e(X)$.

The quadratic model is a special case of the more general *polynomial* model:

$$E(Y) = b_0 + b_1 X + b_2 X^2 + b_3 X^3 + \cdots.$$

This model preserves hierarchy by including all the powers of X up to and including the highest one that is significant—see the optional discussion on page 147 for a mathematical justification for this approach. As we add higher powers of X to the model, we can model more and more complicated curved associations between Y and X. However, just as in Section 3.3.2 on page 96, we should beware of overfitting the sample data so that it reacts to every slight twist and turn in the sample associations between the variables (see Section 5.2.5 on page 212 for an extreme example). A simpler model with fewer powers of X will be preferable *if* it can capture the major, important population associations between the variables without getting distracted by minor, unimportant sample associations. In practice, it is rare to see powers higher than 2 in multiple linear regression models unless there are good theoretical reasons for including those higher powers.

As with the natural logarithm transformation, we can apply polynomial transformations to any (or all) of the predictor variables in a multiple linear regression model. For example, in a dataset with three potential predictors, X_1, X_2, and X_3, we might find that the following model best represents the association between Y and the three predictors:

$$E(Y) = b_0 + b_1 X_1 + b_2 X_1^2 + b_3 X_2 + b_4 X_3 + b_5 X_3^2.$$

In other words, there are quadratic associations between Y and X_1 (holding X_2 and X_3 constant) and between Y and X_3 (holding X_1 and X_2 constant), but a linear association between Y and X_2 (holding X_1 and X_3 constant). A subject-matter theory hypothesizing these associations may have suggested this model, or, when analyzing the data, we may have discovered that the squared terms, X_1^2 and X_3^2, had small, highly significant two-tail p-values.

This model preserves hierarchy because it includes X_1 and X_3 along with X_1^2 and X_3^2. As long as the b_2 and b_5 p-values are small enough to suggest retaining X_1^2 and X_3^2 in the model, we would not usually do hypothesis tests for the regression parameters b_1 and b_4.

When we use polynomial transformations in a multiple linear regression model, interpretation of the regression parameters again becomes more difficult. There are algebraic methods to derive meaningful interpretations for polynomial model regression parameters (see the optional part at the end of this section on page 158). Alternatively, we can construct a predictor effect plot—see Section 5.5—to show the effect of changes in a predictor variable on Y. The curved fitted line in Figure 4.6 is effectively a predictor effect plot for the home prices–age example and shows that *Price* decreases quite steeply as *Age* increases between 0 and 20, levels off for *Age* between 20 and 70, and then increases more steeply as *Age* increases between 70 and 100. Again, quick calculations can quantify these changes more precisely if needed.

In practice, when we use a polynomial transformation for a predictor in a multiple linear regression model, it is common to first *rescale* the values of the predictor to have a mean close to 0 and a standard deviation close to 1 (see Problem 5.3 at the end of Chapter 5 and Section 6.1 for examples of this). This is to avoid numerical problems with estimating the regression model in cases where large values of the predictor would become *very* large if squared or cubed. Rescaling can also reduce multicollinearity problems (between X and X^2, say)—see Section 5.2.3.

It is also possible to apply a power transformation to the response variable—see page 157 for more details.

Optional—mathematical justification for preserving hierarchy when using polynomial transformations.   One reason for preserving hierarchy when using polynomial transformations is that if, for example, we were to include X^2 but not X in a model, then we are making an implicit assumption about the gradient (partial derivative) of the regression equation with respect to X (holding all else constant). In particular, when we differentiate the equation with respect to X, we will have a function with X (multiplied by two times the regression parameter estimate for X^2) and no constant term (which we would have had if we had included X in the model). Then, when we set this partial derivative equal to zero and solve for X, we obtain the solution, $X = 0$. This implies that the gradient of the regression equation in the X-direction is zero at the value $X = 0$ (i.e. there is a local minimum or maximum here). That is quite an assumption to make, so conventionally many regression analysts prefer not to make this assumption. Instead, whenever there is a significant X^2-term in the regression model, they keep X in the model (regardless of its significance). In this way, the data are allowed to decide where there are any local minima or maxima in the regression equation (without forcing them arbitrarily to be at $X = 0$).

For practical purposes, this issue is rarely particularly worrisome. If we have a significant X^2-term (low p-value) coupled with a nonsignificant X-term (relatively high p-value), the data are telling us that there is a local minimum or maximum close to $X = 0$. We could then force the local minimum or maximum to be exactly at $X = 0$ by breaking hierarchy and dropping X from the model. Or, we could just leave X in the model (preserving hierarchy), even though it is not significant. In many cases, both models will be pretty similar in all other respects (regression parameter estimates for the other predictors, model predictions, etc.).

4.1.3   Reciprocal transformation for predictors

The *reciprocal transformation* (sometimes called the *multiplicative inverse transformation*) can be useful in some regression applications where we expect a nonlinear, inverse association between the response variable and a particular predictor variable (adjusting for other predictors in the model). Such circumstances can occur in physical science or engineering applications, for example. To illustrate, consider the fuel efficiency in city miles per gallon (*Cmpg*) and engine size in liters (*Eng*) for 351 new U.S. passenger cars for the 2011 model year. These data, available in the **CARS3** data file, come from a larger dataset (obtained from www.fueleconomy.gov), which is analyzed more fully in a case study in Section 6.2. Figure 4.7 displays the **CARS3** data together with a fitted line from a model that uses the untransformed predictor variable, *Eng* (see computer help #15 and #26 in the software

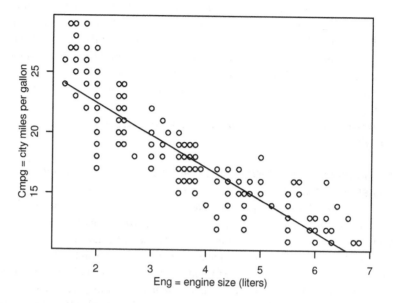

Figure 4.7   Scatterplot of *Cmpg* versus *Eng* for the cars example with a fitted line from E(*Cmpg*) = $b_0 + b_1$ *Eng*.

information files available from the book website):

$$\text{Model 1}: \quad \text{E}(Cmpg) = b_0 + b_1\,Eng.$$

Model 1 seems inappropriate here since the fitted line misses some apparent curvature in the dominant association between the points. Statistical software output for this model (see computer help #25) is:

Model Summary

Model	Sample Size	Multiple R Squared	Adjusted R Squared	Regression Std. Error
1[a]	351	0.7513	0.7506	1.929

[a] Predictors: (Intercept), *Eng*.

Parameters[a]

Model		Estimate	Std. Error	t-stat	Pr($>$ |t|)
1	(Intercept)	27.799	0.298	93.210	0.000
	Eng	−2.672	0.082	−32.474	0.000

[a] Response variable: *Cmpg*.

Contrast this with Figure 4.8, which displays *Cmpg* versus the reciprocal of *Eng* (i.e., 1/*Eng*), together with a fitted line from this model (see computer help #15 and #26):

$$\text{Model 2}: \quad \text{E}(Cmpg) = b_0 + b_1(1/Eng).$$

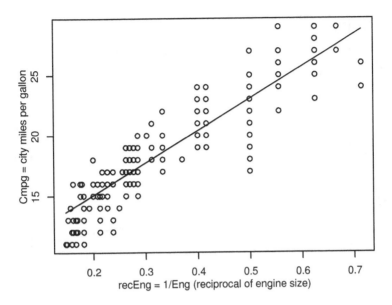

Figure 4.8   Scatterplot of *Cmpg* versus $1/Eng$ for the cars example with a fitted line from $\mathrm{E}(Cmpg) = b_0 + b_1(1/Eng)$.

Statistical software output for this model (see computer help #25) is:

Model Summary

Model	Sample Size	Multiple R Squared	Adjusted R Squared	Regression Std. Error
2[a]	351	0.7975	0.7969	1.741

[a] Predictors: (Intercept), $1/Eng$.

Parameters[a]

Model		Estimate	Std. Error	t-stat	$\Pr(>|t|)$
2	(Intercept)	9.703	0.260	37.302	0.000
	$1/Eng$	26.707	0.720	37.071	0.000

[a] Response variable: *Cmpg*.

We know from Section 3.3 that model 2 (which uses $1/Eng$) provides a more useful description of the association between fuel efficiency and engine size than model 1 (which uses untransformed *Eng*). The regression standard error, *s*, is lower in model 2 (1.741) than in model 1 (1.929), so model 2 will tend to be more accurate when used to predict a car's fuel efficiency from its engine size. Also, R^2 is higher in model 2 (79.7%) than in model 1 (75.1%), so we can explain more of the variation in *Cmpg* using $1/Eng$ in a regression model than using *Eng*. Finally, the (two-tail) p-value for testing whether the regression parameter b_1 could be zero in the population is lower in model 2 than in model 1, so there is more evidence of a nonzero value for b_1 using $1/Eng$ in a regression model than using

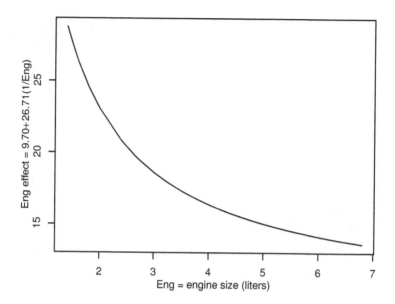

Figure 4.9   Predictor effect plot of *Eng* effect $= 9.70 + 26.71(1/Eng)$ versus *Eng* for the cars example.

Eng. (Although both p-values are 0.000 to three decimal places, we know that model 2 has a lower p-value because its t-statistic is farther from zero—37.1 in model 2 versus -32.5 in model 1.)

Note that the notion of hierarchy does not apply to the reciprocal transformation. In other words, there is no particular reason why we would retain X in a model that includes $1/X$.

As with the other transformations we have considered, we can apply the reciprocal transformation to any (or all) of the predictor variables in a multiple linear regression model. For example, in a dataset with three potential predictors, X_1, X_2, and X_3, we might find that the following model best represents the association between a response variable, Y, and the three predictors:

$$E(Y) = b_0 + b_1(1/X_1) + b_2(1/X_2) + b_3 X_3.$$

In other words, there are curved associations between Y and X_1 (holding X_2 and X_3 constant) and between Y and X_2 (holding X_1 and X_3 constant), but a linear association between Y and X_3 (holding X_1 and X_2 constant). This model may have been suggested by an economic theory that hypothesized these associations, or when analyzing the data we may have noticed that models with $1/X_1$ and $1/X_2$ seemed to fit better than models with X_1 and X_2.

When we use reciprocal transformations in a multiple linear regression model, interpretation of the regression parameters again becomes more difficult. As before, we can construct a predictor effect plot to show the effect on Y of changes in a predictor variable. Figure 4.9 displays a predictor effect plot for the cars example (see computer help #42). We discuss plots such as this more fully in Section 5.5, but for this example it simply displays the association between $\widehat{Cmpg} = 9.70 + 26.71(1/Eng)$ and *Eng*. We can use the axes on the graph to read off approximate answers to questions like "What happens to *Cmpg* when

Eng increases from A to B?" This will be sufficient in circumstances where all we need is a quick visual impression of the association modeled between *Cmpg* and *Eng*. In this case, the plot shows that *Cmpg* decreases steeply as *Eng* increases from 1.4, but the rate of change of the decrease lessens as *Eng* approaches 6.8. Again, we can do some quick calculations to quantify these changes more precisely if needed.

It is also possible to apply a reciprocal transformation to the response variable. We use this approach in the car fuel efficiency examples in Sections 4.3.2 and 6.2, where rather than modeling "city miles per gallon" (*Cmpg*) we instead model "city gallons per hundred miles" (100/*Cmpg*).

4.1.4   Natural logarithm transformation for the response

The *natural logarithm transformation* is sometimes applied to the response variable in a linear regression model. For example, consider the data available in the **WORKEXP** data file, which contains simulated data for the salaries in thousands of dollars (*Sal*) and years of experience (*Exp*) for a sample of 50 workers. Figure 4.10 displays the data together with a fitted line from a model that uses the untransformed *Sal* as the response variable (see computer help #15 and #26 in the software information files available from the book website):

$$\text{Model 1}: \quad \text{E}(Sal) = b_0 + b_1\, Exp.$$

Model 1 seems inappropriate here since the fitted line misses apparent curvature in the association between the points.

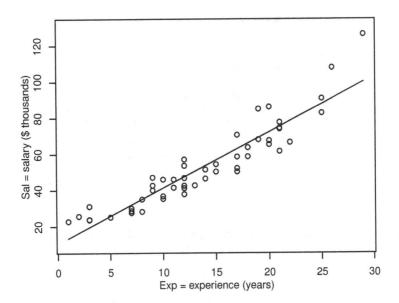

Figure 4.10   Scatterplot of *Sal* versus *Exp* for the work experience example with a fitted line from $\text{E}(Sal) = b_0 + b_1\, Exp$.

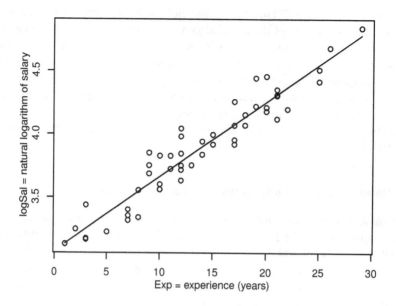

Figure 4.11 Scatterplot of $\log_e(Sal)$ versus Exp for the work experience example with a fitted line from $E(\log_e(Sal)) = b_0 + b_1\,Exp$.

Statistical software output for this model (see computer help #25) is:

Model Summary

Model	Sample Size	Multiple R Squared	Adjusted R Squared	Regression Std. Error
1 [a]	50	0.8647	0.8619	8.357

[a] Predictors: (Intercept), *Exp*.

Parameters [a]

Model		Estimate	Std. Error	t-stat	Pr($> |t|$)
1	(Intercept)	10.323	2.706	3.815	0.000
	Exp	3.094	0.177	17.515	0.000

[a] Response variable: *Sal*.

Contrast this with Figure 4.11, which displays $\log_e(Sal)$ versus Exp, together with a fitted line from this model (see computer help #15 and #26):

$$\text{Model 2}: \quad E(\log_e(Sal)) = b_0 + b_1\,Exp.$$

Statistical software output for this model (see computer help #25) is:

Model Summary

Model	Sample Size	Multiple R Squared	Adjusted R Squared	Regression Std. Error
2^a	50	0.9120	0.9102	0.1248

[a] Predictors: (Intercept), *Exp*.

Parameters[a]

Model		Estimate	Std. Error	t-stat	$\Pr(>|t|)$
2	(Intercept)	3.074	0.040	76.089	0.000
	Exp	0.059	0.003	22.302	0.000

[a] Response variable: $\log_e(Sal)$.

Model 2 (which uses $\log_e(Sal)$ as the response variable) appears to provide a more useful description of the association between salary and experience than model 1 (which uses untransformed *Sal*). The (two-tail) p-value for testing whether the regression parameter b_1 could be zero in the population is lower in model 2 than in model 1, so there is more evidence of a nonzero value for b_1 when using $\log_e(Sal)$ in a regression model than when using *Sal*. (Although both p-values are 0.000 to three decimal places, we know that model 2 has a lower p-value because its t-statistic is farther from zero—22.3 in model 2 versus 17.5 in model 1.)

What about the other two numerical methods we have for evaluating models: the regression standard error, s, and the coefficient of determination, R^2? Suppose that we call the regression standard error s_1 for model 1 and s_2 for model 2. Recall that the regression standard error has the same units as the response variable in a regression model. Thus, for this example, the size of s *cannot* be compared across the two models to see which one will predict more accurately—s_1 is measured in thousands of dollars but s_2 is measured in the natural logarithm of thousands of dollars.

It might seem that we could correct this problem by comparing $\exp(s_2)$ to s_1 so that both measurements are in thousands of dollars. However, this would not be correct either. An approximate 95% prediction interval for *Sal* at a particular value of *Exp* for model 1 is simply $\widehat{Sal} \pm 2s_1$. Similarly, an approximate 95% prediction interval for $\log_e(Sal)$ at the same value of *Exp* for model 2 is $\widehat{\log_e(Sal)} \pm 2s_2$. To find the corresponding 95% prediction interval for *Sal* we exponentiate the endpoints to get $\exp(\widehat{\log_e(Sal)} - 2s_2)$ and $\exp(\widehat{\log_e(Sal)} + 2s_2)$. However, if we tried to compare s_1 to $\exp(s_2)$, this would imply that the endpoints of an approximate 95% prediction interval for *Sal* at the same value of *Exp* for model 2 are $\exp(\widehat{\log_e(Sal)}) - 2\exp(s_2)$ and $\exp(\widehat{\log_e(Sal)}) + 2\exp(s_2)$. If you're not convinced that these two sets of endpoints for model 2 are different, try it for *Exp* = 10 to see. You should find that $\widehat{\log_e(Sal)} = 3.074 + 0.59 = 3.664$, so the correct approximate 95% prediction interval is from $\exp(\widehat{\log_e(Sal)} - 2s_2) = \exp(3.664 - 2(0.1248)) = \exp(3.4144) = 30.4$ to $\exp(\widehat{\log_e(Sal)} + 2s_2) = \exp(3.664 + 2(0.1248)) = \exp(3.9136) = 50.1$. However, the incorrect approximate 95% prediction interval is from $\exp(\widehat{\log_e(Sal)}) - 2\exp(s_2) = \exp(3.664) - 2\exp(0.1248) = 36.8$ to $\exp(\widehat{\log_e(Sal)}) + 2\exp(s_2) = \exp(3.664) + 2\exp(0.1248) = 41.3$.

By contrast, R^2 is unit-free and so can be used *informally* to compare the two models. For example, we can explain 91.2% of the variation in $\log_e(Sal)$ using model 2, but only 86.5% of the variation in *Sal* using model 1. However, since the two numbers measure different things ("proportion of variation in $\log_e(Sal)$ explained by model 2" for the former and "proportion of variation in *Sal* explained by model 1" for the latter), they cannot be compared directly in any *formal* way.

An alternative way to see which model provides a more useful description of the association between salary and experience is to consider the regression assumptions of Section 3.4. For example, the data points in Figure 4.10 tend to be closer to the line at the left of the plot (lower values of *Exp*) than at the right of the plot (higher values of *Exp*). This violates the constant variance assumption for the random errors in a linear regression model, since the variance of the estimated errors in the plot seems to be increasing from left to right. This often happens when using a positively skewed response variable (i.e., with a few response values much higher than the majority of the response values) in a multiple linear regression model. The higher values of the response variable are harder to predict than the lower values, which results in the increasing variance pattern in Figure 4.10.

Recall from Figure 4.4 on page 142 that the natural logarithm transformation works well with positively skewed variables, since it tends to spread out the more common lower values and "pull in" the less common higher values. Thus, regression models for positively skewed response variables, Y, often use $\log_e(Y)$ rather than Y. The use of the natural logarithm transformation in this way often corrects the increasing variance problem, as it does in this case and as illustrated in Figure 4.11 (where the variance of the estimated errors in the plot seems to be approximately constant from left to right). We return to the issue of violation of the constant variance assumption in Section 5.2.1.

The natural logarithm transformation for the response also arises naturally in the context of *multiplicative* models, such as

$$Y = \exp(b_0)\exp(b_1 X_1)\exp(b_2 X_2)\exp(e).$$

Taking the natural logarithm of both sides leads to a multiple linear regression model with response $\log_e(Y)$:

$$\log_e(Y) = b_0 + b_1 X_1 + b_2 X_2 + e.$$

Models like this work well when a unit change in a predictor leads to a *proportional* change in Y rather than a constant change in Y. To see this, consider the expected change in Y when we increase X_1 by 1 unit (and hold X_2 constant) in this model:

$$
\begin{aligned}
\text{expected change in Y} &= \exp(b_0)\exp(b_1(X_1+1))\exp(b_2 X_2) \\
&\quad - \exp(b_0)\exp(b_1 X_1)\exp(b_2 X_2) \\
&= \exp(b_1)\,[\exp(b_0)\exp(b_1 X_1)\exp(b_2 X_2)] \\
&\quad - \exp(b_0)\exp(b_1 X_1)\exp(b_2 X_2) \\
&= [\exp(b_1)-1]\,[\exp(b_0)\exp(b_1 X_1)\exp(b_2 X_2)].
\end{aligned}
$$

We can express this change relative to the expected value of Y *before* we increased X_1 by 1 unit:

$$
\begin{aligned}
\text{expected proportional change in Y} &= \frac{[\exp(b_1)-1]\,[\exp(b_0)\exp(b_1 X_1)\exp(b_2 X_2)]}{\exp(b_0)\exp(b_1 X_1)\exp(b_2 X_2)} \\
&= \exp(b_1)-1.
\end{aligned}
$$

In other words, to interpret b_1 in this model, we first need to exponentiate it and subtract 1. Then the result of this calculation tells us the expected proportional change in Y from increasing X_1 by 1 unit (and holding X_2 constant). Similarly, $\exp(b_2)-1$ would represent the expected proportional change in Y from increasing X_2 by 1 unit (and holding X_1 constant).

For example, in the work experience dataset, our estimated model was

$$\widehat{\log_e(Sal)} = \hat{b}_0 + \hat{b}_1\,Exp$$
$$= 3.074 + 0.059\,Exp.$$

Thus, since $\exp(\hat{b}_1)-1 = \exp(0.059)-1 = 0.0608$, we expect $Sal =$ salary in dollars to increase by a multiplicative factor of 0.0608 (or 6.08%) for each additional year of experience (Exp). In other words, multiply Sal by 0.0608 (or 6.08%) and add, or equivalently, multiply Sal by 1.0608 (or 106.08%).

Another context in which the natural logarithm transformation for the response is common is when response values cannot be negative. For example, consider modeling the weekly sales of an item, $Sales$. By transforming $Sales$ to $\log_e(Sales)$, our regression model will give us predictions for $\log_e(Sales)$, which we can then exponentiate to get predictions for $Sales$. Since the exponentiation function always gives us positive numbers, we won't obtain any nonsensical negative predicted sales.

4.1.5 Transformations for the response and predictors

We can include transformations for the response *and* the predictors in a multiple linear regression model. For example, consider the data available in the **HOMETAX** data file, which contains the annual taxes in $ (Tax) and sale price in $ thousands ($Price$) for a sample of 104 homes sold in Albuquerque, New Mexico in 1993. These data were obtained from the "Data and Story Library" at lib.stat.cmu.edu/DASL/ and have been modified slightly to remove missing values and other problematic observations. This type of data is collected by multiple listing agencies in many cities and is used by realtors as an information base (the original data source in this case is the Albuquerque Board of Realtors). Figure 4.12 displays the data together with a fitted line from a model that uses untransformed Tax and $Price$ (see computer help #15 and #26 in the software information files available from the book website):

$$\text{Model 1}: \quad \text{E}(Tax) = b_0 + b_1\,Price.$$

Model 1 seems inappropriate here since the data points are closer to the line at the left of the plot (for lower values of $Price$) than at the right of the plot (for higher values of $Price$). This violates the constant variance assumption for the random errors in a linear regression model since the variance of the estimated errors in the plot seems to be increasing from left to right.

Contrast this with Figure 4.13, which displays $\log_e(Tax)$ versus $\log_e(Price)$, together with a fitted line from this model (see computer help #15 and #26):

$$\text{Model 2}: \quad \text{E}(\log_e(Tax)) = b_0 + b_1 \log_e(Price).$$

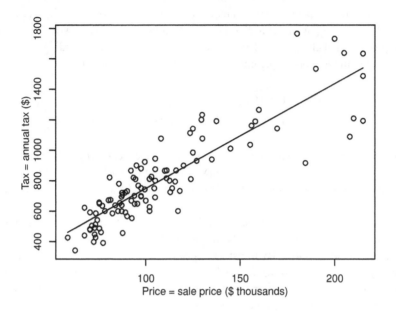

Figure 4.12   Scatterplot of *Tax* versus *Price* for the homes taxes example with a fitted line from $\mathrm{E}(Tax) = b_0 + b_1\, Price$.

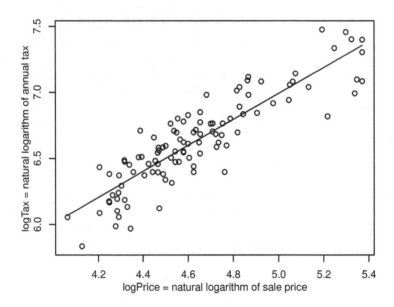

Figure 4.13   Scatterplot of $\log_e(Tax)$ versus $\log_e(Price)$ for the home taxes example with a fitted line from $\mathrm{E}(\log_e(Tax)) = b_0 + b_1 \log_e(Price)$.

Statistical software output for this model (see computer help #25) is:

Model Summary

Model	Sample Size	Multiple R Squared	Adjusted R Squared	Regression Std. Error
2 [a]	104	0.7846	0.7825	0.1616

[a] Predictors: (Intercept), $\log_e(Price)$.

Parameters [a]

Model		Estimate	Std. Error	t-stat	$\Pr(>|t|)$
2	(Intercept)	2.076	0.237	8.762	0.000
	$\log_e(Price)$	0.983	0.051	19.276	0.000

[a] Response variable: $\log_e(Tax)$.

Model 2 is just as easy to use as model 1 for estimating or predicting annual taxes from home sale prices. For example, a home that sold for $100,000 would be expected to have annual taxes of approximately $\exp(2.076 + 0.983 \times \log_e(100)) = \$737$. Similarly, the lower and upper limits of confidence intervals for the mean and prediction intervals for individual values can be exponentiated to obtain confidence and prediction intervals in dollars (rather than log-dollars). These intervals should be more accurate than intervals based on model 1, which will tend to be too wide for low sale prices and too narrow for high sale prices. Experiment with the dataset to see this. For example, you should find that 95% prediction intervals for model 1 at $100k, $150k, and $200k are ($466, $1,032), ($808, $1,377), and ($1,146, $1,727), while the more appropriate intervals for model 2 are ($534, $1,018), ($794, $1,520), and ($1,049, $2,026)—see also Problem 4.2 at the end of the chapter.

A more complicated example for a generic application with response variable Y and predictor variables X_1, X_2, X_3, and X_4 might be

$$E(\log_e(Y)) = b_0 + b_1 X_1 + b_2 X_2 + b_3 X_2^2 + b_4 \log_e(X_3) + b_5(1/X_4).$$

Here, we have transformed Y using the natural logarithm transformation, left X_1 untransformed, transformed X_2 using the quadratic transformation (also including X_2 to preserve hierarchy), transformed X_3 using the natural logarithm transformation, and transformed X_4 using the reciprocal transformation.

We've covered three transformations that are relatively common in practice: logarithms, polynomials, and reciprocals. However, any mathematical function we can think of can be used to transform data. There are even automated methods available for selecting transformations. One such method, proposed by Box and Cox (1964), seeks the best power transformation of the response variable so that the residuals from the regression of the transformed response on the predictors are as close to normally distributed as possible. The selected power does not have to be a whole number, but is typically rounded to something sensible (e.g., if the method determines the best power is 1.8, this might be rounded to 2). The routine is actually based on a "modified power family" that includes nonzero powers such as squares (power 2), square roots (power 0.5), and reciprocals (power −1), but also defines a power of 0 to correspond to a logarithmic transformation. This is sometimes referred to as selecting a transformation from the *ladder of powers* (an orderly way of expressing these modified power transformations proposed by Tukey, 1977). Many statistical software packages have a Box-Cox transformation routine. There is also a multivariate extension of the Box-Cox method (see Velilla, 1993) that can be used to suggest transformations of quantitative predictors. This method seeks to transform a set of variables so that the transformed variables have as strong a linear association with one another as possible. Although the multiple linear regression model does not require predictors to have linear associations (or that at least do not have strong nonlinear associations), it can provide a useful starting point for building regression models. For more details, see Weisberg (2005). Alternatively, Box and Tidwell's (1962) method can be used to suggest modified power transformations of the predictor variables—see Fox and Weisberg (2011) for details.

Another useful tool for identifying appropriate predictor transformations is the *partial residual plot*, also known as the *component plus residual plot*. Cook and Weisberg (1999) provide an excellent introduction to the *Ceres plot* (**c**ombining conditional **e**xpectations and **res**iduals), which is a generalization of the partial residual plot (see also Cook, 1993). Some of the ideas underlying these plots form the basis for *generalized additive modeling*, an iterative model building process based on flexible predictor transformations (see Hastie and Tibshirani, 1990).

Despite the availability of automated methods for selecting transformations, it is better if the transformations in a regression model have been suggested *before* looking at the data from background knowledge about the situation at hand, or from theoretical arguments about why certain transformations make sense in this setting. In the absence of such knowledge, we can use automated methods or try out certain transformations to see if we can find a useful model that fits the data well. One danger with this approach is the potential for overfitting, so that the final model fits the sample data well but generalizes poorly to the wider population. Another danger relates to interpretation of the final model—if there are too many complicated transformations, then a model can be hard to interpret and use. The predictor effect plots suggested in Section 5.5 address this issue, as does the following optional section. Nevertheless, used judiciously, variable transformations provide a useful tool for improving regression models.

Optional—regression parameter interpretations for transformed predictors. One approach to interpreting regression parameters for transformed predictors is to calculate how Y changes as each predictor changes by 1 unit (and all other predictors are held constant). For example, consider the change in Y when we increase X_1 by 1 unit (and all other predictors are held constant) in the model $E(Y) = b_0 + b_1 X_1 + b_2 X_2 + b_3 X_2^2 + b_4 \log_e(X_3) + b_5(1/X_4)$:

$$\text{expected change in Y} = b_0 + b_1(X_1+1) + b_2 X_2 + b_3 X_2^2 + b_4 \log_e(X_3) + b_5(1/X_4)$$
$$- b_0 - b_1 X_1 - b_2 X_2 - b_3 X_2^2 - b_4 \log_e(X_3) - b_5(1/X_4)$$
$$= b_1.$$

This provides an alternative justification of the usual interpretation for a regression parameter for an untransformed predictor: b_1 in this model represents the expected change in Y for a 1-unit increase in X_1 (holding all other predictors constant).

Next, consider the expected change in Y when we increase X_2 by 1 unit in this model (and all other predictors are held constant):

$$\begin{aligned}
\text{expected change in Y} &= b_0 + b_1 X_1 + b_2(X_2+1) + b_3(X_2+1)^2 \\
&\quad + b_4 \log_e(X_3) + b_5(1/X_4) - b_0 - b_1 X_1 \\
&\quad - b_2 X_2 - b_3 X_2^2 - b_4 \log_e(X_3) - b_5(1/X_4) \\
&= b_2 + b_3(2X_2+1) = b_2 + 2b_3\left(\frac{2X_2+1}{2}\right).
\end{aligned}$$

Thus, $b_2 + b_3(2X_2+1)$ represents the expected change in Y for a 1-unit increase in X_2 (holding all other predictors constant). This expected change therefore depends on the starting value of X_2 as well as the values of b_2 and b_3. The quantity $(2X_2+1)/2$ is halfway between X_2 and X_2+1, so this expected change in Y is also equal to the partial derivative of the regression equation with respect to X_2 at this halfway point (i.e., the derivative of the regression equation with respect to X_2, holding all else constant).

The expected change in Y when we increase X_3 by 1 unit in this model (and all other predictors are held constant) is not a simple algebraic expression that we can easily interpret. Instead, consider the expected change in Y when we multiply X_3 by a constant w (and all other predictors are held constant):

$$\begin{aligned}
\text{expected change in Y} &= b_0 + b_1 X_1 + b_2 X_2 + b_3 X_2^2 + b_4 \log_e(wX_3) + b_5(1/X_4) \\
&\quad - b_0 - b_1 X_1 - b_2 X_2 - b_3 X_2^2 - b_4 \log_e(X_3) - b_5(1/X_4) \\
&= b_4 \log_e(w).
\end{aligned}$$

Thus, $b_4 \log_e(w)$ represents the expected change in Y when we multiply X_3 by a constant w (and all other predictors are held constant).

If we use a base-2 logarithmic transformation here rather than the natural logarithm, then b_4 represents the expected change in Y when we multiply X_3 by 2 (and all other predictors are held constant). This is because $\log_2(2) = 1$. Similarly, if we use a base-10 logarithmic transformation, then b_4 represents the expected change in Y when we multiply X_3 by 10 (and all other predictors are held constant), since $\log_{10}(10) = 1$.

Finally, the expected change in Y when we increase X_4 by 1 unit in this model (and all other predictors are held constant) is not a simple algebraic expression that we can easily interpret.

4.2 INTERACTIONS

We can model *interactions* within the multiple linear regression framework when the association between one predictor, X_1, say, and the response variable, Y, depends on (or varies according to) the value of another predictor, X_2, say. We can model the interaction in this situation by including the term $X_1 X_2$ in our model—the value of $X_1 X_2$ for each sample observation is simply the corresponding values of X_1 and X_2 multiplied together.

For example, suppose that for a small luxury goods business we suspect that the association between *Sales* = annual sales (in $m) and *Advert* = annual spending on advertising (in $m) varies according to *Interest* = the prevailing interest rate (in %). We have available the following (simulated) data in the **SALES1** data file to investigate this possibility:

Sales (sales in $m)	4.0	6.0	8.0	2.0	4.5	9.0	4.5	8.0	10.5	5.0	10.0	14.5
Advert (advertising in $m)	3.5	5.5	7.0	1.0	3.0	6.5	2.0	4.0	6.0	1.0	4.0	7.0
Interest (interest rate in %)	5	5	5	4	4	4	3	3	3	2	2	2

To allow for the possibility that the association between *Sales* and *Advert* depends on the value of *Interest*, we calculate the interaction term, $AdvInt = Advert \times Interest$, and include *AdvInt* in an interaction model:

$$\mathrm{E}(Sales) = b_0 + b_1 Advert + b_2 Interest + b_3 AdvInt.$$

One way to see that this model allows the association between *Sales* and *Advert* to depend on the value of *Interest* is to consider the expected change in *Sales* when we increase *Advert* by 1 unit (and hold *Interest* constant):

$$\text{expected change in } Sales = b_0 + b_1(Advert+1) + b_2 Interest + b_3(Advert+1)Interest$$
$$- b_0 - b_1 Advert - b_2 Interest - b_3(Advert \times Interest)$$
$$= b_1 + b_3 Interest.$$

This expected change therefore depends on the value at which *Interest* is being held constant as well as the values of b_1 and b_3. This is essentially the "*Advert* effect" on *Sales*, an example of the *predictor effects* that we consider in Section 5.5. Statistical software output for this model (see computer help #31 in the software information files available from the book website) is:

Model Summary

Model	Sample Size	Multiple R Squared	Adjusted R Squared	Regression Std. Error
1 [a]	12	0.9944	0.9923	0.3075

[a] Predictors: (Intercept), *Advert*, *Interest*, *AdvInt*.

Parameters [a]

Model		Estimate	Std. Error	t-stat	Pr(> |t|)
1	(Intercept)	5.941	0.662	8.979	0.000
	Advert	1.836	0.135	13.611	0.000
	Interest	−1.312	0.197	−6.669	0.000
	AdvInt	−0.126	0.039	−3.261	0.012

[a] Response variable: *Sales*.

The estimated regression equation is therefore

$$\widehat{Sales} = \hat{b}_0 + \hat{b}_1 Advert + \hat{b}_2 Interest + \hat{b}_3 AdvInt$$
$$= 5.941 + 1.836 Advert - 1.312 Interest - 0.126 AdvInt.$$

We estimate the expected change in *Sales* when we increase *Advert* by 1 unit (and hold *Interest* constant) to be

$$\text{estimated expected change in } Sales = \hat{b}_1 + \hat{b}_3 Interest$$
$$= 1.836 - 0.126 Interest.$$

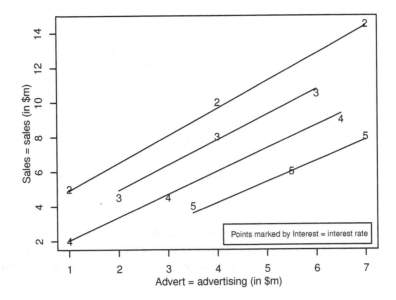

Figure 4.14   Scatterplot of *Sales* versus *Advert* with the points marked according to the value of *Interest* for the sales–advertising–interest rates example with fitted lines from the interaction model $E(Sales) = b_0 + b_1\,Advert + b_2\,Interest + b_3\,AdvInt$. Since the sample fitted lines are relatively far from being exactly parallel, this suggests that an interaction model is appropriate.

So, for example, when the prevailing interest rate is 2%, we expect to increase sales by $1.836 - 0.126(2) = \$1.58\text{m}$ for each additional \$1m we spend on advertising, but when the prevailing interest rate is 5%, we expect to increase sales by only $1.836 - 0.126(5) = \$1.21\text{m}$. As further practice of this concept, calculate the expected increase in sales for each additional \$1m we spend on advertising when the prevailing interest rate is 3% (see also Problem 4.4 at the end of the chapter)—you should find that it is \$1.46m.

We can gain further insight into this particular interaction model by taking advantage of the fact that there are only two predictor variables, and one of them (*Interest*) has only four distinct values—Figure 4.14 illustrates (see computer help #33). Each line represents the estimated linear association between *Sales* and *Advert* at fixed values of *Interest*. For example, the upper line with the steepest slope (which we just calculated to be 1.58) represents the association when *Interest*=2%. Similarly, the lower line with the shallowest slope (which we just calculated to be 1.21) represents the association when *Interest*=5%. Overall, this graph suggests that sales increase as advertising increases and interest rates decrease, but the rate of increase in sales as advertising increases depends on the interest rate.

Geometrically, the presence of interaction in the model results in nonparallel fitted lines in Figure 4.14. However, since we can only graph our sample data, is it possible that in our population the equivalent lines could be exactly parallel instead? With statistics anything is possible, so to investigate this possibility we need to conduct a hypothesis test for the interaction term in the model. Assuming that we have no prior expectation about the sign (positive or negative) of b_3 in the model $E(Sales) = b_0 + b_1\,Advert + b_2\,Interest + b_3\,AdvInt$, we should conduct a two-tail test.

- State null hypothesis: NH: $b_3 = 0$.
- State alternative hypothesis: AH: $b_3 \neq 0$.
- Calculate test statistic: t-statistic $= \dfrac{\hat{b}_3 - b_3}{s_{\hat{b}_3}} = \dfrac{-0.126 - 0}{0.039} = -3.26$.
- Set significance level: 5%.
- Look up t-table:
  - critical value: The 97.5th percentile of the t-distribution with $12 - 3 - 1 = 8$ degrees of freedom is 2.31 (see computer help #8 in the software information files available from the book website); the rejection region is therefore any t-statistic greater than 2.31 or less than -2.31 (we need the 97.5th percentile in this case because this is a two-tail test, so we need half the significance level in each tail).
  - p-value: The sum of the areas to the left of the t-statistic (-3.26) and to the right of the negative of the t-statistic (3.26) for the t-distribution with 8 degrees of freedom is 0.012 (use computer help #9 or observe the p-value of 0.012 in the statistical software output).
- Make decision:
  - Since the t-statistic of -3.26 falls in the rejection region, we reject the null hypothesis in favor of the alternative.
  - Since the p-value of 0.012 is less than the significance level of 0.05, we reject the null hypothesis in favor of the alternative.
- Interpret in the context of the situation: The 12 sample observations suggest that a population regression parameter, b_3, of zero seems implausible—the sample data favor a nonzero value (at a significance level of 5%); in other words, the association between *Sales* and *Advert* does appear to depend on the value of *Interest* (not just in the sample, but in the population, too).

The result of the hypothesis test confirms that we need to model an interaction between *Advert* and *Interest* for these data, since the sample interaction effect is strong enough to suggest a population interaction effect, too.

In this example, we've focused on the association between *Sales* and *Advert* holding *Interest* fixed. However, we could just as easily have focused on the association between *Sales* and *Interest* holding *Advert* fixed. A similar sequence of conclusions and interpretations would have followed—all enabled by including the *AdvInt* interaction term in the model. For example, we estimate the expected change in *Sales* when we increase *Interest* by 1 unit (and hold *Advert* constant) to be $\hat{b}_2 + \hat{b}_3 Advert = -1.312 - 0.126 Advert$. So when we spend \$1m on advertising, we expect to increase sales by $-1.312 - 0.126(1) = -\$1.44$m [or, equivalently, decrease sales by $1.312 + 0.126(1) = \$1.44$m] for each percentage point increase in the prevailing interest rate. However, when we spend \$7m on advertising we expect to decrease sales by $1.312 + 0.126(7) = \$2.19$m for each percentage point increase in the prevailing interest rate. To visualize this we could draw a scatterplot of *Sales* versus *Interest*, with lines representing the regression associations for different values of *Advert*, such as $Advert = 1$ and $Advert = 7$ (try it to see).

For an alternative example, suppose that for a small retail business we suspect that the association between *Sales* = annual sales (in \$m) and *Advert* = annual spending on advertising (in $m) varies according to *Stores* = the number of stores operated. We have collected the following (simulated) data in the **SALES2** data file to investigate this possibility (we looked at this dataset previously on pages 99 and 114):

Sales (sales in $m)	3.8	7.8	7.9	6.5	10.6	13.3	14.7	16.1	18.7	18.8	22.9	24.2
Advert (advertising in $m)	3.5	5.5	7.0	1.0	3.0	6.5	2.0	4.0	6.0	1.0	4.0	7.0
Stores (number of stores)	1	1	1	2	2	2	3	3	3	4	4	4

To allow the association between *Sales* and *Advert* to depend on *Stores*, we calculate the interaction term, $AdvSto = Advert \times Stores$, and include $AdvSto$ in this interaction model:

$$E(Sales) = b_0 + b_1 Advert + b_2 Stores + b_3 AdvSto.$$

Statistical software output for this model (see computer help #31) is:

Model Summary

Model	Sample Size	Multiple R Squared	Adjusted R Squared	Regression Std. Error
1 [a]	12	0.9890	0.9849	0.8112

[a] Predictors: (Intercept), *Advert*, *Stores*, *AdvSto*.

Parameters [a]

Model		Estimate	Std. Error	t-stat	$Pr(> |t|)$
1	(Intercept)	−6.182	1.540	−4.014	0.004
	Advert	1.349	0.297	4.549	0.002
	Stores	6.156	0.519	11.864	0.000
	AdvSto	−0.110	0.102	−1.080	0.312

[a] Response variable: *Sales*.

Figure 4.15 illustrates this model (see computer help #33). Recall that, geometrically, interaction in the model results in nonparallel fitted lines. However, since we can only graph sample data, is it possible that equivalent lines in the population could be exactly parallel instead? It certainly seems possible—more so than in Figure 4.14—so to investigate this possibility we should conduct a hypothesis test for the model interaction term. Assuming that we have no prior expectation about the sign (positive or negative) of b_3 in the model $E(Sales) = b_0 + b_1 Advert + b_2 Stores + b_3 AdvSto$, we should do a two-tail test.

- State null hypothesis: NH: $b_3 = 0$.
- State alternative hypothesis: AH: $b_3 \neq 0$.
- Calculate test statistic: t-statistic $= \dfrac{\hat{b}_3 - b_3}{s_{\hat{b}_3}} = \dfrac{-0.110 - 0}{0.102} = -1.08$.
- Set significance level: 5%.
- Look up t-table:
  - critical value: The 97.5th percentile of the t-distribution with $12 - 3 - 1 = 8$ degrees of freedom is 2.31 (see computer help #8); the rejection region is therefore any t-statistic greater than 2.31 or less than −2.31 (since this is a two-tail test, we need half the significance level in each tail).

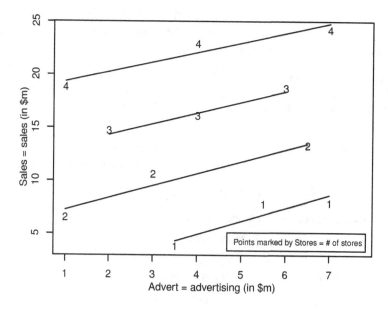

Figure 4.15 Scatterplot of *Sales* versus *Advert* with points marked by *Stores* for the sales–advertising–stores example with fitted lines from the interaction model $E(Sales) = b_0 + b_1 Advert + b_2 Stores + b_3 AdvSto$. The approximately parallel sample fitted lines suggest that the interaction term, *AdvSto*, is not necessary.

- p-value: The sum of the areas to the left of the t-statistic (-1.08) and to the right of the negative of the t-statistic (1.08) for the t-distribution with 8 degrees of freedom is 0.312 (use computer help #9 or observe the p-value of 0.312 in the statistical software output).

- Make decision:

  - Since the t-statistic of -1.08 does not fall in the rejection region, we cannot reject the null hypothesis in favor of the alternative.

  - Since the p-value of 0.312 is more than the significance level of 0.05, we cannot reject the null hypothesis in favor of the alternative.

- Interpret in the context of the situation: The 12 sample observations suggest that a population regression parameter, b_3, of zero is plausible (at a 5% significance level); in other words, the association between *Sales* and *Advert* does *not* appear to depend on *Stores*.

The hypothesis test results suggest that we do not need to include an interaction between *Advert* and *Stores* for these data, since the sample interaction effect is not strong enough to suggest a population interaction effect. In other words, a more appropriate model for these data is

$$E(Sales) = b_0 + b_1 Advert + b_2 Stores.$$

Statistical software output for this model (see computer help #31) is:

Model Summary

Model	Sample Size	Multiple R Squared	Adjusted R Squared	Regression Std. Error
2^a	12	0.9875	0.9847	0.8186

[a] Predictors: (Intercept), *Advert*, *Stores*.

Parameters[a]

Model		Estimate	Std. Error	t-stat	$\Pr(>|t|)$
2	(Intercept)	−4.769	0.820	−5.818	0.000
	Advert	1.053	0.114	9.221	0.000
	Stores	5.645	0.215	26.242	0.000

[a] Response variable: *Sales*.

The estimated regression equation is therefore

$$\widehat{Sales} = -4.77 + 1.05\,Advert + 5.65\,Stores.$$

This model is more appropriate because it excludes the interaction term *AdvSto* (which we determined was unnecessary with the hypothesis test above). Excluding this redundant interaction term allows us to estimate b_1 and b_2 more accurately: The parameter standard errors have decreased from $s_{\hat{b}_1} = 0.297$ and $s_{\hat{b}_2} = 0.519$ for the interaction model to $s_{\hat{b}_1} = 0.114$ and $s_{\hat{b}_2} = 0.215$ for the model without the interaction term.

Note that the term "interaction" in regression means that the association between the response and one of the predictors varies with the value of another predictor. The concept does *not* refer to any association between predictors. Also, the term "no interaction" in regression does *not* mean that there is no association between the response and the predictors. For example, if there is an $X_1 X_2$ interaction in the model, then the association between Y and X_1 can vary with the value of X_2 (and the association between Y and X_2 can vary with the value of X_1). But if there is *no interaction*, there can still be separate associations between the response and each predictor (and these associations are the same whatever the value of the other predictor). For the **SALES1** example illustrated in Figure 4.14 on page 161 there is interaction, so the association between *Sales* and *Advert* varies according to the value of *Interest* (and the association between *Sales* and *Interest* varies according to the value of *Advert*). For the **SALES2** example illustrated in Figure 4.15 on page 164 there is no interaction, so the association between *Sales* and *Advert* is the same for all values of *Stores* (and the association between *Sales* and *Stores* is the same for all values of *Advert*). Figure 3.1 on page 86 is a 3D-scatterplot of a model with two quantitative predictors and no interaction—the slope in the *Party* direction (i.e., the linear association between *Exam* and *Party*) is the same for all values of *Study*, and vice versa. Although there is no interaction (between *Party* and *Study* with respect to their association with *Exam*), there are still separate associations between *Exam* and *Party* and between *Exam* and *Study*. If there were interaction in this context, the regression surface would have a kind of "curved saddle" shape.

The two sales examples demonstrate how interaction terms can improve multiple linear regression modeling (**SALES1**) and when they are unnecessary (**SALES2**). We used the scatterplots in Figures 4.14 and 4.15 to learn about interaction in the context of a model with two quantitative variables, one of which has a small number of distinct whole-number values. However, we can employ interaction terms in more general applications with more than two predictor variables. It is challenging to use scatterplots in such applications to help determine when to include interactions, as we did in the **SALES1** and **SALES2** examples. Instead, to determine if we need interaction terms in most applications, we often combine subject-matter knowledge which suggests that interaction terms might be needed with hypothesis testing to see if those interaction terms are significant. We can even combine the concepts of interaction and transformations (from the preceding section) to produce increasingly sophisticated models capable of capturing quite complex associations between variables.

There are dangers to overusing interactions, just as there are with overusing transformations. At the end of the day, we want to arrive at a model that captures all the important associations in our sample data, but we also need to understand the model results and to generalize the results to the wider population. To balance these goals we need to use interactions judiciously, and they tend to work most effectively if there is some background information that suggests why interactions would be appropriate in a particular application. In the absence of such information, we should use caution when contemplating adding interaction terms to a model.

The notion of hierarchy (see page 145) also applies to interaction terms. Suppose that we have decided to include an interaction term $X_1 X_2$ in a model (and it has a significantly low p-value). Then the principle of hierarchy suggests that there is no need to conduct hypothesis tests for the regression parameters for X_1 or X_2—we retain X_1 and X_2 in the model regardless of their p-values since we have already decided to retain $X_1 X_2$.

In practice, as when using a polynomial transformation for a predictor in a multiple linear regression model, it is common when using interactions first to *rescale* the values of the predictors involved to have means close to 0 and standard deviations close to 1. Again, this can help with numerical estimation problems and multicollinearity issues—see Section 5.2.3.

This section provided an introduction to the topic of interactions in multiple linear regression. For further information, a good place to start is the classic text by Aiken and West (1991).

4.3 QUALITATIVE PREDICTORS

Suppose that we have a *qualitative* (or *categorical*) variable in our sample, for example, gender with values "male" and "female" in a dataset that includes information on annual starting salary and grade point average (GPA) for recent college graduates. If we are interested in modeling the association between salary and GPA, it is reasonable to wonder whether this association differs for males and females. Currently, we have no way of answering this question since our multiple linear regression framework is based on quantitative predictors—variables that take meaningful numerical values and the numbers represent actual quantities of time, money, weight, and so on. Nevertheless, we can easily incorporate qualitative predictor variables into a multiple linear regression model using *indicator* variables (sometimes called *dummy* variables).

4.3.1 Qualitative predictors with two levels

Indicator variables enable us to model differences between categories represented by qualitative variables (also sometimes called categorical variables). To see exactly how this is accomplished, consider the following (simulated) data in the **SALGPA1** data file:

Salary (salary in $ thousands)	59.8	44.2	52.0	62.4	75.4	70.2	59.8	28.6	39.0	49.4
Gpa (grade point average)	3.0	2.5	1.0	3.5	3.5	3.0	2.0	1.5	2.0	1.5
Gender (M=male, F=female)	M	M	F	M	F	F	F	M	M	F
D_F (0=male, 1=female)	0	0	1	0	1	1	1	0	0	1

The variable D_F in this dataset is an example of an indicator variable for a two-level qualitative variable ("gender" has two levels: male and female). We choose one of the levels to be a *reference* level and record values of $D_F = 0$ for observations in this category. It doesn't really matter which level we choose to be the reference level (although more on this later), so we've arbitraily chosen "male" as the reference level—hence the five males in the dataset have values of $D_F = 0$. We then record values of $D_F = 1$ for observations in the other category—hence the five females in the dataset have values of $D_F = 1$. This indicator variable has already been created in the **SALGPA1** data file; if it had not been, we could have used computer help #7 in the software information files available from the book website to do so.

We can think of the indicator variable as providing a method for translating the qualitative information in the variable "gender" into a quantitative form that the computer can understand. The translation dictionary in this case is then:

Gender	D_F
Male	0
Female	1

How can this help us to model different associations for males and females? Well, consider the model

$$E(Salary) = b_0 + b_1 D_F + b_2 Gpa + b_3 D_F Gpa,$$

which includes the indicator variable, D_F, and an interaction between D_F and *Gpa* (calculated by multiplying each observation's values of D_F and *Gpa* together). We can derive two separate equations from this model by plugging in the value $D_F = 0$ to obtain an equation for males and by plugging in the value $D_F = 1$ to obtain an equation for females:

$$E(Salary) \text{ for males} = b_0 + b_1(0) + b_2 Gpa + b_3(0)Gpa \quad (D_F = 0)$$
$$= b_0 + b_2 Gpa,$$
$$E(Salary) \text{ for females} = b_0 + b_1(1) + b_2 Gpa + b_3(1)Gpa \quad (D_F = 1)$$
$$= b_0 + b_1 + b_2 Gpa + b_3 Gpa$$
$$= (b_0 + b_1) + (b_2 + b_3)Gpa.$$

Statistical software output for this model (see computer help #31) is:

Model Summary

Model	Sample Size	Multiple R Squared	Adjusted R Squared	Regression Std. Error
1 [a]	10	0.9664	0.9496	3.218

[a] Predictors: (Intercept), D_F, Gpa, $D_F Gpa$.

Parameters [a]

Model		Estimate	Std. Error	t-stat	Pr(> |t|)
1	(Intercept)	2.600	5.288	0.492	0.640
	D_F	35.614	6.456	5.516	0.001
	Gpa	17.680	2.035	8.687	0.000
	$D_F Gpa$	−7.159	2.559	−2.797	0.031

[a] Response variable: *Salary*.

The estimated regression equation is therefore

$$\widehat{Salary} = 2.60 + 35.61\,D_F + 17.68\,Gpa - 7.16\,D_F\,Gpa.$$

The separate estimated regression equations for males and females are

$$\widehat{Salary} \text{ for males} = \hat{b}_0 + \hat{b}_2 Gpa$$
$$= 2.60 + 17.68\,Gpa,$$
$$\widehat{Salary} \text{ for females} = (\hat{b}_0 + \hat{b}_1) + (\hat{b}_2 + \hat{b}_3)Gpa$$
$$= (2.600 + 35.614) + (17.680 - 7.159)Gpa$$
$$= 38.21 + 10.52\,Gpa.$$

Figure 4.16 illustrates this model (see computer help #33 in the software information files available from the book website). The fitted line for males is below the fitted line for females with a lower intercept (2.60 versus 38.21) but a steeper slope (17.68 versus 10.52).

The regression parameter estimate for D_F (35.61) represents the *difference* in intercept between the level or category represented by $D_F = 1$ (i.e., 38.21 for females) and the reference level represented by $D_F = 0$ (i.e., 2.60 for males). To see this, note that $35.61 = 38.21 - 2.60$. Similarly, the regression parameter estimate for $D_F Gpa$ (−7.16) represents the *difference* in slope (linear association between *Salary* and *Gpa*) between the level or category represented by $D_F = 1$ (i.e., 10.52 for females) and the reference level represented by $D_F = 0$ (i.e., 17.68 for males). To see this, note that $-7.16 = 10.52 - 17.68$.

We can now return to the question of which level we choose to be the reference level. Suppose that instead of male we choose female as the reference level. We'll use the notation D_M for this indicator variable: The five females in the dataset have values of $D_M = 0$ and the five males have values of $D_M = 1$. Now consider the model:

$$\text{E}(Salary) = b_0 + b_1 D_M + b_2 Gpa + b_3 D_M Gpa.$$

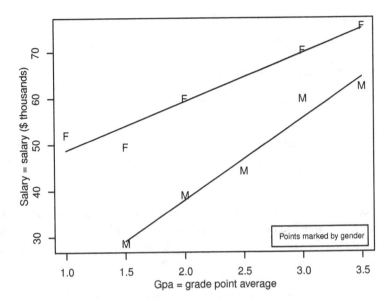

Figure 4.16 Scatterplot of *Salary* versus *Gpa* with the points marked according to gender for the **SALGPA1** data with fitted lines from the model $E(Salary) = b_0 + b_1 D_F + b_2 Gpa + b_3 D_F Gpa$. the nonparallel nature of the sample fitted lines suggests that the interaction term, $D_F Gpa$, is necessary.

We can derive two separate equations from this model by plugging in the value $D_M = 0$ to obtain an equation for females and by plugging in $D_M = 1$ to obtain an equation for males:

$$E(Salary) \text{ for females} = b_0 + b_1(0) + b_2 Gpa + b_3(0)Gpa \quad (D_M = 0)$$
$$= b_0 + b_2 Gpa,$$
$$E(Salary) \text{ for males} = b_0 + b_1(1) + b_2 Gpa + b_3(1)Gpa \quad (D_M = 1)$$
$$= (b_0 + b_1) + (b_2 + b_3)Gpa.$$

Statistical software output for this model (see computer help #31) is:

Model Summary

Model	Sample Size	Multiple R Squared	Adjusted R Squared	Regression Std. Error
2[a]	10	0.9664	0.9496	3.218

[a] Predictors: (Intercept), D_M, Gpa, $D_M Gpa$.

Parameters[a]

Model		Estimate	Std. Error	t-stat	Pr(> |t|)
2	(Intercept)	38.214	3.705	10.314	0.000
	D_M	−35.614	6.456	−5.516	0.001
	Gpa	10.521	1.552	6.780	0.001
	$D_M Gpa$	7.159	2.559	2.797	0.031

[a] Response variable: *Salary*.

The estimated regression equation is therefore

$$\widehat{Salary} = 38.21 - 35.61\,D_M + 10.52\,Gpa + 7.16\,D_M Gpa.$$

The separate estimated regression equations for females and males are

$$
\begin{aligned}
\widehat{Salary} \text{ for females} &= \hat{b}_0 + \hat{b}_2 Gpa \\
&= 38.21 + 10.52\,Gpa, \\
\widehat{Salary} \text{ for males} &= (\hat{b}_0 + \hat{b}_1) + (\hat{b}_2 + \hat{b}_3)Gpa \\
&= (38.214 - 35.614) + (10.521 + 7.159)Gpa \\
&= 2.60 + 17.68\,Gpa.
\end{aligned}
$$

In other words, the separate estimated regression equations for females and males are exactly the same as before and Figure 4.16 illustrates this model also. The values of R^2 and the regression standard error, s, are the same in both models, too. This illustrates the fact that it doesn't really matter for this example which level we use as the reference level—our final model results turn out the same regardless.

We next turn our attention to whether the sample results provide compelling evidence of a difference between males and females in the population. One possibility is that in the population the two lines for males and females are parallel rather than having different slopes. To investigate this possibility we need to conduct a hypothesis test for the interaction term in the model. Since the magnitudes of the estimates and standard errors for the interaction term are equivalent whether we use D_F or D_M, we'll just use the first model with D_F. Assuming that we have no prior expectation about the sign (positive or negative) of b_3 in the first model from page 167, $E(Salary) = b_0 + b_1 D_F + b_2 Gpa + b_3 D_F Gpa$, we should conduct a two-tail test.

- State null hypothesis: NH: $b_3 = 0$.
- State alternative hypothesis: AH: $b_3 \neq 0$.
- Calculate test statistic: t-statistic $= \dfrac{\hat{b}_3 - b_3}{s_{\hat{b}_3}} = \dfrac{-7.159 - 0}{2.559} = -2.80$.
- Set significance level: 5%.
- Look up t-table:
  - critical value: The 97.5th percentile of the t-distribution with $10 - 3 - 1 = 6$ degrees of freedom is 2.45 (see computer help #8—use the 97.5th percentile because a two-tail test needs half the significance level in each tail); the rejection region is therefore any t-statistic greater than 2.45 or less than -2.45.
  - p-value: The sum of the areas to the left of the t-statistic (-2.80) and to the right of the negative of the t-statistic (2.80) for the t-distribution with 6 degrees of freedom is 0.031 (use computer help #9 or observe the p-value of 0.031 in the statistical software output).
- Make decision:
  - Since the t-statistic of -2.80 falls in the rejection region, we reject the null hypothesis in favor of the alternative.
  - Since the p-value of 0.031 is less than the significance level of 0.05, we reject the null hypothesis in favor of the alternative.

- Interpret in the context of the situation: The 10 sample observations suggest that a population regression parameter, b_3, of zero seems implausible—the sample data favor a nonzero value (at a significance level of 5%); in other words, the association between *Salary* and *Gpa* does appear to depend on the value of D_F (i.e., gender).

The result of the hypothesis test confirms that we need to model an interaction between D_F and *Gpa* for these data, since the sample interaction effect is strong enough to suggest a population interaction effect, too. In other words, the slopes of the male and female lines differ (if b_3 were zero, they would be the same and the lines would be parallel).

This model preserves hierarchy because it includes D_F and *Gpa* along with $D_F Gpa$. As long as the b_3 p-value is small enough to suggest retaining $D_F Gpa$ in the model, we would not usually do hypothesis tests for the regression parameters b_1 and b_2.

As an alternative to modeling the two genders together using an indicator variable, we could do a separate analysis for each gender by splitting up the sample and running a separate simple linear regression model for each gender. This would give us two sets of results, one for each gender, which would give us exactly the same estimated regression equations as on page 168 (try this to see). However, this would not be as useful as the single model approach we've taken (for example, it would not be possible then to test for equal slopes, as we just did).

For another example, consider the following (simulated) data in the **SALGPA2** data file:

Salary (salary in $ thousands)	49.4	39.0	41.6	54.6	72.8	65.0	54.6	28.6	33.8	44.2
Gpa (grade point average)	3.0	2.5	1.0	3.5	3.5	3.0	2.0	1.5	2.0	1.5
Gender (M = male, F = female)	M	M	F	M	F	F	F	M	M	F
D_F (0 = male, 1 = female)	0	0	1	0	1	1	1	0	0	1

Consider the model

$$E(Salary) = b_0 + b_1 D_F + b_2 Gpa + b_3 D_F Gpa.$$

Statistical software output for this model (see computer help #31) is:

Model Summary

Model	Sample Size	Multiple R Squared	Adjusted R Squared	Regression Std. Error
1[a]	10	0.9887	0.9830	1.796

[a] Predictors: (Intercept), D_F, *Gpa*, $D_F Gpa$.

Parameters[a]

Model		Estimate	Std. Error	t-stat	$Pr(> |t|)$
1	(Intercept)	7.280	2.951	2.467	0.049
	D_F	20.292	3.603	5.631	0.001
	Gpa	13.520	1.136	11.903	0.000
	$D_F Gpa$	−0.762	1.428	−0.533	0.613

[a] Response variable: *Salary*.

Figure 4.17 illustrates this model (see computer help #33). It seems possible that rather

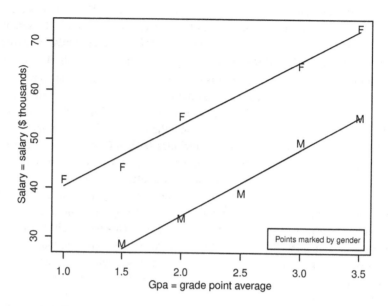

Figure 4.17 Scatterplot of *Salary* versus *Gpa* with the points marked according to gender for the **SALGPA2** with fitted lines from the model $\mathrm{E}(Salary) = b_0 + b_1 D_F + b_2 Gpa + b_3 D_F Gpa$. The parallel nature of the sample fitted lines suggests that the interaction term, $D_F Gpa$, may not be necessary.

than having different slopes, the two population lines for males and females could be parallel. To investigate this possibility we should do a hypothesis test for the interaction term in the model, $\mathrm{E}(Salary) = b_0 + b_1 D_F + b_2 Gpa + b_3 D_F Gpa$. Assuming no prior expectation about the sign (positive or negative) of b_3, we should conduct a two-tail test.

- State null hypothesis: NH: $b_3 = 0$.
- State alternative hypothesis: AH: $b_3 \neq 0$.
- Calculate test statistic: t-statistic $= \dfrac{\hat{b}_3 - b_3}{s_{\hat{b}_3}} = \dfrac{-0.762 - 0}{1.428} = -0.533$.
- Set significance level: 5%.
- Look up t-table:
  - critical value: The 97.5th percentile of the t-distribution with $10 - 3 - 1 = 6$ degrees of freedom is 2.45 (see computer help #8—use the 97.5th percentile because a two-tail test has half the significance level in each tail); the rejection region is any t-statistic greater than 2.45 or less than -2.45.
  - p-value: The sum of the areas to the left of the t-statistic (-0.533) and to the right of the negative of the t-statistic (0.533) for the t-distribution with 6 degrees of freedom is 0.613 (use computer help #9 or observe the p-value of 0.613 in the statistical software output).
- Make decision:
  - Since the t-statistic of -0.533 does not fall in the rejection region, we cannot reject the null hypothesis in favor of the alternative.

- Since the p-value of 0.613 is more than the significance level of 0.05, we cannot reject the null hypothesis in favor of the alternative.

- Interpret in the context of the situation: The 10 sample observations suggest that a population regression parameter, b_3, of zero is plausible (at a significance level of 5%); in other words, the association between *Salary* and *Gpa* does *not* appear to depend on gender.

The result of the hypothesis test suggests that we do not need to model an interaction between D_F and *Gpa* for these data, since the sample interaction effect is not strong enough to suggest a population interaction effect. In other words, the slopes of the male and female lines could be the same in the population (if b_3 were zero, they would be exactly the same and the lines would be parallel). Thus, a more appropriate model for these data is:

$$E(Salary) = b_0 + b_1 D_F + b_2 Gpa.$$

We can derive two equations by plugging in $D_F = 0$ for males and $D_F = 1$ for females:

$$E(Salary) \text{ for males} = b_0 + b_1(0) + b_2 Gpa \quad (D_F = 0)$$
$$= b_0 + b_2 Gpa,$$
$$E(Salary) \text{ for females} = b_0 + b_1(1) + b_2 Gpa \quad (D_F = 1)$$
$$= (b_0 + b_1) + b_2 Gpa.$$

Statistical software output for this model (see computer help #31) is:

Model Summary

Model	Sample Size	Multiple R Squared	Adjusted R Squared	Regression Std. Error
2[a]	10	0.9881	0.9847	1.702

[a] Predictors: (Intercept), D_F, *Gpa*.

Parameters[a]

Model		Estimate	Std. Error	t-stat	Pr($> |$t$|$)
2	(Intercept)	8.484	1.800	4.713	0.002
	D_F	18.471	1.094	16.886	0.000
	Gpa	13.038	0.653	19.980	0.000

[a] Response variable: *Salary*.

The estimated regression equation is therefore

$$\widehat{Salary} = 8.48 + 18.47 D_F + 13.04 Gpa.$$

The separate estimated regression equations for males and females are

$$\widehat{Salary} \text{ for males} = \hat{b}_0 + \hat{b}_2 Gpa$$
$$= 8.48 + 13.04 Gpa,$$
$$\widehat{Salary} \text{ for females} = (\hat{b}_0 + \hat{b}_1) + \hat{b}_2 Gpa$$
$$= (8.484 + 18.471) + 13.04 Gpa$$
$$= 26.96 + 13.04 Gpa.$$

The estimated slopes are both equal to 13.04.

Sometimes a sample comprises a number of different groups, and equivalent regression models may have been fit to the different groups. We can use indicator variables to combine the separate groups into a single dataset, which we can model as in the **SALGPA1** and **SALGPA2** applications. For example, suppose that we have two groups and a single predictor, X. We could define the indicator variable, D, to be 0 for one group and 1 for the other group, and fit the model:

$$E(Y) = b_0 + b_1 D + b_2 X + b_3 DX.$$

We'll get essentially the same results by combining the groups like this as we would get by fitting a simple linear regression model, $E(Y) = b_0 + b_1 X$, to each group separately, but with the added advantage that we have a method to compare slopes (by testing the DX interaction term) and intercepts (by testing the D indicator term). If a nested model F-test suggests that neither of these terms are significant, then both datasets can be modeled by a single common regression line. Otherwise, they are better modeled with different regression lines (that have different intercepts, slopes, or both). Note that even though they are modeled using different lines, both lines can be estimated within a single combined analysis (just as in the **SALGPA1** and **SALGPA2** applications). An alternative approach is to use a *Chow test* (see Chow, 1960), which tests whether sets of parameters in regression models fit to two separate groups are equal.

One concern with combining separate groups into a single regression model relates to the constant variance assumption. Sometimes, residuals in one group will have a very different variance to the residuals in another group, which would violate the constant variance assumption if both groups were modeled together. One way to address this problem uses *weighted least squares* in a way that allows the residual variances in the groups to differ (see page 201).

4.3.2 Qualitative predictors with three or more levels

Indicator variables also enable us to model differences between categories of qualitative variables with more than two levels. For example, consider the **CARS4** data file in Table 4.2, which contains $Eng =$ engine size in liters and $Cmpg =$ city miles per gallon (MPG) for 26 new U.S. passenger cars in 2011. These data come from a larger dataset (obtained from www.fueleconomy.gov), which is analyzed more fully in a case study in Section 6.2. In Section 4.1.3 we modeled the linear association between $Cmpg$ and $1/Eng$. By contrast, as suggested on page 151, here we model the linear association between $Cgphm = 100/Cmpg$ (city gallons per hundred miles) and Eng.

Variables D_{Co} and D_{Sw} in this dataset are indicator variables for the three-level qualitative variable, "Class," with levels: subcompact car (SC), compact car (CO), and station wagon (SW). In the preceding section we used one indicator variable to model differences between two levels (males and females). With three or more levels, apply the following method:

- Choose one of the levels for the reference level. As we saw in the preceding section, for most purposes it doesn't really matter which level is the reference level (although see the discussion of reference levels on page 179). For some applications, it may be practically meaningful to set one of the levels as the reference level (e.g., a placebo drug in a medical study). In other cases the level containing the most sample

Table 4.2 Car data with *Cmpg* = city miles per gallon, *Eng* = engine size in liters, for 7 subcompact cars, 11 compact cars, and 8 station wagons.

Cmpg (city miles per gallon)	*Eng* (engine size in liters)	Class of Car	D_{Co} (indicator for compact car)	D_{Sw} (indicator for station wagon)
29	1.6	SC	0	0
25	1.8	SC	0	0
26	1.6	SC	0	0
17	3.8	SC	0	0
23	2.5	SC	0	0
27	1.8	SC	0	0
29	1.5	SC	0	0
25	1.6	CO	1	0
24	1.4	CO	1	0
22	1.8	CO	1	0
26	1.4	CO	1	0
19	3.6	CO	1	0
25	2.0	CO	1	0
19	3.5	CO	1	0
23	2.4	CO	1	0
22	2.5	CO	1	0
26	1.8	CO	1	0
24	2.5	CO	1	0
22	2.0	SW	0	1
22	2.4	SW	0	1
23	2.0	SW	0	1
20	2.4	SW	0	1
24	2.0	SW	0	1
27	1.6	SW	0	1
21	2.4	SW	0	1
21	2.5	SW	0	1

observations often works well. In this case, we'll simply use the car class listed first in the dataset (i.e., "subcompact cars") to be the reference level.

- Create $L-1$ indicator variables, where L is the number of levels for the qualitative predictor. In this case, since there are three levels for type, we create two indicator variables, D_{Co} and D_{Sw}.

- Give observations in the reference level category values of 0 for all the $L-1$ indicator variables. In this case, $D_{Co} = D_{Sw} = 0$ for the 7 subcompact cars.

- Take each of the other $L-1$ categories in turn and give observations in each category a value of 1 for the corresponding indicator variable, and values of 0 for each of the other $L-2$ indicator variables. In this case, $D_{Co} = 1$ and $D_{Sw} = 0$ for the 11 compact cars, and $D_{Co} = 0$ and $D_{Sw} = 1$ for the 8 station wagons.

Again, D_{Co} and D_{Sw} have already been created in the **CARS4** data file; if they had not been, we could have used computer help #7 in the software information files available from the book website to do so. The indicator variables translate the qualitative information in the variable "Class" into a quantitative form that the computer can understand.

The translation dictionary is:

Class	D_{Co}	D_{Sw}
Subcompact car	0	0
Compact car	1	0
Station wagon	0	1

How can this help us to model different linear associations between *Cgphm* and *Eng* for different car classes? Well, consider the model

$$\mathrm{E}(Cgphm) = b_0 + b_1 D_{Co} + b_2 D_{Sw} + b_3 Eng + b_4 D_{Co}Eng + b_5 D_{Sw}Eng.$$

We can derive three separate equations from this model by plugging in $D_{Co}=D_{Sw}=0$ to obtain an equation for subcompact cars, by plugging in $D_{Co}=1$ and $D_{Sw}=0$ to obtain an equation for compact cars, and by plugging in $D_{Co}=0$ and $D_{Sw}=1$ to obtain an equation for station wagons.

$$\mathrm{E}(Cgphm) \text{ for subcompact cars} = b_0 + b_1(0) + b_2(0) + b_3 Eng + b_4(0)Eng + b_5(0)Eng$$
$$= b_0 + b_3 Eng \quad [D_{Co} = D_{Sw} = 0],$$
$$\mathrm{E}(Cgphm) \text{ for compact cars} = b_0 + b_1(1) + b_2(0) + b_3 Eng + b_4(1)Eng + b_5(0)Eng$$
$$= (b_0 + b_1) + (b_3 + b_4)Eng \quad [D_{Co} = 1, D_{Sw} = 0],$$
$$\mathrm{E}(Cgphm) \text{ for station wagons} = b_0 + b_1(0) + b_2(1) + b_3 Eng + b_4(0)Eng + b_5(1)Eng$$
$$= (b_0 + b_2) + (b_3 + b_5)Eng \quad [D_{Co}=0, D_{Sw}=1].$$

We can easily estimate the model by creating the interaction terms $D_{Co}Eng$ and $D_{Sw}Eng$ and fitting the model as usual. The values of the interaction terms for each sample observation are simply the corresponding values of D_{Co} (or D_{Sw}) and *Eng* multiplied together. Statistical software output for this model (see computer help #31 in the software information files available from the book website) is:

Model Summary

Model	Sample Size	Multiple R Squared	Adjusted R Squared	Regression Std. Error
1 [a]	26	0.8899	0.8623	0.2177

[a] Predictors: (Intercept), D_{Co}, D_{Sw}, Eng, $D_{Co}Eng$, $D_{Sw}Eng$.

Parameters [a]

Model		Estimate	Std. Error	t-stat	Pr($> |t|$)
1	(Intercept)	1.994	0.239	8.338	0.000
	D_{Co}	1.070	0.319	3.356	0.003
	D_{Sw}	−0.062	0.624	−0.100	0.921
	Eng	1.008	0.108	9.362	0.000
	$D_{Co}Eng$	−0.425	0.140	−3.029	0.007
	$D_{Sw}Eng$	0.170	0.285	0.596	0.558

[a] Response variable: *Cgphm*.

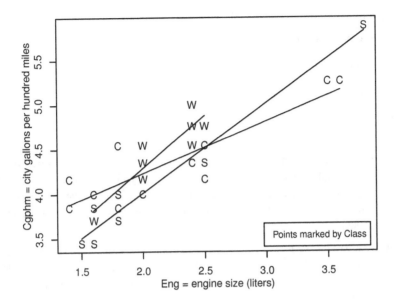

Figure 4.18   Scatterplot of *Cgphm* versus *Eng* with the points marked according to car class (S=subcompact cars, C=compact cars, W=station wagons) for the car example with fitted lines from the model $E(Cgphm) = b_0 + b_1 D_{Co} + b_2 D_{Sw} + b_3 Eng + b_4 D_{Co}Eng + b_5 D_{Sw}Eng$. Since the sample fitted lines for subcompact cars and station wagons are approximately parallel, this suggests that the interaction term $D_{Sw}Eng$ may be unnecessary.

The estimated regression equation is therefore

$$\widehat{Cgphm} = 1.994 + 1.070\,D_{Co} - 0.062\,D_{Sw} + 1.008\,Eng - 0.425\,D_{Co}Eng + 0.170\,D_{Sw}Eng.$$

We can use this equation to write out three regression equations—one for each car class—by plugging in $D_{Co}=D_{Sw}=0$ for subcompact cars, $D_{Co}=1$ and $D_{Sw}=0$ for compact cars, and $D_{Co}=1$ and $D_{Sw}=0$ for station wagons. As an alternative to modeling the three car classes together using indicator variables, we could do a separate analysis for each car class by splitting up the sample and fitting a simple linear regression model to each class. This would give us three sets of results, one for each car class, which would give us exactly the same estimated regression equations as we just obtained using a single model (try this to see). However, this would not be as useful as the single model approach we've taken (for example, it would not be possible then to test for equal slopes, as we will now do).

Figure 4.18 illustrates the current model (see computer help #33). It seems possible that in the population the line for subcompact cars and the line for station wagons could be parallel rather than having different slopes. To investigate this possibility we need to conduct a hypothesis test for the relevant interaction term in the model. Since D_{Sw} is the indicator variable that has a value of 1 for station wagons (and the reference level category is subcompact cars), the relevant interaction term is $D_{Sw}Eng$. Assuming that we have no prior expectation about the sign (positive or negative) of b_5 in the model $E(Cgphm) = b_0 + b_1 D_{Co} + b_2 D_{Sw} + b_3 Eng + b_4 D_{Co}Eng + b_5 D_{Sw}Eng$, we should conduct a two-tail test.

- State null hypothesis: NH: $b_5 = 0$.
- State alternative hypothesis: AH: $b_5 \neq 0$.
- Calculate test statistic: t-statistic $= \dfrac{\hat{b}_5 - b_5}{s_{\hat{b}_5}} = \dfrac{0.170 - 0}{0.285} = 0.596$.
- Set significance level: 5%.
- Look up t-table:

  - critical value: The 97.5th percentile of the t-distribution with $26 - 5 - 1 = 20$ degrees of freedom is 2.09 (see computer help #8 in the software information files available from the book website); the rejection region is therefore any t-statistic greater than 2.09 or less than -2.09 (we need the 97.5th percentile in this case because this is a two-tail test, so we need half the significance level in each tail).

  - p-value: The sum of the areas to the right of the t-statistic (0.596) and to the left of the negative of the t-statistic (-0.596) for the t-distribution with 20 degrees of freedom is 0.558 (use computer help #9 or observe the p-value of 0.558 in the statistical software output).

- Make decision:

  - Since the t-statistic of 0.596 does not fall in the rejection region, we cannot reject the null hypothesis in favor of the alternative.

  - Since the p-value of 0.558 is more than the significance level of 0.05, we cannot reject the null hypothesis in favor of the alternative.

- Interpret in the context of the situation: The 26 sample observations suggest that a population regression parameter, b_5, of zero seems plausible (at a significance level of 5%); in other words, the linear association between *Cgphm* and *Eng* does *not* appear to differ for subcompact cars and station wagons.

The result of the hypothesis test suggests that we do not need to model an interaction between D_{Sw} and *Eng* for these data, since the sample interaction effect is not strong enough to suggest a population interaction effect. In other words, the slopes of the subcompact car and station wagon lines could be the same in the population (if b_5 were zero, they would be exactly the same and the lines would be parallel). Thus, a more appropriate model for these data is

$$\mathrm{E}(Cgphm) = b_0 + b_1\,D_{Co} + b_2\,D_{Sw} + b_3\,Eng + b_4\,D_{Co}Eng.$$

Statistical software output for this model (see computer help #31) is:

Model Summary

Model	Sample Size	Multiple R Squared	Adjusted R Squared	Regression Std. Error
2[a]	26	0.8879	0.8666	0.2143

[a] Predictors: (Intercept), D_{Co}, D_{Sw}, Eng, $D_{Co}Eng$.

Parameters[a]

Model		Estimate	Std. Error	t-stat	$\Pr(> |t|)$
2	(Intercept)	1.944	0.220	8.827	0.000
	D_{Co}	1.120	0.303	3.703	0.001
	D_{Sw}	0.303	0.111	2.727	0.013
	Eng	1.032	0.098	10.515	0.000
	$D_{Co}Eng$	−0.449	0.132	−3.398	0.003

[a] Response variable: $Cgphm$.

The estimated regression equation is therefore

$$\widehat{Cgphm} = 1.944 + 1.120 D_{Co} + 0.303 D_{Sw} + 1.032 Eng - 0.449 D_{Co}Eng.$$

The separate estimated regression equations for subcompact cars, compact cars, and station wagons are:

$$\widehat{Cgphm} \text{ for subcompact cars} = \hat{b}_0 + \hat{b}_3 Eng$$
$$= 1.94 + 1.03 Eng,$$
$$\widehat{Cgphm} \text{ for compact cars} = (\hat{b}_0 + \hat{b}_1) + (\hat{b}_3 + \hat{b}_4)Eng$$
$$= (1.944 + 1.120) + (1.032 - 0.449)Eng$$
$$= 3.06 + 0.58 Eng,$$
$$\widehat{Cgphm} \text{ for station wagons} = (\hat{b}_0 + \hat{b}_2) + \hat{b}_3 Eng$$
$$= (1.944 + 0.303) + 1.032 Eng$$
$$= 2.25 + 1.03 Eng.$$

The estimated slopes for subcompact cars and station wagons are both equal to 1.03.

The reference level selected for a set of indicator variables plays a subtle role in multiple linear regression modeling. If we change the reference level, regression results might appear to have changed drastically, but if we write out each estimated regression equation (representing the different levels of the qualitative predictor), we will find that they are the same whatever the reference level is. In other words, we can fit equivalent models in different ways using different sets of indicator variables. However, the individual p-values in equivalent models will differ, and this can cause difficulties if we're not careful. For a hypothetical example, suppose that a qualitative variable has three levels ("A," "B," and "C") and the average response values for levels "B" and "C" are significantly different from the average response value for level "A" but not significantly different from each other. If we use D_B and D_C as indicator variables for levels "B" and "C" (with respect to reference level "A"), we'll get significant p-values for D_B and D_C. By contrast, if we use D_A and D_C as indicator variables for levels "A" and "C" (with respect to reference level "B"), we'll get a significant p-value for D_A but not for D_C (since the average response value for level "C" is not significantly different from the average response value for the reference level "B").

It gets more complicated to think this through when we add interactions between indicator variables and quantitative predictors to the mix, but the same principles apply. Continuing the hypothetical example, suppose that there is a quantitative predictor, X, in addition to the qualitative predictor with three levels. Also, suppose that the intercepts for levels "B" and "C" are significantly different from the intercept for level "A" but are not significantly different from each other. Finally, suppose that all three slopes are significantly different from zero, and that the slopes for levels "A" and "C" are significantly different from the slope for level "B" but are not significantly different from each other. If we fit a model with D_B, D_C, X, $D_B X$, and $D_C X$ (so that level "A" is the reference level), then D_B, D_C, X, and $D_B X$ will all be significant but $D_C X$ will not (since we said that in terms of slopes, levels "A" and "C" are not significantly different from each other). By contrast, if we fit a model with D_A, D_C, X, $D_A X$, and $D_C X$ (so that level "B" is the reference level), then D_A, X, $D_A X$, and $D_C X$ will all be significant but D_C will not (since we said that in terms of intercepts, levels "B" and "C" are not significantly different from each other).

To see how this plays out in a real example, try using the **CARS4** data file to fit a model with D_{Sc}, D_{Co}, Eng, $D_{Sc} Eng$, and $D_{Co} Eng$ (so "station wagons" is the reference level). You should find that $D_{Sc} Eng$ is not significant (agreeing with our previous finding that the linear association between $Cgphm$ and Eng does not appear to differ for station wagons and subcompact cars). If you drop $D_{Sc} Eng$ from this model, the remaining terms are all significant and you'll have a model equivalent to model 2 on page 179. However, if instead you fit a model with D_{Sc}, D_{Sw}, Eng, $D_{Sc} Eng$, and $D_{Sw} Eng$ (so that "compact cars" is the reference level), you should find that both interaction terms are significant. This tells us that the linear association between $Cgphm$ and Eng appears to differ for compact cars and subcompact cars and also for compact cars and station wagons. However, we cannot tell from this model whether the linear association between $Cgphm$ and Eng differs significantly for subcompact cars and station wagons. The only way to do this with an individual t-test is to fit one of the previous models (with either "subcompact cars" or "station wagons" as the reference level). Alternatively, it is possible to use a nested model F-test to test whether the regression parameters for $D_{Sc} Eng$ and $D_{Sw} Eng$ could be the same—see the optional discussion at the end of this section for details.

The key thing to remember when working with indicator variables is that the corresponding regression parameters measure *differences* between one level and the reference level. So, with two levels there is only one difference, so only one indicator variable is needed; with three levels there are two differences, so only two indicator variables are needed; and so on. To see why there are only two differences needed for three levels, consider a hypothetical example that models differences between level 1 and level 2 (call this diff$_{12}$) and between level 1 and level 3 (diff$_{13}$). "What about the difference between level 2 and level 3 (diff$_{23}$)," you might say? Well, we could derive that by calculating "the difference of the other two differences" (i.e., diff$_{23}$ = diff$_{13}$ − diff$_{12}$).

Tempting as it might be, we *should not* convert a qualitative variable to a quantitative one using numbers such as 1, 2, 3, . . . to represent the levels. If we did this, we would restrict the estimated regression model so that the difference between each level is the same (it would be equal to the estimated regression parameter for the new variable). For example, if this estimated regression parameter was equal to 2, this would imply that the difference between level 2 and level 1 is 2, and the difference between level 3 and level 2 is also 2. We should not even assume that level 3 is larger than level 2 and that level 2 is larger than level 1, let alone that the differences between the levels are all the same.

To convince yourself of all this, you should experiment using statistical software to see what happens. For example, see what happens when you try to use two indicator variables when your qualitative variable only has two levels (or try three indicator variables for a three-level qualitative variable). Also, see what happens in a three-level qualitative variable example when you try to use a quantitative variable with values 1, 2, and 3 in place of two indicator variables.

Another common misunderstanding with indicator variables is the belief that dropping an indicator variable from a model excludes the observations represented by that indicator variable from the analysis. This is incorrect; rather, what happens is that the observations represented by that indicator variable are now included as part of the reference level. For example, if we dropped the D_{Sw} indicator variable from the **CARS4** analysis and simply included the D_{Co} indicator variable in our models, the reference level would consist of subcompact cars *and* station wagons (since both those vehicle classes would have the value zero for D_{Co}, which is the only indicator variable left in the analysis). Take a look at Problem 4.9 at the end of this chapter if you're still not sure about this.

The examples we have considered in this section illustrate how to use indicator variables to model differences between two levels/categories of a qualitative predictor (e.g., male and female) and between three levels/categories (e.g., subcompact cars, compact cars, and station wagons). However, we can employ indicator variables in more general applications with qualitative predictors that have any number of levels/categories (within reason). For example, consider an application with one quantitative predictor (X) and two qualitative predictors, one with two levels ("A" and "B," say) and the other with four levels ("C," "D," "E," and "F"). If the linear association between Y and X could differ for the two levels of the first qualitative predictor, then we need to create an indicator variable—D_B, say (with reference level "A")—and include it in the model together with a $D_B X$ interaction. If the linear association between Y and X could also differ for the four levels of the second qualitative predictor, then we need to create three additional indicator variables—D_D, D_E, and D_F, say (with reference level "C")—and include them in the model together with $D_D X$, $D_E X$, and $D_F X$ interactions. To find out whether there are significant differences in linear association between Y and X for all these different levels, we can fit a series of multiple linear regression models that include $D_B X$, $D_D X$, $D_E X$, and $D_F X$, systematically removing those interactions that are not significant. We may even need to repeat this "model building" more than once using different sets of indicator variables with different reference levels to make sure that we've considered all the level comparisons we want to.

The examples in this section had only one quantitative predictor variable, which meant that we could illustrate results in scatterplots with that quantitative predictor on the horizontal axis. We then used these plots to guide our use of hypothesis tests for slope equality. However, in more complex applications with more than one quantitative predictor variable, we instead have to rely on the model building guidelines from Section 5.3 to determine which interactions between indicator variables and quantitative predictors we need in a model.

The indicator variables considered in this section are not the only way to code qualitative predictor variables for use in multiple linear regression models. Two other approaches include "deviation coding" and "Helmert coding"—see Fox and Weisberg (2011) for details.

Survey data often include variables measured on a multipoint scale, often 5-point or 7-point *Likert-type* scales. It is possible to include such variables as regular quantitative predictor variables in a multiple linear regression model (e.g., a 5-point scale variable could simply have the values 1–5). Whether this makes sense or not depends on the extent to which it is reasonable to assume that, for example, a "5" represents five times as much of whatever is being measured as a "1." If this does not seem reasonable, then it might be better to consider defining indicator variables for the five categories represented in the variable. However, this could soon add up to an awful lot of indicator variables if there are many such variables, which is why treating them quantitatively is often preferable in cases where it is justified.

If the response variable is also measured on a multipoint scale, this presents more of a challenge, however. For example, suppose that the response variable is measured on a 5-point scale. The multiple linear regression model assumes that the response variable is quantitative, and thus produces predictions that are continuous on some range (i.e., the predictions would not be restricted to whole numbers between 1 and 5). One way to address this would be to map predictions to whole numbers between 1 and 5 (e.g., a prediction between 3.5 and 4.5 maps to "4," a prediction above 4.5 maps to "5," etc.). Alternatively, more complex "generalized linear models" could be used (see Section 7.1). Two references for these types of models are McCullagh and Nelder (1989) and Agresti (2002).

Optional—testing equality of regression parameters with the nested model test. On page 180 we discussed using the **CARS4** data file to fit a model with D_{Sc}, D_{Sw}, Eng, $D_{Sc}Eng$, and $D_{Sw}Eng$ (so "compact cars" is the reference level). We could not tell from this model whether the linear association between $Cgphm$ and Eng differs significantly for subcompact cars and station wagons. One way to answer this question is to fit a model with a different set of indicator variables (with either "subcompact cars" or "station wagons" as the reference level) and use an individual t-test (recall that the absolute value of the resulting t-statistic was 0.596 with p-value 0.558). Alternatively, it is possible to use a nested model F-test to test whether the regression parameters for $D_{Sc}Eng$ and $D_{Sw}Eng$ could be equal. To do this, we define a new predictor term equal to $(D_{Sc}+D_{Sw})Eng$ and fit a model with D_{Sc}, D_{Sw}, Eng, and $(D_{Sc}+D_{Sw})Eng$. We then conduct a nested model F-test where this is the "reduced model" and the previous model (with D_{Sc}, D_{Sw}, Eng, $D_{Sc}Eng$, and $D_{Sw}Eng$) is the "complete model." The resulting F-statistic is 0.355 with p-value 0.558, again confirming the previous finding that the linear association between $Cgphm$ and Eng does not differ significantly for subcompact cars and station wagons. These results also demonstrate the results from page 113 that this F-statistic is equal to the square of the corresponding t-statistic, $0.355 = 0.596^2$, and the p-values for the two tests are identical (0.558).

We use this approach of testing equality of regression parameters with the nested model test for the case study in Section 6.2.

4.4 CHAPTER SUMMARY

The major concepts that we covered in this chapter relating to multiple linear regression model building are as follows:

Transformations are mathematical functions applied to data values that can sometimes allow us to model the association between a response variable, Y, and predictor variables, $(X_1, X_2, \ldots, X_k)$, more effectively. Background knowledge about the regression application often motivates the use of certain transformations, but in other applications trying certain transformations out during model building can reveal whether they might be useful. Particular transformations common in regression applications include the following.

Natural logarithm transformations can help to make certain kinds of skewed data more symmetric and normal. Linear regression models often fit better when the quantitative predictors are not too highly skewed. The logarithm transformation can also be applied to the response variable if it too is highly skewed, or if a multiplicative model is warranted (when changes in predictor values are associated with proportional changes in the response rather than additive changes).

Polynomial transformations include adding terms like X^2 to a multiple linear regression model. These transformations can help in situations where the association between the response variable and a predictor variable (holding all else equal) is expected to be a curved rather than a linear association (e.g., if high and low values of the predictor are associated with low values of the response, but medium predictor values are associated with high response values).

Reciprocal transformations relate to replacing a predictor, X, with its reciprocal (or inverse), $1/X$, in a multiple linear regression model. This can be helpful in situations where a negative association between the response variable and a predictor variable (holding all else equal) starts off quite strong (say) at low values of the predictor, but then becomes weaker at high values of the predictor.

Other transformations include any mathematical function we can think of (and that we can use computer software to calculate). For example, polynomial transformations include positive whole-number powers $(2, 3, \ldots)$ of a variable, but it is also possible to apply fractional power transformations if this might prove useful. For example, a half-power transformation, $X^{0.5}$, also called a square root transformation (since $X^{0.5} = \sqrt{X}$), is sometimes helpful in cases where a variable is skewed but a logarithm transformation is too "strong." Problem 4.3 at the end of this chapter provides an example of the use of square root transformations.

Interactions are terms like $X_1 X_2$ (predictors X_1 and X_2 multiplied together) in a multiple linear regression model. This allows the association between the response variable, Y, and the X_1 predictor to depend on (or vary according to) the value of the X_2 predictor. (Similarly, the association between Y and X_2 then depends on the value of X_1.)

Qualitative predictors are categorical variables in which values represent the category to which an observation belongs (e.g., male or female, or small, medium, and large), rather than specific quantities (e.g., dollars or pounds of weight). To incorporate the information in a qualitative predictor into a multiple linear regression model, we need to translate the categories into numerical codes that a computer can understand. Indicator variables accomplish this by taking the value 1 to represent a particular category (or level) of the qualitative predictor, and 0 to represent all other levels. We need one fewer indicator variable than there are levels, since one level is selected to be the "reference level" (which has the value 0 for all the indicator variables). Each indicator variable then represents average response differences between the level represented by a 1 and the reference level. Interactions between indicator variables and quantitative predictors represent differences in that predictor's association with the response between the level represented by a 1 and the reference level.

PROBLEMS

- "Computer help" refers to the numbered items in the software information files available from the book website.

- There are *brief* answers to the even-numbered problems in Appendix E.

4.1   This problem is adapted from one in McClave et al. (2005). The **NYJUICE** data file contains data on demand for cases of 96-ounce containers of chilled orange juice over 40 sale days at a warehouse in New York City that has been experiencing a large number of out-of-stock situations. To better understand demand for this product, the company wishes to model the number of cases of orange juice, *Cases*, as a function of sale day, *Day*.

   (a) Construct a scatterplot for these data, and add a quadratic regression line to your scatterplot [computer help #15 and #32].

   (b) Fit a simple linear regression model to these data, that is, use statistical software to fit the model with *Day* as the single predictor [computer help #25] and write out the resulting fitted regression equation.

   (c) Fit a quadratic model to these data, that is, use statistical software to fit the model with both *Day* and Day^2 as predictors and write out the resulting fitted regression equation. You will first need to create the Day^2 term in the dataset [computer help #6 and #31].
   Hint: To square a variable, X, in statistical software you may need to write it as "X∗∗2" or "X∧2."

   (d) Based on the scatterplot in part (a), it appears that a quadratic model would better explain variation in orange juice demand than a simple linear regression model. To show this formally, do a hypothesis test to assess whether the Day^2 term is statistically significant in the quadratic model (at a 5% significance level). If it is, the more complicated quadratic model is justified. If not, the simpler simple linear regression model would be preferred.
   Hint: Understanding the example on page 145 will help you solve this part.

4.2   Investigate the claim on page 157 that prediction intervals based on the transformed model 2 for home taxes are more appropriate than intervals based on the untransformed model 1. The data are in the **HOMETAX** data file. Confirm the values given on page 157 [computer help #30], and draw the intervals by hand onto Figure 4.12 to see which set of intervals are best supported by the data.

4.3   Recall Problem 2.1 from page 78 in which you fit a simple linear regression model to data for 212 countries with response *Int* (percentage of the population that are Internet users) and predictor *Gdp* (GDP per capita in US$ thousands). This model is a reasonable one, but it is possible to improve it by transforming both the response and predictor variables. Investigate the use of transformations for this application using the data in the **INTERNET** data file. In particular, investigate natural logarithm and square root transformations for both the response and predictor variables. Which transformations seem to provide the most useful model for understanding any association between *Int* and *Gdp*? Write up your results in a short report (no more than two pages) that compares and contrasts the different models that you fit. Include a few paragraphs describing your conclusions with respect to the various ways of comparing models and perhaps some scatterplots.

Hint: Compare the following three models: (1) response Int and predictor Gdp; (2) response $\log_e(Int)$ and predictor $\log_e(Gdp)$; (3) response $\sqrt{Int}$ and predictor $\sqrt{Gdp}$. Use the methods from the example in Section 4.1.4 to guide your comparison.

4.4 As suggested on page 161 for the **SALES1** example, calculate the expected increase in sales for each additional $1m we spend on advertising when the prevailing interest rate is 3%.

4.5 This problem extends the home prices example used previously to 76 homes (Section 6.1 contains a complete case study of these data). We wish to model the association between the price of a single-family home (*Price*, in $ thousands) and the following predictors:

> $Floor$ = floor size (thousands of square feet)
> Lot = lot size category (from 1 to 11—see page 89)
> $Bath$ = number of bathrooms (with half-bathrooms counting as "0.1")
> Bed = number of bedrooms (between 2 and 6)
> Age = age (standardized: (year built $-$ 1970)/10)
> Gar = garage size (0, 1, 2, or 3 cars)
> D_{Ac} = indicator for "active listing" (rather than pending or sold)
> D_{Ed} = indicator for proximity to Edison Elementary
> D_{Ha} = indicator for proximity to Harris Elementary

Consider the following model, which includes an interaction between *Bath* and *Bed*:

$$E(Price) = b_0 + b_1 \, Floor + b_2 \, Lot + b_3 \, Bath + b_4 \, Bed + b_5 \, BathBed$$
$$+ b_6 \, Age + b_7 \, Age^2 + b_8 \, Gar + b_9 \, D_{Ac} + b_{10} \, D_{Ed} + b_{11} \, D_{Ha}.$$

The regression results for this model are:

Predictor variable	Parameter estimate	Two-tail p-value
Intercept	332.47	0.00
Floor	56.72	0.05
Lot	9.92	0.01
Bath	−98.15	0.02
Bed	−78.91	0.01
BathBed	30.39	0.01
Age	3.30	0.30
Age^2	1.64	0.03
Gar	13.12	0.12
D_{Ac}	27.43	0.02
D_{Ed}	67.06	0.00
D_{Ha}	47.27	0.00

*Hint: Understanding the **SALES1** example beginning on page 159 will help you solve this problem.*

    (a) Test whether the linear association between home price (*Price*) and number of bathrooms (*Bath*) depends on number of bedrooms (*Bed*), all else equal (use significance level 5%).

    (b) How does the linear association between *Price* and *Bath* vary with *Bed*? We can investigate this by isolating the part of the model involving just *Bath*: The

"*Bath* effect" on *Price* is given by $b_3 Bath + b_5 BathBed = (b_3 + b_5 Bed) Bath$. For example, when $Bed = 2$, this effect is estimated to be $(-98.15 + 30.39(2)) Bath = -37.37 Bath$. Thus, for two-bedroom homes, there is a negative linear association between home price and number of bathrooms (for each additional bathroom, the sale price drops by \$37,370, all else being equal—perhaps adding extra bathrooms to two-bedroom homes is considered a waste of space and so has a negative impact on price). Use similar calculations to show the linear association between *Price* and *Bath* for three-bedroom homes, and also for four-bedroom homes.

4.6 A consumer products company wishes to focus its marketing efforts on current customers who are likely to be more profitable in the future. Using past data on customer transactions it has calculated profitability scores (variable *Score*) for 200 customers based on their purchasing behavior over the last five years. Scores range from 0 (customers with few purchases who have provided little return to the company) to 10 (very profitable customers with many purchases). The company would like to predict future profitability scores for customers acquired within the last year using two potential predictor variables: purchase frequency in the last year (variable *Freq*) and average purchase amount (in dollars) in the last year (variable *Amt*). The idea is that the company could focus marketing efforts on recently acquired customers with a high predicted (long-term) profitability score, rather than wasting resources on customers with low predicted profitability scores. Two nested models were estimated from the data for the 200 long-term customers. Statistical software output for the (complete) model, $E(Score) = b_0 + b_1 Freq + b_2 Amt + b_3 FreqAmt + b_4 Freq^2 + b_5 Amt^2$, is:

ANOVA [a]

Model		Sum of Squares	df	Mean Square	Global F-stat	Pr(>F)
1	Regression	636.326	5	127.265	254.022	0.000[b]
	Residual	97.194	194	0.501		
	Total	733.520	199			

[a] Response variable: *Score*.
[b] Predictors: (Intercept), *Freq*, *Amt*, *FreqAmt*, *Freq²*, *Amt²*.

Statistical software output for the (reduced) model, $E(Score) = b_0 + b_1 Freq + b_2 Amt$, is:

ANOVA [a]

Model		Sum of Squares	df	Mean Square	Global F-stat	Pr(>F)
2	Regression	609.037	2	304.519	481.916	0.000[b]
	Residual	124.483	197	0.632		
	Total	733.520	199			

[a] Response variable: *Score*.
[b] Predictors: (Intercept), *Freq*, *Amt*.

(a) Write down the null and alternative hypotheses to test whether at least one of the predictor terms in the complete model—*Freq*, *Amt*, *FreqAmt*, *Freq²*, and *Amt²*—is linearly associated with *Score*.

(b) Do the hypothesis test from part (a) using a significance level of 5% (use the fact that the 95th percentile of the F-distribution with 5 numerator degrees of freedom and 194 denominator degrees of freedom is 2.26—see computer help #8). Remember to draw an appropriate conclusion from the results of the test.

(c) Write down the null and alternative hypotheses to test whether *FreqAmt*, *Freq²*, and *Amt²* provide useful information about *Score* beyond the information provided by *Freq* and *Amt* alone.

(d) Do the hypothesis test from part (c) using a significance level of 5% (use the fact that the 95th percentile of the F-distribution with 3 numerator degrees of freedom and 194 denominator degrees of freedom is 2.65—see computer help #8). Remember to draw an appropriate conclusion from the results of the test.

(e) Which of the two models would be more appropriate to use to predict *Score*?

4.7   Consider the data available in the **HOMES5** data file—these data are for 40 single-family homes in south Eugene, Oregon in 2005 and were provided by Victoria Whitman, a realtor in Eugene. For the sake of illustration, here we investigate whether any linear association between *Floor* (floor size in thousands of square feet) and *Price* (sale price in thousands of dollars) differs for two particular neighborhoods (defined by closest elementary school) in this housing market—we ignore the other predictors (lot size, age, etc.) for now. (We'll revisit this application in Section 6.1 when we consider a complete analysis of the whole dataset.) In the **HOMES5** data file there are 26 homes whose closest elementary school is "Redwood" ($D_{Ha}=0$) and 14 homes whose closest elementary school is "Harris" ($D_{Ha}=1$). (*Hint: Understanding the **SALGPA2** example beginning on page 171 will help you solve this problem.*)

(a) Write the equation of a model relating sale price (*Price*) to floor size (*Floor*) and neighborhood (D_{Ha}) that allows for different slopes and intercepts for each neighborhood.
Hint: You should have regression parameters (b's) and predictor terms in your equation, but no numbers.

(b) Draw a scatterplot that illustrates the model in part (a). Include two regression lines, one for each neighborhood, on your plot [computer help #15, #17, and #33].

(c) Use statistical software to fit the model from part (a) to the data and write out the resulting estimated regression equation. You will first need to create the $D_{Ha}Floor$ interaction term in the dataset [computer help #6 and #31].
Hint: This is where you replace the regression parameters in your equation from part (a) with numbers obtained using statistical software.

(d) Conduct a hypothesis test to determine whether the slopes associated with the two neighborhoods are significantly different. Use significance level 5%.

(e) Based on the results from part (d), fit a new model to the data, and write two separate equations (with actual numbers) for predicting *Price* from *Floor*, one for the Redwood neighborhood and the other for the Harris neighborhood.

4.8   Consider a multiple linear regression model for a response Y, with one quantitative predictor variable, X, and one qualitative predictor variable with three levels.

(a) Write the equation for a simple linear regression model that relates the mean response $E(Y)$ to X.

(b) Add two terms to the model that will allow the mean response to differ for the three levels of the qualitative predictor variable. Specify the indicator variable coding scheme you use.

(c) Add two further terms to the model that will allow the association between X and Y to differ for the three levels of the qualitative predictor variable (these terms will be interaction terms).

(d) Under what circumstance will the regression associations represented by the model in part (c) be three parallel lines?

(e) Under what circumstance will the regression associations represented by the model in part (c) be a single straight line?

4.9 The following example illustrates that dropping an indicator variable from a model does not drop the corresponding observations from the analysis. Consider the following dataset:

Y	X	Manufacturer	D_A	D_B
3.4	2.3	A	1	0
3.8	3.0	A	1	0
3.5	2.9	B	0	1
3.9	3.8	B	0	1
3.5	2.3	C	0	0
3.9	3.2	C	0	0

(a) Fit a multiple linear regression model with D_A, D_B, and X as the predictors and write out the fitted regression equation [computer help #31].

(b) Plug in 1's and 0's for the indicator variables to obtain an equation for each of the three manufacturers.
Hint: Manufacturer C represents the reference level ($D_A = D_B = 0$).

(c) You should have found in part (a) that although we have significant p-values for D_B and X (at the 5% significance level), we have an insignificant p-value for D_A. So, drop D_A from the model, fit a multiple linear regression model with D_B and X as the predictors, and write out the fitted regression equation [computer help #31].

(d) Plug in 1's and 0's for the indicator variables to obtain an equation for each of the three manufacturers.
Hint: Manufacturers A and C should now have the same equations since they both now represent the reference level ($D_B = 0$).

CHAPTER 5

REGRESSION MODEL BUILDING II

This chapter covers additional model building topics that should be addressed in any regression analysis. In Section 5.1 we consider the role that individual data points can play in a multiple linear regression model, particularly overly influential points. As with any mathematical model that attempts to approximate a potentially complex reality, there are a number of pitfalls that can cause problems with a multiple linear regression analysis. We outline a few of the major pitfalls in Section 5.2 and suggest some remedies. Adding transformations, interactions, and qualitative predictors to our toolbox (as discussed in Chapter 4) creates a very flexible methodology for using multiple linear regression modeling; Section 5.3 provides some guidelines and strategies for employing these methods successfully in real-life applications. In Section 5.4 we discuss computer-automated model selection methods, which are used to aid predictor selection for large datasets. Finally, in Section 5.5 we propose some graphical methods for understanding and presenting the results of a multiple linear regression model.

5.1 INFLUENTIAL POINTS

5.1.1 Outliers

Informally, an *outlier* is a data value that lies a long way from the majority of the other data values. In terms of regression, we think of outliers as sample observations whose *response* Y-value lies a long way from the predicted $\hat{Y}$-value from the model. In other words, they

Applied Regression Modeling, Second Edition. By Iain Pardoe
Copyright © 2012 John Wiley & Sons, Inc.

are observations that have a residual ($\hat{e}_i = Y_i - \hat{Y}_i$) with a large magnitude. Since the measurement scale for residuals is the same as the measurement scale for the response variable, we need a method for standardizing residuals so that we can identify potential outliers for any application. One simple method to accomplish this is to subtract from each residual their sample mean (which will actually be zero) and divide by the regression standard error, s (which will be approximately equal to the sample standard deviation of the residuals). A slightly more sophisticated method takes into account the distance that each observation's predictor values are from the sample mean of the predictors—this produces *standardized residuals.* An even more sophisticated method adjusts each residual according to the regression standard error for a model fit to all the other observations—this produces *studentized residuals.* We can use statistical software to calculate the studentized residuals for a multiple linear regression model (although a formula for studentized residuals is provided at the end of this section for interest). Beware that some statistical software calls the studentized residuals that we are talking about here something different, such as "deleted studentized residuals" or "externally studentized residuals" (computer help #35 in the information files at the book website provides details for the major software packages).

When the four multiple linear regression assumptions (zero mean, constant variance, normality, and independence) are satisfied, the studentized residuals should have an approximate standard normal distribution with a mean of 0 and a variance close to 1. Thus, if we identify an observation with a studentized residual outside the range -3 to $+3$, then we've either witnessed a very unusual event (one with probability less than 0.002) or we've found an observation with a response value that doesn't fit the pattern in the rest of the dataset *relative to its predicted value from the model* (i.e., an outlier). So, for the purposes of this book, we formally define a potential outlier to be an observation with studentized residual less than -3 or greater than $+3$. (Note that other textbooks and statisticians may define outliers slightly differently, although the basic principle will be the same.)

Suppose that we have fit a multiple linear regression model to a dataset and calculated the resulting studentized residuals (and perhaps graphed them in a histogram). If no observations have studentized residuals less than -3 or greater than $+3$, then we can conclude that there are probably no outliers in the dataset, and we can go on to evaluate and interpret the model. If we do have one or more studentized residuals less than -3 or greater than $+3$, then we can conclude that we have one or more potential outliers in the dataset. We should then go on to investigate *why* the potential outliers have such unusual response values (Y) relative to their predicted values from the model ($\hat{Y}$). Some possible reasons include the following.

- There has been a data input mistake and some values have been recorded incorrectly in the dataset. Remedy: If possible, identify and correct the mistake(s) and reanalyze the data.

- An important predictor has been omitted from the model. Remedy: If possible, identify any potentially useful predictors that have not been included in the model and reanalyze the data by including them.

- One or more regression assumptions have been violated. Remedy: Try reformulating the model (using transformations or interactions, say) to correct the problem.

- The potential outliers really do represent observations that differ substantively from the other sample observations. Remedy: Remove the observations from the dataset and reanalyze the remainder of the dataset separately.

Table 5.1 Car data with $Cmpg$ = city miles per gallon, Eng = engine size (liters), Cyl = number of cylinders, Vol = interior passenger and cargo volume (hundreds of cubic feet).

Car	Cmpg	Eng	Cyl	Vol	Car	Cmpg	Eng	Cyl	Vol
Aston Martin DB9	13	5.9	12	0.83	Hyundai Genesis Coupe 3.8L	17	3.8	6	0.99
BMW 128Ci Convertible	18	3.0	6	0.86	Infiniti EX35	17	3.5	6	1.11
BMW 128i	18	3.0	6	0.96	Infiniti G25	20	2.5	6	1.13
BMW 328Ci	18	3.0	6	1.00	Infiniti G37 Convertible	17	3.7	6	0.88
BMW 328Ci Convertible	18	3.0	6	0.93	Infiniti M37	18	3.7	6	1.19
BMW 328i	18	3.0	6	1.05	Jaguar XJ	15	5.0	8	1.20
BMW 328i Sports Wagon	18	3.0	6	1.18	Lexus HS 250h	35	2.4	4	1.02
BMW 335Ci Convertible	19	3.0	6	0.93	Lexus LS 460	16	4.6	8	1.17
BMW 335is Convertible	17	3.0	6	0.93	Mercedes C 350	17	3.5	6	1.00
BMW 335is Coupe	17	3.0	6	1.00	Mercedes E 350 Convertible	17	3.5	6	0.86
BMW 550i Gran Turismo	15	4.4	8	1.22	Mercedes E 350 Coupe	17	3.5	6	0.92
BMW 740i	17	3.0	6	1.20	Mercedes E 550	15	4.7	8	1.11
BMW 740Li	17	3.0	6	1.29	Mercedes E 550 Convertible	15	5.5	8	0.86
BMW 750i	15	4.4	8	1.20	Mercedes E 550 Coupe	15	5.5	8	0.92
Cadillac CTS 3.6L	18	3.6	6	1.14	Mercedes GLK 350	16	3.5	6	1.26
Cadillac CTS 3L	18	3.0	6	1.14	Mercedes ML 350	16	3.5	6	1.48
Cadillac CTS Wagon 3.6L	18	3.6	6	1.26	Mercedes S 550	15	5.5	8	1.25
Cadillac CTS Wagon 3L	18	3.0	6	1.26	Nissan 370Z	19	3.7	6	0.59
Chevrolet Camaro	18	3.6	6	1.04	Nissan 370Z Roadster	18	3.7	6	0.56
Chrysler 300 FFV	18	3.6	6	1.22	Porsche 911 Carrera	19	3.6	6	0.75
Dodge Challenger FFV	18	3.6	6	1.07	Porsche 911 Carrera Cab.	19	3.6	6	0.73
Dodge Challenger SRT8	14	6.4	8	1.07	Porsche 911 Carrera S Cab.	19	3.8	6	0.73
Dodge Charger FFV	18	3.6	6	1.20	Porsche 911 Turbo Coupe	17	3.8	6	0.75
Hyundai Genesis	18	3.8	6	1.25	Porsche Panamera	18	3.6	6	1.23
Hyundai Genesis Coupe 2L	20	2.0	4	0.99	Rolls-Royce Ghost	13	6.6	12	1.25

One general approach is to exclude the sample observation with the largest magnitude studentized residual from the dataset, refit the model to the remaining sample observations, and see whether the results change substantially. If they do (say, if regression parameter estimates change their significance dramatically or change signs), then the observation in question should probably be analyzed separately from the remainder of the sample. If there is little change (say, if regression parameter estimates change only marginally), then the observation in question can probably remain in the main analysis and we should merely note the fact that it represents a potential outlier.

For example, consider the **CARS5** data file in Table 5.1, which contains Eng = engine size (liters), Cyl = number of cylinders, Vol = interior passenger and cargo volume (hundreds of cubic feet), and $Cmpg$ = city miles per gallon (MPG) for 50 new U.S. passenger cars in 2011. These data come from a larger dataset (obtained from www.fueleconomy.gov), which is analyzed more fully in a case study in Section 6.2. As in Section 4.3.2, we use $Cgphm = 100/Cmpg$ (city gallons per hundred miles) as our response variable. Consider the following multiple linear regression model:

$$E(Cgphm) = b_0 + b_1\, Eng + b_2\, Cyl + b_3\, Vol.$$

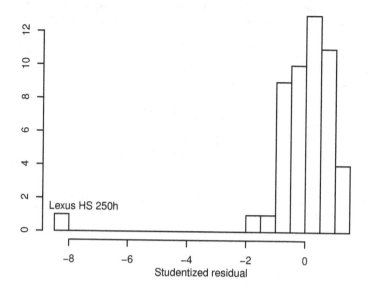

Figure 5.1   Histogram of studentized residuals from the model fit to all the vehicles in the **CARS5** dataset (model fit 1).

Statistical software output for this model is:

Parameters[a]

Model		Estimate	Std. Error	t-stat	Pr($>$|t|)
1	(Intercept)	2.542	0.365	6.964	0.000
	Eng	0.235	0.112	2.104	0.041
	Cyl	0.294	0.076	3.855	0.000
	Vol	0.476	0.279	1.709	0.094

[a] Response variable: *Cgphm*.

The results are a little surprising since we might have expected all three predictors to be highly significant, but the p-value for *Vol* is 0.094. However, on looking at the histogram of studentized residuals in Figure 5.1, we can see that these results are not to be trusted—there is a clear outlier with much lower response value than predicted (see computer help #35 and #14 in the software information files available from the book website).

If we look back at the dataset in Table 5.1, we can see that the city MPG for the Lexus HS 250h is higher than might be expected relative to the pattern of values for the entire sample (and consequently, its value of *Cgphm* is surprisingly low)—this is the outlier that we can see in Figure 5.1. On investigation, it turns out that this vehicle is a hybrid car, so it is not surprising that it fails to fit the dominant pattern of the rest of the (standard gasoline-powered) vehicles.

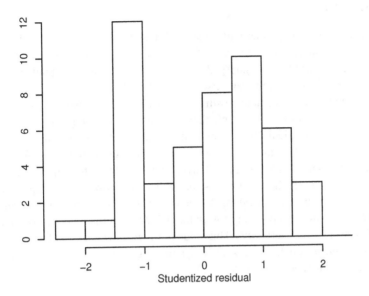

Figure 5.2   Histogram of studentized residuals from the model fit to the **CARS5** dataset excluding the Lexus HS 250h (model fit 2).

We therefore remove this car from the dataset (see computer help #19) and refit the model to obtain the following results:

Parameters^a

Model		Estimate	Std. Error	t-stat	$\Pr(>|t|)$
2	(Intercept)	2.905	0.237	12.281	0.000
	Eng	0.259	0.071	3.645	0.001
	Cyl	0.228	0.049	4.634	0.000
	Vol	0.492	0.177	2.774	0.008

[a] Response variable: *Cgphm*.

The regression parameter estimates for b_1, b_2, and b_3 have changed quite dramatically, are all now highly significant, and have smaller standard errors (so model results are now more precise for the remaining 49 cars). Also, there are no longer any apparent outliers in the histogram of studentized residuals in Figure 5.2 (see computer help #35 and #14).

Optional—formulas for standardized and studentized residuals. The ith observation's standardized residual is $r_i = \frac{\hat{e}_i}{s\sqrt{1-h_i}}$, where $\hat{e}_i$ is the ith observation's residual, s is the regression standard error, and h_i is the ith observation's leverage (defined in the next section). The ith observation's studentized residual is $t_i = r_i\sqrt{\frac{n-k-2}{n-k-1-r_i^2}}$, where r_i is the ith observation's standardized residual, n is the sample size, and k is the number of predictor terms in the model (excluding the constant intercept term).

5.1.2   Leverage

Outliers are sample observations with "unusual" response values. What about sample observations with unusual predictor values? For example, suppose that we have a dataset with two predictor variables, X_1 and X_2, that we would like to use to explain and predict Y in a multiple linear regression model. Also suppose that the sample observations for these two variables tend to be moderately positively associated, that is, when X_1 is low, then X_2 also tends to be low, and when X_1 is high, then X_2 also tends to be high. If there is a sample observation that has a particularly low value of X_1 paired with a particularly high value of X_2 (or vice versa), then this unusual observation may have an exaggerated influence on the regression model. Since this point lies a relatively long way from the majority of the sample observations, it can "pull" the fitted regression model close toward its Y-value, so that the fitted model becomes biased toward this single observation and away from the majority.

One way to measure this potential for undue influence on the model is to calculate the *leverage* of each sample observation. This value represents the potential for an observation to have undue influence on the regression model, ranging from 0 (low potential influence) to 1 (high potential influence). Leverages are higher for observations with unusual predictor values.

We can use statistical software to calculate the leverages for each sample observation (although for interest a formula for leverage is provided at the end of this section). Then we can note any that have a leverage greater than $3(k+1)/n$, where k is the number of predictor terms in the model (excluding the constant intercept term) and n is the sample size, or that are very isolated with a leverage greater than $2(k+1)/n$. This is particularly easy if we plot the leverages on the vertical axis of a scatterplot (with any other convenient sample variable on the horizontal axis). Then we can identify the observation with the highest leverage and see if this leverage is greater than $3(k+1)/n$. If it is, or if it is very isolated and has a leverage greater than $2(k+1)/n$, we should investigate what happens to the model results when the observation in question is removed from the analysis. If results change substantially (say, if regression parameter estimates change their significance dramatically or change signs), then the observation in question should probably be analyzed separately from the remainder of the sample. If there is little change (say, if regression parameter estimates change only marginally), then the observation in question can probably remain in the main analysis and we should merely note the fact that it represents a high-leverage observation.

The thresholds $2(k+1)/n$ and $3(k+1)/n$ are simple rules of thumb that usually work well provided that n is not too small (say, < 25). In these small sample size situations it is prudent to investigate what happens when the observation with the largest leverage is removed, and then repeat this process if permanent removal of this observation is warranted.

Recall the multiple linear regression model fit to the **CARS5** dataset excluding the Lexus HS 250h—see model fit 2 on page 193. Figure 5.3 shows the leverages for this model plotted versus ID number (see computer help #37). The horizontal dashed lines show the warning thresholds, $2(k+1)/n = 2(3+1)/49 = 0.16$ and $3(k+1)/n = 3(3+1)/49 = 0.24$, with three vehicles exceeding the upper threshold. We use ID number as the horizontal axis variable simply to be able to identify observations in the dataset. It is really only the vertical axis in this plot that conveys useful information about influence, so it does not matter what we use on the horizontal axis, as long as it "spreads the points out horizontally" so that it is easy to tell where the highest values are vertically—ID number does this very

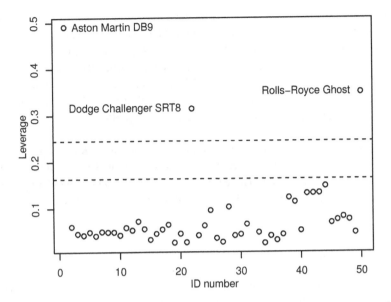

Figure 5.3   Scatterplot of leverage versus ID number for the model fit to the **CARS5** dataset excluding the Lexus HS 250h (model fit 2). The horizontal dashed lines show the warning thresholds, $2(k+1)/n$ and $3(k+1)/n$.

conveniently in this case. Some statistical software also allows the user to interactively identify individual points in a scatterplot—see computer help #18.

Looking back at Table 5.1, we see that the Aston Martin DB9 has a large engine and the most cylinders in the sample (12), but a relatively small interior passenger and cargo volume. The Rolls-Royce Ghost has the largest engine in the sample and also 12 cylinders, but has quite a large volume. The Dodge Challenger SRT8 has a large engine and 8 cylinders, but has a medium volume. Since the Aston Martin DB9 has the highest leverage of 0.49, which exceeds the threshold value $3(k+1)/n = 0.24$, we'll also remove this car from the dataset (see computer help #19) and refit the model to see what impact its removal has:

Parameters[a]

Model		Estimate	Std. Error	t-stat	Pr($>$|t|)
3	(Intercept)	2.949	0.247	11.931	0.000
	Eng	0.282	0.080	3.547	0.001
	Cyl	0.201	0.064	3.139	0.003
	Vol	0.532	0.188	2.824	0.007

[a] Response variable: *Cgphm*.

We can assess how the regression parameter estimates change from model fit 2 on page 193 (which just excluded the Lexus HS 250h), to these results for model fit 3 (which excludes both the Lexus HS 250h and the Aston Martin DB9). The relative changes in parameter estimates are less than we noted previously, where the changes from model fit 1 on page 192 (including all data) to model fit 2 were quite dramatic. Furthermore, the regression parameter estimate standard errors have increased in size from model fit 2 to model fit 3, so that removing the Aston Martin DB9 has not improved the precision of the model results for the remaining 48 cars. Leverages tell us the *potential* for undue influence, but in this case, while there was the potential for the Aston Martin DB9 to strongly influence the results, in the end it did not really do so. Thus, we can probably safely leave this car in the analysis. You can confirm for yourself that removal of the Rolls-Royce Ghost or the Dodge Challenger SRT8 also influences the results only weakly.

Optional—formula for leverages.   Define the predictor matrix $\mathbf{X}$ as on page 92. Then calculate the following matrix, $\mathbf{H} = \mathbf{X}(\mathbf{X}^\mathrm{T}\mathbf{X})^{-1}\mathbf{X}^\mathrm{T}$. The n diagonal entries of this matrix are the leverages for each observation. There is an alternative formula for calculating leverages from Cook's distances and studentized residuals at the end of the next section.

5.1.3 Cook's distance

Outliers are sample observations with unusual response values, while sample observations with unusual predictor values have high leverage. We can consider both concepts together in a combined measure called *Cook's distance* (Cook, 1977). An observation with a large Cook's distance can be an outlier, a high-leverage observation, or both. We can use statistical software to calculate Cook's distances and identify all such observations (again, for interest a formula for Cook's distance is provided at the end of this section).

In particular, we can plot Cook's distances on the vertical axis of a scatterplot (with any other convenient sample variable on the horizontal axis), and then identify the observation with the highest Cook's distance—see computer help #18. How high would a Cook's distance have to be for us to be concerned? As with most aspects of complex regression modeling, there are no hard and fast rules but, rather, general guidelines that work well in most situations. A useful rule of thumb is to consider investigating further if the highest Cook's distance is greater than 0.5, particularly if it is greater than 1. Observations with a Cook's distance less than 0.5 are *rarely* so influential that they should be removed from the main analysis. Those with a Cook's distance between 0.5 and 1 are *sometimes* sufficiently influential that they should be removed from the main analysis. Those with a Cook's distance greater than 1 are *often* sufficiently influential that they should be removed from the main analysis.

Recall the multiple linear regression model fit to the **CARS5** dataset, including all the vehicles—see model fit 1 on page 192. Figure 5.4 shows Cook's distances for this model plotted versus ID number, with the Lexus HS 250h clearly having the highest Cook's distance (see computer help #38). Cook's distance for the Lexus HS 250h is greater than the 0.5 warning threshold (shown by the horizontal dashed line), suggesting further investigation to see if it is an outlier or high-leverage point that might need to be removed from the analysis. We saw in Section 5.1.1 that this car is indeed an outlier that should probably be removed from the analysis—results are then as in model fit 2 on page 193.

Figure 5.5 shows Cook's distances for model fit 2 (see computer help #38) plotted versus ID number. Since none of the Cook's distances are greater than 0.5, there is probably no need to investigate further vehicles. We saw in Section 5.1.2 that while the Aston Martin DB9, Rolls-Royce Ghost, and Dodge Challenger SRT8 each have a high leverage and thus the potential to strongly influence the results, in the end none of them really does so.

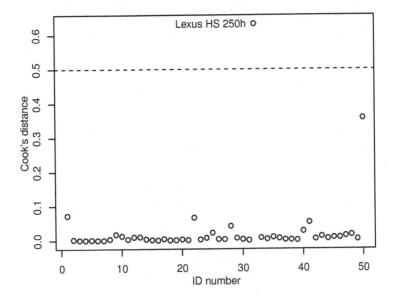

Figure 5.4   Scatterplot of Cook's distance versus ID number for the model fit to all the vehicles in the **CARS5** dataset (model fit 1). The horizontal dashed line shows the warning threshold at 0.5.

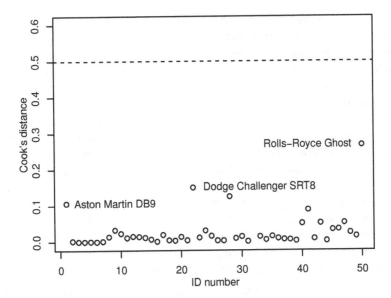

Figure 5.5   Scatterplot of Cook's distance versus ID number for the model fit to the **CARS5** dataset excluding the Lexus HS 250h (model fit 2). The horizontal dashed line shows the warning threshold at 0.5.

Since Cook's distances help to identify both outliers and high-leverage observations, one approach in linear regression modeling is to look first at a particular model's Cook's distances to identify potentially influential points, and then to investigate the nature of the potentially influential points in terms of their outlyingness and leverage. Remember that studentized residuals, leverages, and Cook's distances are related but measure different things:

- Studentized residuals measure outlyingness of response values relative to their predicted values.

- Leverage considers whether an observation has the potential to have a large impact on an estimated regression model because it has an unusual combination of predictor values.

- Cook's distance is a composite measurement that draws our attention to observations that are influential either because they are outliers or have high leverage, or both.

An observation can be an outlier but not have high leverage, or can have high leverage but not be an outlier, or be neither an outlier nor have high leverage, or be both an outlier and have high leverage! In the **CARS5** example, the Lexus HS 250h is an outlier (but does not have high leverage), while the Aston Martin DB9 has high leverage (but is not an outlier). Cook's distance flags the Lexus HS 250h (because of its high studentized residual) but not the Aston Martin DB9 (because, despite its high leverage, it does not strongly influence the results). The Rolls-Royce Ghost actually has the next-highest Cook's distance after the Lexus HS 250h—its combination of studentized residual and leverage is potentially more of a problem than the Aston Martin DB9's combination of studentized residual and leverage, but it too is below the 0.5 threshold and doesn't have much of an impact on the model fit.

If we find an influential point that has an adverse effect on the fit of a model, there are a number of potential remedies, such as the ones listed on page 190. We should never remove an observation arbitrarily just to get a better-fitting model—this could be considered manipulating the analysis to suit our purposes. The goal is to find the most appropriate and useful model for the data we have, and determining what to do about influential observations is a major decision not to be taken lightly.

The presence of outliers or high-leverage observations can indicate the need to transform one or more variables. For example, sometimes a logarithmic transformation can make extreme values less extreme and reduce their influence. Other possibilities include *binning* (in which the range of a variable is divided into intervals and each data value is replaced by a number that is representative of that interval) and *Winsorizing* (in which any data values beyond, say, the 5th and 95th percentiles, are replaced by those percentiles). However, employing transformations need not resolve issues relating to influential points, so we should still look at Cook's distances, residuals, and leverages in models with transformations.

Collectively, residuals, leverages, and Cook's distances are known as *regression diagnostics*. There are many other types of diagnostic, including:

- The *dfbetas* for the ith observation and jth regression parameter is the change in the estimate of the jth regression parameter when we drop the ith observation from the analysis (standardized by the standard error of the parameter estimate).

- The *dffits* for the ith observation is the change in predicted response when we drop the ith observation from the analysis (standardized by the regression standard error).

- The *press residual* or *predicted residual* for the ith observation is the residual for the ith observation when we estimate the regression parameters without the ith observation in the analysis.

In many applications, if we investigate residuals, leverages, and Cook's distances thoroughly (and follow up any observations flagged as a result), it is relatively rare that these additional diagnostics add much to the analysis. However, since they address influence from different points of view, they can prove useful in some applications. For more details, see Belsley et al. (2004), Cook and Weisberg (1982), Draper and Smith (1998), Kutner et al. (2004), Ryan (2008), and Weisberg (2005).

An alternative—or complementary—approach to individually investigating influential points through the use of regression diagnostics is *robust regression*. This is a set of techniques designed to mitigate problems that can result when highly influential points are present in a dataset (e.g., M-estimation and quantile regression). For more details, see Andersen (2007), Fox (2008), Huber and Ronchetti (2009), and Koenker (2005).

Optional—formula for Cook's distances. One formula for the ith observation's Cook's distance is

$$D_i = \frac{r_i^2 h_i}{(k+1)(1-h_i)},$$

where r_i is the ith observation's standardized residual, h_i is the ith observation's leverage, and k is the number of predictor terms in the model (excluding the constant intercept term). Rearranging this formula provides another way to calculate the leverage for the ith observation:

$$h_i = \frac{(k+1)D_i}{(k+1)D_i + r_i^2},$$

where D_i is the ith observation's Cook's distance.

5.2 REGRESSION PITFALLS

In this section we describe some of the pitfalls that can cause problems with a regression analysis.

5.2.1 Nonconstant variance

One of the four multiple linear regression assumptions states that the probability distribution of the random error at each set of predictor values has constant variance. When this assumption is violated, the error variance is said to be *nonconstant* or *heteroscedastic*.

For example, consider the following dataset obtained from www.gapminder.org (a "non-profit venture promoting sustainable global development and achievement of the United Nations Millennium Development Goals"). These data, available in the **FERTILITY** data file, contain the following variables for 169 localities (countries for the most part) for the year 2008:

> Fer = total fertility, children per women
> Inc = gross domestic product per capita, $ thousands (2005 $)
> Edu = mean years in school (women of reproductive age, 15 to 44)

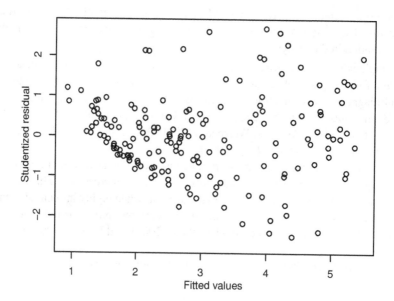

Figure 5.6   Scatterplot of studentized residuals versus fitted values for the model $E(Fer) = b_0 + b_1 \log_e(Inc) + b_2 Edu$ for the fertility dataset.

Here, *Inc* is based on purchasing power parities to account for between-country differences in price levels. Consider the following illustrative model of fertility (inspired by a similar example in Weisberg, 2005):

$$E(Fer) = b_0 + b_1 \log_e(Inc) + b_2 Edu.$$

Figure 5.6 displays a residual plot for this model (see computer help #35 and #15 in the software information files available from the book website). This plot shows clear evidence of nonconstant variance since the variation of the residuals increases from the left of the plot to the right. This violates the constant variance assumption for linear regression models.

To complement the visual impression of the residual variance in a residual plot, there are a number of test procedures available for diagnosing nonconstant variance.

- Breusch and Pagan (1979) and Cook and Weisberg (1983) independently developed a nonconstant variance test based on modeling the variance as a function of the fitted values from the model (or a selected subset of the predictor variables or a variable representing a quantity such as spatial location or time). If we apply this test to the fertility example using the fitted values from the model, $E(Fer) = b_0 + b_1 \log_e(Inc) + b_2 Edu$, we obtain a highly significant p-value of 0.000.

- Other tests we don't discuss further here include those proposed by White (1980) and Goldfeld and Quandt (1965).

Excessive nonconstant variance creates some technical difficulties with the multiple linear regression model. For example, if the residual variance increases with the value of a particular predictor, then prediction intervals will tend to be wider than they should be for low values of that predictor and narrower than they should be for high values of that predictor. We can employ a variety of remedies for refining a model that appears to exhibit excessive nonconstant variance, such as the following:

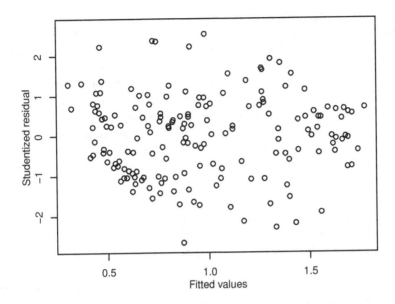

Figure 5.7   Scatterplot of studentized residuals versus fitted values for the model $E(\log_e(Fer)) = b_0 + b_1 \log_e(Inc) + b_2 Edu$ for the fertility dataset.

- Sometimes a nonconstant variance problem can be addressed by an appropriate "variance-stabilizing" transformation of the response variable, Y. One common variance-stabilizing transformation that we saw in Sections 4.1.4 and 4.1.5 is the log-arithmic transformation. This is sometimes used when errors tend to be proportional to the response variable rather than additive. Two similar variance-stabilizing trans-formations are the square root transformation (which can be effective if a logarithmic transformation is "too strong") and the reciprocal transformation (which can be ef-fective if a logarithmic transformation is "too weak"). Finally, the "arcsine square root" is another variance-stabilizing transformation, which is sometimes used when the response variable is constrained to a limited range (e.g., a proportion between 0 and 1). If we apply a natural logarithm transformation to the response variable in the fertility example and fit the model, $E(\log_e(Fer)) = b_0 + b_1 \log_e(Inc) + b_2 Edu$, we ob-tain the residual plot in Figure 5.7. The residuals in this plot appear to approximately satisfy the constant variance assumption. The Breusch-Pagan/Cook-Weisberg test supports this visual impression, with a nonsignificant p-value of 0.891.

- An alternative approach is to relax the constant variance assumption in the following way. Rather than assuming that the regression errors have constant variance at each set of values $(X_1, X_2, \ldots, X_k)$, it can make more sense to *weight* the variances so that they can be different for each set of values $(X_1, X_2, \ldots, X_k)$. This leads to the use of *weighted least squares* (or WLS), in which the data observations are given different weights when estimating the linear regression model. For example, if observations of a response variable represent group averages, then observations from larger groups will be less variable than observations from smaller groups, so the former are given relatively more weight when estimating the linear regression model. Or, if observations of a response variable represent group totals, then observations from larger groups will be more variable than observations from smaller groups, so the former are given relatively less weight when estimating the linear regression model. It is even possible to weight observations according to the value of one of the predictors (e.g., if the response variable tends to be more variable as the value of a particular predictor increases, then observations with low values of that predictor can be given relatively more weight than observations with high values), or according to the probability of selection in a complex survey design. When all observations are weighted equally (as for all the examples considered in this book), the method used to estimate the linear regression model is called simply *ordinary least squares* (or OLS). Weisberg (2005) includes a good discussion of applying weighted least squares in practice.

- A generalization of weighted least squares is to allow the regression errors to be correlated with one another in addition to having different variances. This leads to *generalized least squares* (GLS), in which various forms of nonconstant variance can be modeled. The details of GLS lie beyond the scope of this book, but some useful references include: Draper and Smith (1998), Thisted (1988), Pinheiro and Bates (2000), and Greene (2011) (who gives a thorough discussion of the econometric approaches of White, 1980, and Newey and West, 1987). Long and Ervin (2000) propose additional adjustments to White's classic correction for nonconstant variance.

- For some applications, we can explicitly model the variance as a function of the mean, $E(Y)$. This approach uses the framework of generalized linear models, which we discuss briefly in Section 7.1.

5.2.2 Autocorrelation

One of the four multiple linear regression assumptions states that the value of the random error for one observation should be independent of the value of the random error for any other observation (in other words, knowing the value of one random error gives us no information about the value of another one). One common way for this assumption to be violated is when the sample data have been collected over time and the regression model fails to effectively capture any time trends. In such a circumstance, the random errors in the model are often positively correlated over time; that is, each random error is more likely to be similar to the previous random error than it would be if the random errors were independent of one another. This phenomenon is known as *autocorrelation* or *serial correlation* and can sometimes be detected by plotting the model residuals (on the vertical axis of a scatterplot) versus time (on the horizontal axis).

For example, consider the following macroeconomic dataset obtained from www.economagic.com (a comprehensive collection of economic time series developed by University of Alabama economists). These annual data, available in the **UNEMP** data file, contain the following variables for the 52 years between 1959 and 2010:

Unr = U.S. unemployment rate, %
Bms = broad money supply (M2), $ billions
Ipd = implicit price deflator for GDP, 2005 = 100
Gdp = gross domestic purchases, $ billions (2005 $)
Exp = exports of goods and services, $ billions (2005 $)
$Time$ = time period

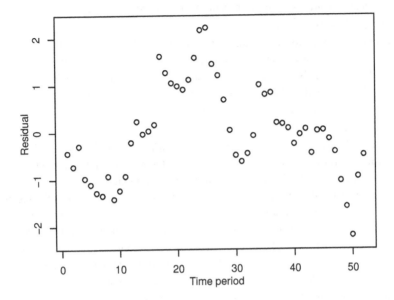

Figure 5.8   Scatterplot of the residuals from the model $E(Unr) = b_0 + b_1 \log_e(Bms/Ipd) + b_2 \log_e(Gdp) + b_3 \log_e(Exp) + b_4\,Time$ versus *Time* for the unemployment example.

Consider the following illustrative model of unemployment (based on an example in Rea, 1983):

$$E(Unr) = b_0 + b_1 \log_e(Bms/Ipd) + b_2 \log_e(Gdp) + b_3 \log_e(Exp) + b_4\,Time.$$

Figure 5.8 displays a residual plot for this model (see computer help #35 and #15 in the software information files available from the book website). This plot shows clear evidence of autocorrelation since each residual follows the previous residual much more closely than would be expected if the residuals were random. This violates the independent errors assumption for linear regression models.

The simplest form of autocorrelation is based on *first-order autoregressive* or AR(1) errors, $e_i = re_{i-1} + v_i$, where index i keeps track of the time order of the sample observations, e_i are the model errors, v_i are independent, normally distributed random variables with zero mean and constant variance, and r is the first-order autocorrelation (a number between -1 and 1). A value of r close to 1 indicates strong, positive autocorrelation, while a value close to 0 corresponds to the usual "uncorrelated errors" situation. It is also possible, although less common, to have negative autocorrelation, indicated by a value of r close to -1. There are more complex forms of autocorrelation that involve sophisticated time series methods, but many situations involving autocorrelation can be handled effectively within an AR(1) framework.

The pattern in a residual plot need not be as clear as in Figure 5.8 for there to be an auto-correlation issue. Thus it can be prudent to calculate the sample first-order autocorrelation of the residuals for any model where autocorrelation is suspected. One rule of thumb is to investigate further if the autocorrelation is greater than 0.3 (or less than -0.3 for negative autocorrelation). The residuals in Figure 5.8 have a first-order autocorrelation of 0.87. We can also look at higher-order autocorrelations if these are likely to be a problem. If further investigation seems warranted, there are a number of more formal test procedures available for diagnosing autocorrelation.

- One particularly simple procedure is the Wald and Wolfowitz (1940) *runs test*, which in this context looks at sequences of residuals that are all positive or all negative (called runs). This nonparametric test detects when there are longer runs than would be expected if the residuals were random, indicating that the residuals are autocorrelated. For the residuals in Figure 5.8, the runs test results in a highly significant p-value of 0.000.

- A historically important test is that of Durbin and Watson (1950, 1951, 1971). The test statistic is approximately $2(1 - \hat{r})$, where $\hat{r}$ is the sample first-order autocorrelation of the residuals. Thus, if the Durbin-Watson test statistic is close to 0, this indicates (positive) first-order autocorrelation, while if it is close to 2, this indicates no first-order autocorrelation. For the residuals in Figure 5.8, the Durbin-Watson test results in a highly significant test statistic of 0.24 with a p-value of 0.000. The Durbin-Watson test is a little tricky to apply since it can give inconclusive results. Also, it cannot be used for models in which the lagged response variable is used as a predictor (see below), which is common in many time series applications. Thus, the following alternative tests are generally preferred in practice.

- The Breusch (1978) and Godfrey (1978) test works by regressing the residuals on the original predictors plus the lag-1 residuals. Informally, a significant regression parameter for the lag-1 residuals indicates first-order autocorrelation. The test actually uses the F or χ^2 distributions (the χ^2 or *chi-squared* distribution, like the F-distribution, is a positively skewed probability distribution that takes only positive values). For the residuals in Figure 5.8, the Breusch-Godfrey test results in a highly significant p-value of 0.000. The test can also be generalized to test for higher-order autoregressive or moving average errors.

- Other tests we don't discuss here include the Durbin (1970) h statistic, the Box and Pierce (1970) Q *test*, and the Ljung and Box (1979) refinement of the Q test.

Excessive autocorrelation creates some technical difficulties with the multiple linear regression model, including, for example, underestimated standard errors. We can employ a variety of remedies for refining a model that appears to exhibit excessive autocorrelation.

- Sometimes autocorrelation arises when the model is misspecified through the omission of potential predictor variables. In particular, a dataset collected over time may have a time trend that requires inclusion of a predictor variable representing time (such as *Time* in the **UNEMP** dataset) or an appropriate nonlinear transformation of the time variable (if the trend is nonlinear). In addition, cyclical or seasonal patterns can often be effectively modeled by including indicator variables to represent appropriate time periods (e.g., months or quarters)—see Problem 5.2 at the end of this chapter for an example of this approach. While inclusion of time trend and seasonal variables often alleviates autocorrelation problems, in some cases there will be remaining autocorrelation that may still need to be dealt with.

- Although we cannot include the response variable (or transformations of the response variable) directly as a predictor in a multiple linear regression model, in some time series applications lagged response variables are used as predictors. The lag-1 response variable is defined so that its value in time period i is equal to the value of the response variable in the previous time period, $i-1$. When including the lag-1

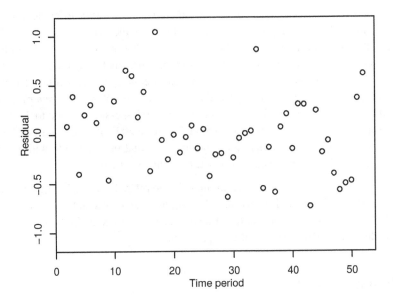

Figure 5.9   Scatterplot of the residuals versus *Time* for the unemployment model using first-differenced variables.

response variable as a predictor, we lose the first observation in the time series (since there is no previous time period), so the sample size is reduced by 1. We can also include lag-2 (and higher) response variables as predictors, in which case the sample size would be reduced by 2 (or higher). Again, as with inclusion of time trend and seasonal variables, adding lagged response variables as predictors may alleviate autocorrelation problems, but in some cases additional model refinement may be needed. For example, for the **UNEMP** dataset, including lag-1 unemployment as a predictor in the model on page 203 removes much of the autocorrelation evident in Figure 5.8, but the Breusch-Godfrey test still results in a significant p-value of 0.001.

- A common approach in time series applications is to "first difference" the data by subtracting the value of each variable in the previous time period (for the response variable and predictor variables alike). For example, $Y_i^{diff} = Y_i - Y_{i-1}$ and $X_i^{diff} = X_i - X_{i-1}$ (for each predictor). We then use these first-differenced variables in our multiple linear regression model in place of the original variables. As with using the lag-1 response variable as a predictor, we lose the first observation in the time series. In addition, we must estimate the model without an intercept term. If we apply this method to the **UNEMP** dataset, replacing the variables in the model on page 203 with their first-differenced versions, we obtain the residual plot in Figure 5.9. The residuals in this plot do not exhibit any strong autocorrelation patterns and appear to satisfy the independent errors assumption. The Breusch-Godfrey test supports this visual impression, with a nonsignificant p-value of 0.995.

- A generalization of the first-differencing procedure is to "quasi difference" the data by subtracting r times the value of each variable in the previous time period, where r is the first-order autocorrelation of the model errors. Since r is an unknown population parameter, we have to estimate it in practice with, for example, $\hat{r}$, the sample first-order autocorrelation of the residuals. We then use these quasi-differenced variables in our multiple linear regression model in place of the original variables. This is called the Cochrane and Orcutt (1949) procedure (omitting the first observation in the time series) or the Prais and Winsten (1954) procedure (including the first observation using an appropriate transformation). We can iterate either procedure if excessive autocorrelation remains. These procedures are examples of *feasible generalized least squares* (FGLS). If we apply the Cochrane-Orcutt single-step method to the **UNEMP** dataset, replacing the variables in the model on page 203 with their quasi-differenced versions, we obtain a residual plot very similar to that in Figure 5.9, with $\hat{r} = 0.874$ and a nonsignificant Breusch-Godfrey test p-value of 0.938.

- Other methods we don't discuss here include the related approaches of Hildreth and Lu (1960) and Beach and MacKinnon (1978), the corrected standard errors of Newey and West (1987), and other more sophisticated time series methods. Greene (2011) provides a good discussion of these approaches.

5.2.3 Multicollinearity

With multiple linear regression modeling we are attempting to quantify how a response variable, Y, is associated with the values of two or more predictor variables, $X_1, X_2, \ldots,$ in a model such as $\mathrm{E}(Y) = b_0 + b_1 X_1 + b_2 X_2 + \cdots$. This is easiest to do if the sample associations between Y and each of the predictors are far stronger than any of the sample associations among the predictors themselves. Recall that the interpretation of a regression parameter (e.g., b_1 measuring the association between X_1 and Y) involves the notion of "keeping the values of the other predictors constant." Thus, we can estimate b_1 most accurately if there is plenty of variation in X_1 when we keep the values of the other predictors constant. Conversely, if (say) X_1 and X_2 are highly correlated in the sample, then there is little variation in X_1 when we keep X_2 constant, so it is much harder to estimate b_1.

This phenomenon, known as *multicollinearity* or *collinearity*, can lead to unstable models and inflated standard errors (making the models less accurate and useful). Methods for identifying multicollinearity (when there is excessive correlation between quantitative predictor variables) include the following.

- Construct a scatterplot matrix of the response variable, Y, and all the quantitative predictors; there is a potential multicollinearity problem if any of the scatterplots involving just predictors look very highly correlated.

- Calculate the absolute values of the correlations between Y and each of the quantitative predictors, and between each pair of quantitative predictors; there is a potential multicollinearity problem if any of the (absolute) correlations between each pair of predictors is greater than the highest (absolute) correlation between Y and each of the predictors.

- Calculate *variance inflation factors* (Marquardt, 1970) for the regression parameter estimates for each of the quantitative predictors in the model; these provide an estimate of how much larger the variance of each parameter estimate becomes when the corresponding predictor is included in the model; a useful rule of thumb is that there is a potential multicollinearity problem if the variance inflation factor for a quantitative predictor is greater than 10.

We can use all three methods to look at multicollinearity since they can all help enrich the analysis and each comes at the issue from a different perspective. For example, suppose that for a small high-tech business we suspect that *Sales* = sales (in $m) depends on *Trad* = traditional advertising on TV and in newspapers (in $m) and *Int* = advertising using the Internet (in $m). We have available the following (simulated) data in the **SALES3** data file to investigate this possibility (we looked at this dataset previously on pages 100 and 116):

Sales (sales in $m)	5.0	7.0	7.0	5.0	9.0	11.0	12.0	13.0	15.0	17.0	18.0	19.0
Trad (advertising in $m)	1.0	2.0	2.5	1.5	3.5	4.0	5.0	6.0	6.5	8.0	8.0	7.5
Int (Internet in $m)	2.0	2.5	3.0	1.5	4.0	4.5	5.0	5.5	6.0	7.5	8.0	7.0

We obtain the following results from this model (see computer help #31 and #41 in the software information files available from the book website):

$$E(Sales) = b_0 + b_1 \, Trad + b_2 \, Int.$$

Model Summary

Model	Sample Size	Multiple R Squared	Adjusted R Squared	Regression Std. Error
1[a]	12	0.9740	0.9682	0.8916

[a] Predictors: (Intercept), *Trad*, *Int*.

Parameters[a]

Model		Estimate	Std. Error	t-stat	Pr(> |t|)	VIF
1	(Intercept)	1.992	0.902	2.210	0.054	
	Trad	1.275	0.737	1.730	0.118	49.541
	Int	0.767	0.868	0.884	0.400	49.541

[a] Response variable: *Sales*.

On the one hand, the relatively high p-value for b_1 (0.118) suggests that there is no linear association between *Sales* and *Trad* (holding *Int* constant). On the other hand, the relatively high p-value for b_2 (0.400) suggests that there is no linear association between *Sales* and *Int* (holding *Trad* constant). However, there does appear to be some kind of association between *Sales* and (*Trad*, *Int*) together because the value of R^2, 97.4%, is pretty high.

The problem is multicollinearity between *Trad* and *Int*—the variance inflation factors (VIF in the statistical software output) are each 49.5, much larger than our rule-of-thumb threshold of 10. Also, the correlation between *Trad* and *Int* of 0.990 is greater than the correlation between *Sales* and *Trad* (0.986) and between *Sales* and *Int* (0.983)—values obtained using statistical software, but software output not shown (see computer help #40).

To address this problem, we can:

- Attempt to collect new data with a lower correlation between the collinear predictors; this is rarely going to be possible with observational data.
- If possible, combine the collinear predictors together to form one new predictor; this usually makes sense only if the predictors in question have the same measurement scale (as they happen to in this example).

- Remove one of the collinear predictors from the model, for example, the one with the highest p-value; we necessarily lose information in the dataset with this approach, but it may be the only remedy if the previous two approaches are not feasible.

For the sales data, since *Trad* and *Int* are both measured in $m, we can therefore try the second approach using this model:

$$E(Sales) = b_0 + b_1(Trad + Int).$$

Model Summary

Model	Sample Size	Multiple R Squared	Adjusted R Squared	Regression Std. Error
2[a]	12	0.9737	0.9711	0.8505

[a] Predictors: (Intercept), *Trad + Int*.

Parameters[a]

Model		Estimate	Std. Error	t-stat	Pr(> |t|)
2	(Intercept)	1.776	0.562	3.160	0.010
	Trad + Int	1.042	0.054	19.240	0.000

[a] Response variable: *Sales*.

R^2 is essentially unchanged with this model, while we now have a highly significant regression parameter estimate for b_1 (p-value is 0.000).

We previously also tried the third approach by removing *Int* from the analysis and fitting a model using just *Trad* as a single predictor—see page 116 for the results of that approach.

The quantitative predictors in a regression model will nearly always be correlated to some degree (one exception being in designed experiments with orthogonal predictors). We should therefore be pragmatic when dealing with multicollinearity. If multicollinearity is extreme enough to be causing problems (e.g., disproportionately large standard errors for some of the regression parameter estimates, which causes unstable model results), then we probably need to address it (by combining the offending predictors if logically possible, or by removing one or more predictors if not). Otherwise, there is probably no need to worry about it. Sometimes, variance inflation factors can signal a potential multicollinearity problem that turns out not to be worth worrying about. For example, models with quadratic transformations or interactions involving quantitative predictors often have large variance inflation factors for the predictor terms involved. *Rescaling* the values of those predictor terms to have a mean close to 0 and a standard deviation close to 1 can mitigate such problems. However, if inclusion of the quadratic or interaction terms does not result in regression parameter standard errors becoming alarmingly large, again there is probably no need to worry. See Problem 5.3 at the end of the chapter to see an example of this.

This section has focused on multicollinearity problems with quantitative predictor variables. However, it is also possible for multicollinearity problems to arise with indicator variables for two or more qualitative predictors. Variance inflation factors can again help to signal when further investigation is warranted, but scatterplot matrices are not particularly helpful for indicator variables. We can calculate a regular correlation for indicator variables, but there are other methods that can be more appropriate. The association between two binary variables is typically summarized in a two-by-two table (the simplest example of a "contingency table" or "cross-tab"). Such a table shows the number of times each combination of categories occurs. There are many summary measures of association for contingency tables, including the "Phi Coefficient" and "Kendall's Tau." It turns out that for two-by-two tables, both these measures are numerically equivalent to the usual (Pearson product-moment) correlation. Both the Phi Coefficient and Kendall's Tau measure the degree to which the column and row values tend to match (i.e., both are 0 or both are 1). Most statistical software packages will compute them (among various other measures of association for contingency tables). For further information, see Agresti (2002).

Optional—formula for variance inflation factors. One way to think about the variance inflation factor (VIF) for a predictor is to note the regression parameter estimate standard errors for all predictors except the one in question. Then take that predictor out of the model and look at how the parameter estimate standard errors for all predictors left in the model have changed. Now, summarize that change using some complicated formula. That is essentially what the VIF does for each predictor in turn. As an example, the actual formula for VIF for X_1 in a multiple linear regression model is $1/(1-R_1^2)$, where R_1^2 is the coefficient of determination for the regression of X_1 on the other predictors in the model. Similarly, the VIF for X_2 is $1/(1-R_2^2)$, where R_2^2 is the coefficient of determination for the regression of X_2 on the other predictors in the model, and so on.

5.2.4 Excluding important predictor variables

When model building, we don't want to include unimportant predictors whose presence can overcomplicate things unnecessarily and can increase our uncertainty about the magnitudes of the effects for the important predictors (particularly if some of those predictors are highly collinear). However, there is potentially greater risk from excluding important predictors than from including unimportant ones. For example, consider the data in Table 5.2 (available in the **PARADOX** data file), which contains data for a simulated production process in

Table 5.2 Computer component data.

Quality	Speed	Angle	Quality	Speed	Angle
0.86	3.01	1	1.41	2.46	1
1.48	3.96	2	2.11	5.86	3
3.21	0.55	1	2.83	3.10	2
3.11	4.08	3	3.51	7.12	4
4.02	2.15	2	4.22	4.17	3
3.60	5.77	4	3.69	7.93	5
4.69	2.65	3	4.93	4.62	4
5.08	6.47	5	5.62	1.77	3
6.52	3.85	4	5.79	6.04	5
5.62	8.22	6	7.13	3.31	4
7.07	4.86	5	6.70	6.65	6
7.32	9.23	7	7.95	4.29	5
7.83	5.56	6	8.44	8.21	7
9.21	6.97	7			

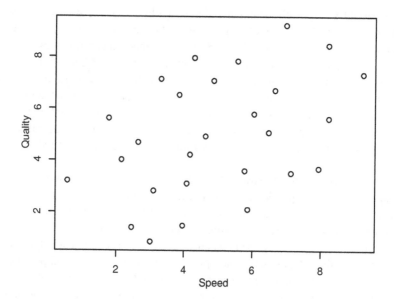

Figure 5.10 Scatterplot of *Quality* versus *Speed* for the computer components example.

which we have measurements on $n=27$ high-precision computer components with *Quality* potentially depending on two controllable machine factors: *Speed* and *Angle*.

Figure 5.10 suggests that there is a positive linear association between *Quality* and *Speed* when we ignore *Angle* (see computer help #15 in the software information files available from the book website). This is confirmed with a simple linear regression model:

$$E(Quality) = b_0 + b_1\,Speed.$$

Parameters[a]

Model		Estimate	Std. Error	t-stat	Pr($>|t|$)
1	(Intercept)	2.847	1.011	2.817	0.009
	Speed	0.430	0.188	2.288	0.031

[a] Response variable: *Quality*.

This suggests that as machine speed increases (by 1 unit) component quality increases (by 0.430 unit). However, this analysis ignores any potential information about the process that results from changing the value of *Angle*. Figure 5.11 suggests that there is actually a *negative* linear association between *Quality* and *Speed* when we account for *Angle* (see computer help #17). In particular, when we hold *Angle* fixed (at any value), there is a negative linear association between *Quality* and *Speed*. This is confirmed with a multiple linear regression model:

$$E(Quality) = b_0 + b_1\,Speed + b_2\,Angle.$$

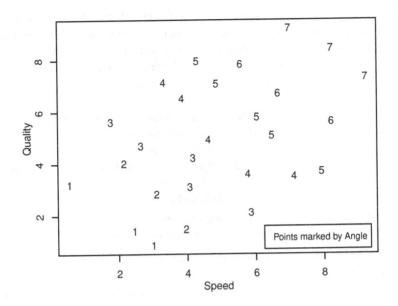

Figure 5.11 Scatterplot of *Quality* versus *Speed* with points marked by *Angle* for the computer components example.

Parameters[a]

Model		Estimate	Std. Error	t-stat	Pr(> |t|)
2	(Intercept)	1.638	0.217	7.551	0.000
	Speed	−0.962	0.071	−13.539	0.000
	Angle	2.014	0.086	23.473	0.000

[a] Response variable: *Quality*.

This suggests that when we hold machine angle fixed (at any value), as speed increases (by 1 unit) component quality decreases (by 0.962 unit). We can also see that when we hold speed fixed (at any value), as the angle increases (by 1 unit) component quality increases (by 2.014 unit).

If we used the simple linear regression model (which excludes the important predictor *Angle*), we would make the *incorrect* decision to increase machine speed to improve component quality. If we used the multiple linear regression model (which includes the important predictor *Angle*), we would make the *correct* decision to decrease machine speed and increase machine angle simultaneously to improve component quality. The result that a linear association between two variables (i.e., *Quality* and *Speed*) *ignoring* other relevant variables (i.e., *Angle*) can have a different direction from the association that *conditions on* (controls for) other relevant variables is called *Simpson's paradox* (Simpson, 1951; Yule, 1903).

In other applications, the linear association between two variables *ignoring* other relevant variables may differ only in magnitude (but not direction) from the association that *conditions on* (controls for) other relevant variables. Nevertheless, whether results differ in magnitude alone or magnitude *and* direction, this is potentially a huge problem. Whereas the potential cost of *including unimportant predictors* might be increased difficulty with interpretation and reduced prediction accuracy, the potential cost of *excluding important predictors* can be a completely meaningless model containing misleading associations. Results can vary considerably depending on whether such predictors are (inappropriately) excluded or (appropriately) included. These predictors are sometimes called *confounding variables* or *lurking variables*, and their absence from a model can lead to incorrect conclusions and poor decision-making.

Thus, it is often more prudent to include all potentially important predictors (and transformations and interactions) even when they are not quite statistically significant (at a 5% significance level, say). In other words, the usual 5% significance threshold is just a guideline, and a good final model for a particular context may include one or more predictors with p-values higher than this threshold. There are other factors that can help us to make the decision about which predictors to include and exclude:

- Does inclusion of a high p-value predictor result in far-worse prediction accuracy (i.e., does the regression standard error, s, increase greatly)?

- Does inclusion of a high p-value predictor cause multicollinearity problems with other included predictors?

- Do overall conclusions about the effects of other predictors on the response variable remain essentially unchanged whether the high p-value predictor is included or excluded?

Answering yes to any of these questions provides more compelling evidence to remove the high p-value predictor than just the fact that it has a high p-value.

5.2.5 Overfitting

Although in the preceding section we suggested that excluding important predictors is often more costly than including unimportant predictors, the latter strategy can lead to problematic overfitting. For an extreme example, consider adding higher and higher powers of X to a model for predicting Y for the data shown in Figure 5.12. These data were generated from the model $E(Y) = b_0 + b_1 X$ (left-hand plot), but consider what happens when we fit

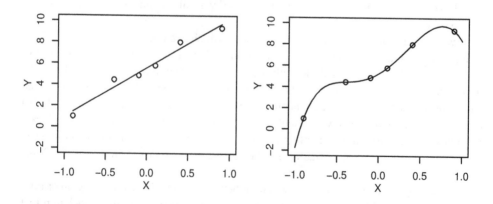

Figure 5.12   Scatterplots of Y versus X with the fitted line from the correct model used to generate the data, $E(Y) = b_0 + b_1 X$ (left), and the fitted line from an overfit, incorrect model, $E(Y) = b_0 + b_1 X + b_2 X^2 + b_3 X^3 + b_4 X^4 + b_5 X^5$ (right).

the more complicated model $E(Y) = b_0 + b_1 X + b_2 X^2 + b_3 X^3 + b_4 X^4 + b_5 X^5$ (right-hand plot). This more complicated model fits the sample data perfectly (going through all the points), but it seems unlikely that this wiggly line represents the true, underlying association in the population from which this sample came. (In this simulated example, we know this is not the case since we generated the data from a simple linear regression model.) The wiggly line overfits the sample data and provides a poor understanding of the underlying population association. In other words, the complicated model contains unimportant predictors (X^2, X^3, X^4, and X^5), which appear to improve prediction of the sample response values but which actually overcomplicate things, and lead to poor predictions of population response values not included in the sample. For example, imagine predicting Y at $X = -0.6$. The complicated model on the right in Figure 5.12 has an odd "bump" here pointing to a prediction for Y of about 4. The simple linear regression model on the left in Figure 5.12 correctly predicts a value for Y closer to 3.

In more complicated examples with more predictor variables, it is not as easy to "see" overfitting like this using graphs. However, overfitting can always occur the more complicated a model becomes, and the more predictor variables, transformations, and interactions are added to the model. It is always prudent to apply a "sanity check" to any model being used to make decisions. Models should always make sense, preferably grounded in some kind of background theory or sensible expectation about the types of associations allowed between variables. Furthermore, any predictions the model gives should be reasonable (overly complicated models can give quirky results that may not reflect reality).

5.2.6 Extrapolation

A multiple linear regression model can be a very powerful method for explaining the association between a response variable, Y, and predictor variables, $X_1, X_2, \ldots$, and for predicting future values of the response at particular values of the predictors. However, the model should only be applied in situations similar to that represented by the sample dataset. This is one of the principles of statistical inference—we use sample data to tell us about expected patterns of behavior in a population by assuming that the sample is representative of the population.

In other words, we should *not* apply a multiple linear regression model to predictor values that are very different from those in our sample. To do so would be to *extrapolate* our model results to settings in which they may or may not hold (where we have no data to support our conclusions).

For example, consider the simulated data represented in Figure 5.13. These data represent a hypothetical production situation, where a simple linear regression model model and a quadratic model have been fit to 30 sample points—these points are represented by circles in Figure 5.13. The lines represent the fitted model predictions—these lines are dashed beyond the sample data region to indicate that they are extrapolations there. The two crosses represent future values that were not available when the models were fit.

In the case of the future value at *Labor* = 6, there is no problem, and both the simple linear regression model and the quadratic model provide a good prediction for the resulting value of *Production*. However, predicting *Production* at the future value of *Labor* = 12 is more difficult, because our sample data only go up to *Labor* = 8 or so. If we were to use the simple linear regression model to predict *Production* at *Labor* = 12, we would clearly overshoot and predict much too high. Perhaps we can get around this problem by using the

Figure 5.13   Scatterplot of *Production* = production in hundreds of units versus *Labor* = labor hours in hundreds for simulated production data where a simple linear regression model and a quadratic model have been fit to the sample points (circles). The lines represent the fitted model predictions—these lines are dashed beyond the sample data region to indicate that they are extrapolations there. The two crosses represent future values that were not available when the models were fit.

quadratic model in an attempt to model the apparent curvature in the sample data. However, although this quadratic model fits the sample data better, it has the curious property that *Production* is predicted to start going down as *Labor* increases above 8. This doesn't make any practical sense, and indeed if we were to use the quadratic model to predict *Production* at *Labor* = 12, we would clearly undershoot and predict much too low.

In both cases (the simple linear regression model and the quadratic model), the underlying problem is the same—attempting to use the model to predict response values well outside the sample range of predictor values. With more predictors, this pitfall can become more subtle since it can be difficult to draw a picture like Figure 5.13 to avoid making such a mistake. Furthermore, it is not enough just to ensure that predictions will be made only for predictor values within their sample ranges. For example, suppose that two predictors, X_1 and X_2, have sample ranges of 0–10, but that they are reasonably highly correlated so that there are no sample values close to the region of $X_1 = 2$ and $X_2 = 8$. Then, it is unwise to use a model based on the sample values of X_1 and X_2 to predict a response value at $X_1 = 2$ and $X_2 = 8$, even though $X_1 = 2$ is within the sample range of X_1-values and $X_2 = 8$ is within the sample range of X_2-values.

That said, it is sometimes reasonable to extrapolate "just beyond" the sample data region if we have reason to expect that the regression model results extend to here. We saw in Section 3.6 that the prediction interval for an individual response value becomes wider (reflecting increased uncertainty) as the particular predictor values at which we are predicting get farther away from their sample means. However, beware of using this approach a long way from the sample data region—results here will be very unreliable.

5.2.7 Missing data

Real-world datasets frequently contain *missing values*, so that we do not know the values of particular variables for some of the sample observations. For example, such values may be missing because they were impossible to obtain during data collection. Dealing with missing data is a challenging task, and there are many issues and approaches to consider that we are unable to get into too deeply here (a classic reference in this area is Little and Rubin, 2002). Nevertheless, with a few simple ideas we can mitigate some of the problems caused by missing data.

To frame these ideas, consider the simulated dataset in Table 5.3, which is available in the **MISSING** data file. A sample observation may have useful information for all potential predictor variables except one, but if the latter predictor is used in a regression model then all that information is ignored. For example, if we fit a multiple linear regression model that contains all four predictors, X_1, X_2, X_3, and X_4, five observations out of the total sample of 30 will be excluded. If we fit a model that does not contain X_2 (i.e., it uses X_1, X_3, and X_4 only), one observation out of the total sample of 30 will be excluded. If we fit a model that contains neither X_2 nor X_3 (i.e., it uses X_1 and X_4 only), all 30 observations will be analyzed.

Missing data therefore has the potential to affect a regression analysis adversely by reducing the total usable sample size. The best solution to this problem is to try extremely hard to avoid having missing data in the first place—in situations where sample sizes are small enough that they can be drastically affected by missing data, it is well worth spending a great deal of time and effort minimizing the total number of missing values. When there are missing values that are impossible (or too costly) to avoid, one approach is to replace the missing values with plausible estimates—this is known as *imputation* and lies beyond the scope of this book.

Another approach to dealing with missing data is to consider only models that contain predictors with no (or few) missing values. This can be a little extreme, however, since even

Table 5.3 Simulated dataset containing missing values.

Y	X_1	X_2	X_3	X_4	Y	X_1	X_2	X_3	X_4
13	5	7	7	6	11	4	—	7	3
16	7	7	7	8	10	7	5	4	7
9	1	2	3	5	8	5	4	3	5
16	6	5	9	6	6	3	—	1	4
8	3	3	6	2	16	7	6	7	8
7	4	—	—	3	11	4	5	4	6
8	3	3	5	3	8	1	0	3	5
14	8	7	7	5	12	9	7	3	7
14	5	8	7	6	11	7	7	1	8
3	2	4	1	2	12	6	6	4	5
16	6	5	8	8	3	2	4	1	2
10	6	6	4	4	6	4	3	2	4
9	3	—	6	2	15	7	10	6	7
13	4	6	5	6	11	4	—	6	4
8	5	4	2	5	7	4	4	4	2

predictor variables with a large number of missing values can contain useful information. Rather than discarding such predictor variables solely on the basis of their containing many missing values, we should factor this into our usual strategy for identifying whether predictor variables are useful or redundant. For example, in the simulated dataset in Table 5.3, we might be more inclined to remove X_2 from a multiple linear regression model than to remove X_1, X_3, or X_4 (since it has the largest amount of missing data), but we would still need to be convinced that the information available in its nonmissing values was redundant before actually removing it.

In particular, consider the results for a model that contains all four predictors (and that has a usable sample size of 25):

$$E(Y) = b_0 + b_1 X_1 + b_2 X_2 + b_3 X_3 + b_4 X_4.$$

Model Summary

Model	Sample Size	Multiple R Squared	Adjusted R Squared	Regression Std. Error
1 [a]	25	0.9586	0.9503	0.8649

[a] Predictors: (Intercept), X_1, X_2, X_3, X_4.

Parameters [a]

Model		Estimate	Std. Error	t-stat	$\Pr(> |t|)$
1	(Intercept)	−0.219	0.558	−0.392	0.699
	X_1	0.070	0.144	0.485	0.633
	X_2	0.245	0.131	1.872	0.076
	X_3	0.925	0.086	10.802	0.000
	X_4	0.969	0.122	7.917	0.000

[a] Response variable: Y.

X_1 seems redundant here (since it has such a large p-value), but there is a question mark about X_2 as well—its p-value is quite a bit smaller than that of X_1, but it is larger than the usual 5% significance level. Furthermore, removing X_2 would increase the usable sample size to 29, whereas removing X_1 would keep the usable sample size at 25. Nevertheless, since the p-value for X_1 is so much larger than that for X_2, we first consider the results for a model that does not contain X_1 (and that has a usable sample size of 25):

$$E(Y) = b_0 + b_2 X_2 + b_3 X_3 + b_4 X_4.$$

Model Summary

Model	Sample Size	Multiple R Squared	Adjusted R Squared	Regression Std. Error
2 [a]	25	0.9581	0.9521	0.8490

[a] Predictors: (Intercept), X_2, X_3, X_4.

Parameters[a]

Model		Estimate	Std. Error	t-stat	$\Pr(>|t|)$
2	(Intercept)	−0.224	0.548	−0.409	0.687
	X_2	0.286	0.099	2.875	0.009
	X_3	0.926	0.084	11.024	0.000
	X_4	0.995	0.108	9.252	0.000

[a] Response variable: Y.

Contrast these results with those for a model that does not contain X_2 (and that has a usable sample size of 29):

$$E(Y) = b_0 + b_1 X_1 + b_3 X_3 + b_4 X_4.$$

Model Summary

Model	Sample Size	Multiple R Squared	Adjusted R Squared	Regression Std. Error
3 [a]	29	0.9529	0.9472	0.8524

[a] Predictors: (Intercept), X_1, X_3, X_4.

Parameters[a]

Model		Estimate	Std. Error	t-stat	$\Pr(>|t|)$
3	(Intercept)	0.209	0.493	0.424	0.675
	X_1	0.237	0.109	2.172	0.040
	X_3	0.966	0.074	13.054	0.000
	X_4	0.943	0.111	8.480	0.000

[a] Response variable: Y.

Both three-predictor models fit well (and X_2 is significant in one, while X_1 is significant in the other). However, the (X_1, X_3, X_4) model fits slightly less well than the (X_2, X_3, X_4) model: The former model's R^2 of 0.9529 is 0.5% smaller than the latter model's R^2 of 0.9581, while the former model's s of 0.8524 is 0.4% larger than the latter model's s of 0.8490. Ordinarily, then, we would probably favor the (X_2, X_3, X_4) model. However, the (X_1, X_3, X_4) model applies to much more of the sample than the (X_2, X_3, X_4) model: The former model's usable sample size of 29 is 16% larger than the latter model's usable sample size of 25. Thus, in this case, we would probably favor the (X_1, X_3, X_4) model (since R^2 and s are roughly equivalent, but the usable sample size is much larger).

5.2.8 Power and sample size

In small datasets, a lack of observations can lead to poorly estimated models with a large regression standard error and large regression parameter standard errors. Such models are said to lack statistical *power* because there is insufficient data to be able to detect significant associations between the response and the predictors. That begs the question, "How much data do we need to conduct a successful regression analysis?" A common rule of thumb is that 10 data observations per predictor variable is a pragmatic lower bound for sample size. However, it is not so much the number of data observations that determines whether a regression model is going to be useful, but rather whether the resulting model satisfies the four regression model assumptions (see Section 3.4). So, in some circumstances, a model applied to fewer than 10 data observations per predictor variable might be perfectly fine (if, say, the model fits the data really well and the regression assumptions seem fine), while in other circumstances a model applied to a few hundred data points per predictor variable might be pretty poor (if, say, the model fits the data badly and one or more regression assumptions are seriously violated).

For another example, in general we'd need more data to model interaction compared to a similar model without the interaction. However, it is difficult to say exactly how much data would be needed. It is possible that we could adequately model interaction with a relatively small number of observations if the interaction effect was pronounced and there was little statistical error. Conversely, in datasets with only weak interaction effects and relatively large statistical error, it might take a much larger number of observations to have a satisfactory model. In practice, we have methods for assessing the regression assumptions, so it is possible to consider whether an interaction model approximately satisfies the assumptions on a case-by-case basis.

In conclusion, there is not really a good standard for determining sample size given the number of predictors, since the only truthful answer is, "It depends." In many cases, it soon becomes pretty clear when working on a particular dataset if we are trying to fit a model with too many predictor terms for the number of sample observations (results can start to get a little odd and standard errors greatly increase). From a different perspective, if we are designing a study and need to know how much data to collect, then we need to get into sample size and power calculations, which rapidly become quite complex and beyond the scope of this book. Some statistical software packages will do sample size and power calculations, and there is even some software specifically designed to do just that. When designing a large, expensive study, it is recommended that such software be used or to get advice from a statistician with sample size expertise.

5.3   MODEL BUILDING GUIDELINES

It is easy to become overwhelmed by possibilities when faced with modeling a large dataset with many potential predictors. Our goal is to come up with a useful model for understanding the association between a response variable, Y, and k predictor variables, $(X_1, X_2, \ldots, X_k)$. Yet there is unlikely to be a single "best model" that we'll be able to discover as long as we work at it long enough. Chances are there are many good models, however, and all we need to do is find one of these good ones. While differing slightly with respect to which predictors have been included and which transformations have been used, the good models usually yield similar overall interpretations and predictions. Problem 5.6 at the end of the chapter illustrates this point.

There are many strategies for attacking a regression problem, but the following approach tends to be quite successful most of the time:

1. Carefully frame your questions and identify the data that will help answer the questions.

2. Collect and organize the data—this will usually be the most time-consuming part of the whole process (we haven't said much about this step in the book because our focus has been elsewhere).

3. Organize the data for analysis—this includes checking for mistakes and coding qualitative variables using indicator variables (see page 174).

4. Graph the data (using a scatterplot matrix, say) and perhaps calculate some summary statistics to get a feel for the dataset—this can also alert you to potential data entry mistakes and might suggest some variable transformations (e.g., highly skewed variables will often reveal themselves in graphs and are often best transformed to natural logarithms before further analysis).

5. Fit an initial model using each of the potential quantitative predictors and indicator variables for the qualitative predictors. Use untransformed Y as the response variable unless a transformed Y is suggested beforehand [e.g., highly skewed Y is often better analyzed as $\log_e(Y)$]. If there is background knowledge which suggests that particular transformations or interactions might be important, then include them in this initial model.

6. Check the four regression assumptions for this initial model using residual plots. If there is a strong suggestion that one or more of the assumptions is violated, then proceed to step 7; otherwise, if everything checks out, proceed to step 8.

7. Include interactions and transformations (if not included in the model already) to attempt to improve the model so that the four regression assumptions do check out. Start "simple" and try more complicated models only as needed (once the model is adequate and the regression assumptions check out, proceed to step 8). Ideally, the question of which interactions and transformations to try should be guided by background information about the particular application. In the absence of such information, a pragmatic approach is to proceed as follows:

   - First try adding interactions between indicator variables and quantitative predictors, for example, DX_1 and DX_2.
   - If the model is still inadequate, next try adding/replacing transformations of the quantitative predictors, for example, add X_1^2, or replace X_1 with $\log_e(X_1)$, or replace X_1 with $1/X_1$.
   - If the model is still inadequate, next try adding interactions between quantitative predictors, for example, $X_1 X_2$ and $X_1 X_3$.

8. Simplify the model in stages (aim for a parsimonious model that can capture the major, important population associations between the variables without getting distracted by minor, unimportant sample associations):

   - Evaluate each model that you fit to identify which predictors, interactions, and transformations to retain and which to remove—methods include adjusted R^2, the regression standard error (s), and hypothesis tests for the regression parameters (global usefulness test, nested model tests, individual tests).

- Remove all redundant predictors, interactions, and transformations (remember to preserve hierarchy while doing this, e.g., retain X_1 if including X_1^2)—note that this step usually proceeds just a few predictor terms at a time.

9. Evaluate the final model and confirm that the regression assumptions still hold.

10. During the whole process identify any outliers or other influential points, and look out for potential pitfalls (nonconstant variance, autocorrelation, multicollinearity, excluding important predictors, overfitting, missing data)—address any problems as they occur.

11. Interpret the final model, including understanding predictor effects on Y, and estimating expected values of Y and predicting individual values of Y at particular values of the predictors (taking care not to extrapolate well outside the sample data region).

12. Interpret the final model using graphics—see the next section.

There are some examples of this model building process in Section 5.5, in Problems 5.5 and 5.7 at the end of this chapter, and in the case studies of Chapter 6.

Model building can also be framed in terms of the regression errors, $e = Y - E(Y)$. Recall that these "errors" are not mistakes, but rather represent the part of Y that cannot be explained by the predictors in the model. It is possible that this unexplained part could be partly explained with additional predictor variables (or using the existing predictor variables in a different way in the model through transformations or interactions, for example). Thus one rationale for model building is that we iteratively fit a series of models to explain as much of the variation in Y as we can (or, equivalently, reduce the errors as much as we can). At some point, however, we are unable to improve the model beyond a certain point, and the errors for this final model could be considered to represent some kind of "natural random variation" that is left over once the effects of all the predictors have been accounted for.

By trying to explain as much of the variation in Y as we can, we are aiming to obtain a "high" value of R^2. However, it is not possible to tell just from the value of R^2 whether a model is a success or not, since it depends on the context and the underlying variability in the data. In some instances, a model with an R^2 of 30% could be considered a success (if the assumptions are satisfied and there are underlying conditions which mean that 70% of the variation in the data is always going to be unexplainable), while in others a model with an R^2 of 95% could be considered poor (if the assumptions are strongly violated, say).

One final, brief thought on the ethics of statistical modeling. Throughout any regression analysis, we should use background knowledge to guide our model building (e.g., to suggest transformations). However, we should dispassionately assess the evidence provided by the data in drawing any conclusions, and resist any temptation to manipulate the modeling process to fit preconceived ideas. In other words, we should give the data a chance to speak for itself, to reveal what is important and what is not. After all, scientific breakthroughs are often made when unexpected things happen.

Given this, how then should we handle predictor variables that we expect to include in a regression model but that do not turn out to be particularly significant? This is where regression modeling becomes as much an art as a science. Ideally, we want the best model we can find that explains as much of the variation in Y as possible (high R^2), predicts Y as accurately as possible (low s), contains only significant predictor terms (those with low p-values), and that satisfies the four linear regression assumptions. However, we should also take into account the goals of the analysis, whether it is important to have a relatively simple model or not, background information about the specific context, etc. If we have particular predictors that we believe could be important, we could first try them in the model untransformed. If the resulting model is unsatisfactory, we could next try transforming any quantitative predictors (either using background knowledge of which transformations make sense in this context or to make the predictors closer to normally distributed) and also creating interactions (either with indicator variables representing qualitative information or with other quantitative predictors). If none of the terms involving the specific predictor variables that we are interested in end up being significant, then we can reasonably conclude that they do not appear to be associated with the response variable (after accounting for the associations of the other predictors in the model). We could then either drop these predictors (and/or transformations and interactions) from our final model, or, if their presence does not cause multicollinearity problems or unstable results, just leave them in. The disadvantage of this second approach is that results will be a little messier and predictions could be a little less accurate (s will invariably be higher if the model contains a bunch of insignificant predictor terms). On the other hand, the advantage of this second approach is that someone looking at the results can see for themselves that the predictor terms in question are insignificant (otherwise they would have to trust that they had been determined to be insignificant during the analysis).

5.4 MODEL SELECTION

The model building guidelines in Section 5.3 generally work well in applications with a reasonable number of potential predictors (say, less than 10) and/or strong supplementary background information about model specification. In other cases where the number of potential predictors is prohibitively large, *computer-automated* model selection methods can provide a useful complement to the approach in Section 5.3. One possibility is to use automated methods to make an initial pass through the data to identify a manageable set of potentially useful predictors that we can then evaluate more carefully manually.

If we have f predictor terms as possible candidates for inclusion in a linear regression model, then there are 2^f possible models (since each predictor term can either be included or excluded for a particular model). The predictor terms can include indicator variables, transformations, and interactions. For example, the simulated data in the **SIMULATE** data file contain 100 observations of seven predictor variables (an indicator variable, D_1, and six quantitative predictors, $X_2, \ldots, X_7$) and a response variable, Y. If we consider models that could contain interactions between D_1 and each of the quantitative predictors, as well as quadratic (squared) terms for each of the quantitative predictors, then there are $2^{(1+6+6+6)} = 2^{19} \approx 500{,}000$ possible models. Requiring that we only consider models that preserve hierarchy (e.g., retain X_1 if including X_1^2) reduces this number somewhat, but still leaves us with an overwhelming number of models to evaluate.

Since multiple linear regression models are estimated using the method of least squares (see Section 3.2), the "full model" with f predictor terms will predict best within sample (in terms of smallest residual sum of squares). However, this model will generally not be the best at predicting out of sample since it tends to "overfit" the sample data so that it doesn't generalize to the population particularly well. A model with fewer predictor terms will predict better out of sample, but the problem is identifying the predictor terms to include using only the sample data. Thus, predictor selection can be framed as a trade-off between the ability of a model to predict within sample as well as possible (i.e., small residuals) and model complexity (i.e., the number of predictor terms in the model). Since the most appropriate way to do this is not clear, many model selection criteria have been proposed, including:

- Regression standard error, $s = \sqrt{\frac{\text{RSS}}{n-k-1}}$, where RSS is the residual sum of squares, n is the sample size, and k is the number of predictor terms (excluding the intercept) in the model under consideration. Smaller values of s generally indicate better-fitting models.

- Adjusted $R^2 = 1 - \left(\frac{n-1}{n-k-1}\right)\left(\frac{\text{RSS}}{\text{TSS}}\right)$, where TSS is the total sum of squares. Larger values of adjusted R^2 generally indicate better-fitting models.

- The Akaike Information Criterion, $AIC = n\log_e\left(\frac{\text{RSS}}{n}\right) + 2(k+1)$. Smaller (or more negative) values of AIC generally indicate better-fitting models (Sakamoto et al., 1987).

- Schwarz's Bayes Information Criterion, $BIC = n\log_e\left(\frac{\text{RSS}}{n}\right) + (k+1)\log_e(n)$. Smaller (or more negative) values of BIC generally indicate better-fitting models (Schwarz, 1978).

- Mallows' $C_p = \frac{\text{RSS}}{s_f^2} + 2(k+1) - n$, where s_f is the regression standard error for the full f-predictor term model. Smaller (or more negative) values of C_p generally indicate better-fitting models (Mallows, 1973).

If we consider a subset of models with a fixed number of predictor terms, k (e.g., all 7-predictor term models), then all the criteria will agree that the "best" model is the one with the smallest value of RSS. However, the criteria need not agree about which model is "best" overall since each criterion penalizes the number of predictor terms differently. In other words, there is no single, definitive "best model." Rather, in most applications there will be many "good models," which, while differing slightly with respect to how many and which predictor terms have been included, will generally yield similar overall interpretations and predictions (see Problem 5.6 at the end of the chapter). We can thus use a particular model selection criterion to identify a set of potentially good models that have broadly equivalent criterion values. Often, the sets of potentially good models identified by each model selection criterion will be similar, but if they differ greatly they can always be combined into a larger set of potentially good models that can then be investigated further.

If computer power and time allow, some software packages allow an *exhaustive search* through all possible models to identify a set of potentially good models based on a particular model selection criterion (clever computer algorithms can avoid the need to search through every single model). Otherwise, there are a number of approaches to efficiently searching through the set of possible models for potentially good ones. Examples include *forward selection* (predictor terms are added sequentially to an initial zero-term model), *backward elimination* (predictors are excluded sequentially from the full f-predictor model), and a combined *stepwise* method (which can proceed forward or backward at each stage). Most statistical software with such automated model selection (or predictor or variable selection) methods allow the user to specify particular predictor terms to be included in all models under consideration. In this way, the user can restrict attention to models that preserve hierarchy and to ensure that sets of indicator variables (e.g., two indicator variables used

Table 5.4 Some automated model selection results for the **SIMULATE** data file.

Predictor Terms	k	RSS	s	Adj. R^2	AIC	BIC	C_p
Full $(D_1, X_2–X_7, D_1X_2–D_1X_7, X_2^2–X_7^2)$	19	86.0	1.037	0.740	24.88	77.0	20.00
$D_1, X_2, X_3, X_4, X_5, X_6, D_1X_2, X_3^2$	8	87.4	0.980	0.768*	4.57	28.0	−0.63
$D_1, X_2, X_3, X_4, X_5, X_7, D_1X_2, X_3^2$	8	87.9	0.983	0.766	5.07	28.5	−0.23
$D_1, X_2, X_3, X_4, X_5, D_1X_2, X_3^2$	7	88.2	0.979*	0.768*	3.45	24.3	−1.92
$D_1, X_2, X_3, X_6, D_1X_2, X_3^2$	6	89.3	0.980	0.768*	2.63*	20.9*	−2.94*

to represent a three-level qualitative predictor) can only be included or excluded from a model together.

To illustrate, Table 5.4 displays some results for the **SIMULATE** data file, where we considered models including interactions between D_1 and each of the quantitative predictors, as well as quadratic (squared) terms for each of the quantitative predictors. The stars in the table indicate the optimal values of each model selection criterion—s, adjusted R^2, AIC, BIC, and C_p—for the displayed models. Each of the model selection criteria suggest that the full 19-predictor term model fits less well than the other four models in the table, but do not agree which of the four is "best." In this case, since the data were generated from the 7-predictor term model $(D_1, X_2, X_3, X_4, X_5, D_1X_2, X_3^2)$, we actually know the best model. Of course in real applications, all we can hope is that we find a good, useful model that picks up all the genuine patterns and associations in the population without overemphasizing spurious patterns in the sample data.

How can we make sense of the results in Table 5.4? Although s and adjusted R^2 favor the 7-predictor term model $(D_1, X_2, X_3, X_4, X_5, D_1X_2, X_3^2)$, AIC, BIC, and C_p all favor the 6-predictor term model $(D_1, X_2, X_3, X_6, D_1X_2, X_3^2)$. In this case, X_6 is approximately equal to $X_4 + X_5$, so both these models are essentially equivalent and by investigating the models further it becomes clear that either model could be used effectively (see Problem 5.6 at the end of the chapter). What then of the 8-predictor term model with X_4, X_5, and X_6? The model selection criteria all indicate that this should be a good model, but it turns out that since X_4, X_5, and X_6 are collinear, this is not a very effective model (again, see Problem 5.6). This illustrates that automated model selection can find apparently good models that are problematic for one reason or another. Thus, automated model selection results should never be relied on in isolation, and all potential models should be investigated more thoroughly using the methods described in the rest of this book. Finally, there is another 8-predictor term model in Table 5.4, which includes X_7 rather than X_6. Although X_7 was not used in generating Y in this dataset, including it in a model is not particularly harmful (again, see Problem 5.6). However, the model selection criteria all suggest that this model is worse than the other 8-predictor term model (the one with X_6, which we just noted is a problematic model). Again, this reinforces the notion that care must be taken when using automated model selection methods.

An alternative to the automated methods described so far, which attempt to evaluate model fit *within sample* by penalizing model complexity, is to *cross-validate* model results *out of sample*. One simple way to do this is to use the *predicted residuals* introduced on page 199. The predicted residual for the ith observation for a particular model is the residual for the ith observation when we estimate that model without the ith observation in the analysis. The predicted residual sum of squares (*PRESS*) is the sum of squares of these predicted residuals, which can be used as a model selection criterion without the need for a model complexity correction.

A more extensive way to cross-validate model results in applications with plenty of available data is to partition the dataset into three sets, "training," "validation," and "test," prior to analysis. The idea is that we can estimate all our models using the training data. Then to get a fair, unbiased assessment of which model fits "best," we assess the fit of each model with the validation data—this avoids the problem of overfitting the models to the training data. Finally, we use the test data to get a fair, unbiased assessment of how well our final selected model will work for predicting new observations. Typically, more observations would be used for the training set than for the validation set, which in turn would have more observations than the test set, so 50%/30%/20% are common proportions. Sometimes, the dataset is partitioned into just training and test sets. The proportions then might be 60%/40% or 70%/30%. Partitioning is a topic that is generally covered in more detail in "data mining" courses and textbooks, for applications where there are typically a lot of data available.

The main difference between automated model selection methods and the approach described in Section 5.3 is that the former is computer-driven while the latter is human-driven. A human analyst can try things out, see where there are problems (e.g., by looking at residual plots), and attempt to correct the problems in an interactive, iterative way. Furthermore, a human analyst can try out specific interactions and transformations as needed. A computer using automated model selection methods can only select from the available predictors those that fit best according to a particular model selection criterion. In addition, a good regression analysis should be driven at least as much by subject matter knowledge as by statistical testing. Human analysts with good knowledge about the data and situation at hand can use this to inform their model building. A computer can only use the information it has (i.e., just the data values). In applications with an unmanageable number of potential predictor terms, automated model selection methods can be useful, but more as a screening device than as a rigorous way to build useful models. In other words, automated methods should be used as the starting point of further manual model building, not the endpoint.

All of the methods discussed in this section have considered model fit purely from the perspective of accuracy of prediction. However, it could be argued that to really assess how well a model fits, we need to ask ourselves what are the benefits of making good predictions (and conversely, the costs of making bad predictions) and compare models using some kind of cost-benefit analysis. This approach quickly becomes very involved and lies beyond the scope of this book.

5.5   MODEL INTERPRETATION USING GRAPHICS

We discovered in Sections 4.1 and 4.2 that while transformations and interactions can enhance our ability to model multiple linear regression associations, they can make the resulting models more difficult to understand and interpret. One approach to interpreting such models is to use graphs to plot how Y changes as each predictor changes (and all other predictors are held constant). We discuss these *predictor effect plots* in this section.

Remember that estimated regression parameters cannot usually be given *causal* interpretations. The regression modeling described in this book can really only be used to quantify linear associations and to identify whether a change in one variable is associated with a change in another variable, not to establish whether changing one variable "causes" another to change. We use the term "predictor effect" in this section to indicate how a regression model expects Y to change as each predictor changes (and all other predictors are held constant), but without suggesting at all that this is some kind of causal effect.

To motivate the use of predictor effect plots, consider the data in Table 5.5 (available in the **CREDITCARD** data file). This simulated dataset contains $Bal =$ average outstanding monthly balance (in dollars) for 50 individual credit card accounts. A credit card company wishes to investigate these accounts to predict the creditworthiness of future potential customers. Possible predictors include $Purch =$ average monthly purchases (in hundreds of dollars), $House =$ average monthly housing payments (in hundreds of dollars), type of housing (whether the account holder rents accommodations or is a homeowner), and gender (male or female). D_R is an indicator variable that has value 1 for renters and 0 for homeowners, while D_M is an indicator variable that has value 1 for males and 0 for females.

We first propose the following model, which allows the association between Bal and $(Purch, House)$ to be curved and to vary according to renting status and gender:

$$E(Bal) = b_0 + b_1 D_R + b_2 D_M + b_3 Purch + b_4 House + b_5 D_R Purch + b_6 D_R House$$
$$+ b_7 D_M Purch + b_8 D_M House + b_9 Purch^2 + b_{10} House^2.$$

Here is statistical software output that displays some results for this model (see computer help #31 in the software information files available from the book website):

Parameters[a]

Model		Estimate	Std. Error	t-stat	Pr($>|t|$)
1	(Intercept)	62.053	36.372	1.706	0.096
	D_R	−39.541	35.891	−1.102	0.277
	D_M	−37.700	25.421	−1.483	0.146
	$Purch$	18.353	5.891	3.115	0.003
	$House$	−15.290	8.525	−1.794	0.081
	$D_R Purch$	3.537	4.230	0.836	0.408
	$D_R House$	17.789	4.113	4.325	0.000
	$D_M Purch$	2.994	3.387	0.884	0.382
	$D_M House$	2.325	3.362	0.692	0.493
	$Purch^2$	−1.364	0.570	−2.392	0.022
	$House^2$	0.673	0.703	0.958	0.344

[a] Response variable: Bal.

Since many of the individual regression parameter p-values are on the large side, some of the terms in this model may be redundant. To preserve hierarchy, we first focus on the question of whether we can remove any of the interactions or the squared terms. There is a suggestion from the larger p-values in the results that perhaps ($D_R Purch$, $D_M Purch$, $D_M House$, $House^2$) do not provide useful information about the response, Bal, beyond the information provided by (D_R, D_M, $Purch$, $House$, $D_R House$, $Purch^2$). To test this formally, we do a nested model test (at significance level 5%) of the following hypotheses:

- NH: $b_5 = b_7 = b_8 = b_{10} = 0$.
- AH: at least one of b_5, b_7, b_8, or b_{10} is not equal to zero.

Table 5.5 Credit card data to illustrate model interpretation using predictor effect plots.

Bal	Purch	House	Type	D_R	Gender	D_M
174	5.00	9.11	R	1	M	1
67	1.45	5.19	R	1	M	1
143	5.31	5.73	R	1	M	1
113	5.33	4.93	R	1	M	1
186	9.48	8.51	R	1	M	1
143	6.31	5.68	R	1	F	0
83	5.61	3.92	R	1	M	1
96	4.49	3.28	H	0	F	0
83	7.76	2.19	H	0	M	1
138	4.95	6.49	R	1	F	0
100	4.59	4.78	R	1	M	1
58	3.79	4.34	H	0	F	0
64	4.14	6.17	H	0	F	0
34	4.13	5.92	H	0	M	1
161	4.92	7.79	R	1	F	0
79	1.23	6.68	R	1	F	0
65	4.41	4.67	H	0	M	1
114	4.31	4.81	R	1	M	1
105	7.50	3.94	R	1	F	0
134	6.27	5.06	R	1	M	1
98	2.62	3.05	R	1	F	0
92	3.45	4.77	R	1	F	0
121	3.57	4.70	R	1	F	0
58	3.08	4.09	H	0	F	0
112	3.12	4.68	R	1	M	1
57	4.83	5.75	H	0	M	1
80	4.67	2.44	R	1	M	1
56	4.98	6.15	H	0	F	0
85	1.15	6.47	R	1	F	0
17	3.48	7.04	H	0	M	1
137	5.00	5.82	R	1	M	1
120	4.60	6.91	R	1	M	1
153	6.97	6.84	R	1	F	0
166	6.69	5.10	R	1	F	0
98	3.75	5.34	R	1	F	0
60	5.74	3.86	H	0	F	0
114	5.13	4.83	R	1	M	1
149	5.08	5.49	R	1	F	0
0	0.93	9.40	H	0	F	0
157	4.05	8.74	R	1	M	1
26	4.48	5.16	H	0	F	0
136	4.41	5.85	R	1	F	0
0	2.89	5.01	H	0	M	1
51	4.52	6.33	H	0	F	0
112	5.45	5.97	R	1	M	1
50	4.69	5.68	H	0	F	0
34	5.23	6.51	H	0	M	1
173	9.28	6.69	R	1	M	1
141	4.11	5.63	R	1	M	1
116	4.32	6.41	R	1	M	1

Here is the output produced by statistical software that displays the results of this nested model test (see computer help #34):

Model Summary

Model	R Squared	Adjusted R Squared	Regression Std. Error	Change Statistics F-stat	df1	df2	Pr($>$F)
2^a	0.8972	0.8828	16.03				
1^b	0.9052	0.8809	16.16	0.830	4	39	0.515

^{a} Predictors: (Intercept), D_R, D_M, Purch, House, D_RHouse, Purch^2.
^{b} Predictors: (Intercept), D_R, D_M, Purch, House, D_RHouse, Purch^2, D_RPurch, D_MPurch,
    D_MHouse, House^2.

Since the p-value of 0.515 is more than our significance level (0.05), we cannot reject the null hypothesis in favor of the alternative hypothesis. Thus, a more appropriate model for these data is

$$E(Bal) = b_0 + b_1 D_R + b_2 D_M + b_3 Purch + b_4 House + b_6 D_R House + b_9 Purch^2.$$

Statistical software output for this model (see computer help #31) is:

Model Summary

Model	Sample Size	Multiple R Squared	Adjusted R Squared	Regression Std. Error
2^a	50	0.8972	0.8828	16.03

^{a} Predictors: (Intercept), D_R, D_M, Purch, House, D_RHouse, Purch^2.

Parameters^a

Model		Estimate	Std. Error	t-stat	Pr($>$ |t|)
2	(Intercept)	19.094	21.176	0.902	0.372
	D_R	−23.919	18.531	−1.291	0.204
	D_M	−10.677	4.683	−2.280	0.028
	Purch	18.627	4.791	3.888	0.000
	House	−5.335	2.595	−2.056	0.046
	D_RHouse	17.579	3.289	5.344	0.000
	Purch^2	−0.922	0.457	−2.019	0.050

^{a} Response variable: Bal.

The estimated regression equation is therefore

$$\widehat{Bal} = 19.09 - 23.92\,D_R - 10.68\,D_M + 18.63\,Purch - 5.33\,House$$
$$+ 17.58\,D_R House - 0.92\,Purch^2.$$

First consider how Bal changes as Purch changes. Since Purch is not included in any interaction terms, we can isolate this change in Bal when we hold House, housing status, and gender constant. A convenient value at which to hold House constant is the sample mean of House (which comes to 5.598). Convenient values at which to hold qualitative variables constant are their reference category levels, that is, "homeowner" ($D_R = 0$) for

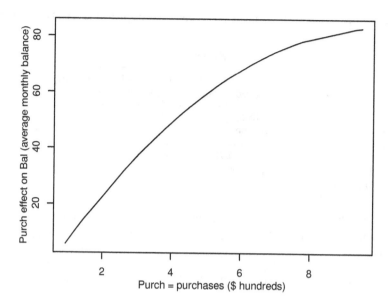

Figure 5.14 Predictor effect plot for *Purch* in the credit card example. *Purch* effect $= -10.77 + 18.63\,Purch - 0.92\,Purch^2$ is on the vertical axis, while *Purch* is on the horizontal axis.

housing status and "female" ($D_M = 0$) for gender. Then the "*Purch* effect on *Bal*" for female homeowners whose value of *House* is equal to the sample mean is

$$Purch \text{ effect} = 19.094 - 23.92(0) - 10.68(0) + 18.63\,Purch - 5.335(5.598)$$
$$+ 17.58(0)(5.598) - 0.92\,Purch^2$$
$$= 19.094 + 18.63\,Purch - 29.865 - 0.92\,Purch^2$$
$$= -10.77 + 18.63\,Purch - 0.92\,Purch^2.$$

We can then construct a line plot with this *Purch* effect on the vertical axis and *Purch* on the horizontal axis—Figure 5.14 illustrates (see computer help #42). This predictor effect plot for *Purch* shows how the predicted values of *Bal* from the model vary according to the value of *Purch* for female homeowners whose value of *House* is equal to the sample mean. Thus, high-spending female homeowners with average monthly housing expenses tend to carry a higher balance on their credit card than do low spenders, but the balance increases become smaller the more that is spent each month. To quantify the effect more precisely, when *Purch* increases from $200 to $300, we expect *Bal* to increase by

$$(-10.77 + 18.63(3) - 0.92(3^2)) - (-10.77 + 18.63(2) - 0.92(2^2)) \approx \$14.$$

However, when *Purch* increases from $800 to $900, we expect *Bal* to increase by

$$(-10.77 + 18.63(9) - 0.92(9^2)) - (-10.77 + 18.63(8) - 0.92(8^2)) \approx \$3.$$

In this example, these increases also hold for other values of *House*, D_R, and D_M. For example, let us repeat these calculations for male renters whose value of *House* is equal to

4. When *Purch* increases from \$200 to \$300, we expect *Bal* to increase by

$$(19.09 - 23.92(1) - 10.68(1) + 18.63(3) - 5.33(4) + 17.58(1)(4) - 0.92(3^2))$$
$$- (19.09 - 23.92(1) - 10.68(1) + 18.63(2) - 5.33(4) + 17.58(1)(4) - 0.92(2^2))$$
$$\approx \$14.$$

However, when *Purch* increases from \$800 to \$900, we expect *Bal* to increase by

$$(19.09 - 23.92(1) - 10.68(1) + 18.63(9) - 5.33(4) + 17.58(1)(4) - 0.51(9^2))$$
$$- (19.09 - 23.92(1) - 10.68(1) + 18.63(8) - 5.33(4) + 17.58(1)(4) - 0.51(8^2))$$
$$\approx \$3.$$

Thus, while the scale of the vertical axis in Figure 5.14 refers to female homeowners whose value of *House* is equal to the sample mean, the vertical differences between points along the line represent changes in *Bal* as *Purch* changes for *all* individuals in the population.

Next, consider how *Bal* changes as *House* changes. Since *House* is included in an interaction term, $D_R House$, we need to take into account housing status when calculating how *Bal* changes as *House* changes. We will hold the other predictors constant at convenient values, that is, the sample mean of *Purch* (which comes to 4.665) for *Purch* and "female" ($D_M = 0$) for gender. Then the "*House* effect on *Bal*" for females whose value of *Purch* is equal to the sample mean is

$$\begin{aligned} \textit{House effect} &= 19.094 - 23.92 D_R - 10.68(0) + 18.627(4.665) - 5.33 \textit{House} \\ &\quad + 17.58 D_R \textit{House} - 0.922(4.665^2) \\ &= 19.094 - 23.92 D_R + 86.895 - 5.33 \textit{House} + 17.58 D_R \textit{House} - 20.065 \\ &= 85.92 - 23.92 D_R - 5.33 \textit{House} + 17.58 D_R \textit{House}. \end{aligned}$$

We can then construct a line plot with this *House* effect on the vertical axis, *House* on the horizontal axis, and two separate lines for homeowners and renters—Figure 5.15 illustrates (see computer help #42). This predictor effect plot for *House* shows how the predicted values of *Bal* from the model vary according to the value of *House*, for female homeowners whose value of *Purch* is equal to the sample mean, and for female renters whose value of *Purch* is equal to the sample mean. We can derive the equations for the two separate lines by plugging in the value $D_R = 0$ to obtain an equation for homeowners and by plugging in the value $D_R = 1$ to obtain an equation for renters:

$$\textit{House effect for homeowners } (D_R = 0)$$
$$= 85.92 - 23.92(0) - 5.33 \textit{House} + 17.58(0)\textit{House}$$
$$= 85.92 - 5.33 \textit{House},$$
$$\textit{House effect for renters } (D_R = 1)$$
$$= 85.92 - 23.92(1) - 5.33 \textit{House} + 17.58(1)\textit{House}$$
$$= 62.00 + 12.25 \textit{House}.$$

Thus, renters tend to carry a higher balance on their credit card than do homeowners, and renters' balances increase as housing expenses increase, whereas homeowners' balances decrease as housing expenses increase. In particular, an additional \$100 in housing expenses

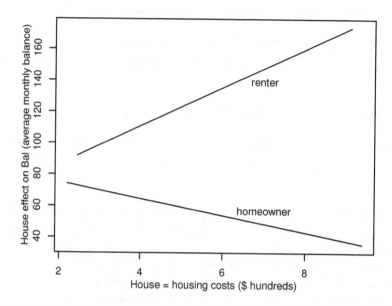

Figure 5.15   Predictor effect plot for *House* in the credit card example. *House* effect $= 85.92 - 23.92\,D_R - 5.33\,House + 17.58\,D_R House$ is on the vertical axis, while *House* is on the horizontal axis. Since the *House* effect depends on D_R, there are two separate lines—one for homeowners $(D_R = 0)$ and one for renters $(D_R = 1)$.

tends to decrease the monthly balance carried by a homeowner by approximately $5, but it tends to increase the monthly balance carried by a renter by approximately $12.

Next, consider the expected difference in *Bal* between renters $(D_R = 1)$ and homeowners $(D_R = 0)$:

$$\begin{aligned}
& (19.09 - 23.92(1) - 10.68\,D_M + 18.63\,Purch \\
& \quad - 5.33\,House + 17.58(1)House - 0.92\,Purch^2) \\
& - (19.09 - 23.92(0) - 10.68\,D_M + 18.63\,Purch \\
& \quad - 5.33\,House + 17.58(0)House - 0.92\,Purch^2) \\
& = -23.92 + 17.58\,House.
\end{aligned}$$

This represents the vertical distance between the two lines in Figure 5.15—renters tend to carry higher balances on their credit cards and the difference becomes greater as *House* = average monthly housing expenses increases.

Finally, consider the expected difference in *Bal* between males $(D_M = 1)$ and females $(D_M = 0)$:

$$\begin{aligned}
& (19.09 - 23.92\,D_R - 10.68(1) + 18.63\,Purch \\
& \quad - 5.33\,House + 17.58\,D_R House - 0.92\,Purch^2) \\
& - (19.09 - 23.92\,D_R - 10.68(0) + 18.63\,Purch \\
& \quad - 5.33\,House + 17.58\,D_R House - 0.92\,Purch^2) \\
& = -10.68.
\end{aligned}$$

Thus, males tend to carry lower balances on their credit cards than do women (approximately $11 lower on average).

We can employ similar strategies to interpret and present the results for each predictor effect in any multiple linear regression model. Predictor effects can be more difficult to interpret when there are interactions between quantitative variables in the model. One approach generalizes the method we used for the *House* effect in the credit card example. Suppose that there was a *PurchHouse* interaction in our model rather than the $D_R House$ interaction that we did have. We can calculate the *House* effect as before, but now it will depend on *Purch* (rather than D_R). To display this dependence in the predictor effect plot we can plot separate lines for different values of *Purch*, say, the sample minimum, 25th percentile, 50th percentile, 75th percentile, and maximum.

There are further examples of the use of predictor effect plots in Problems 5.5, 5.6, and 5.7 at the end of this chapter and in the case studies in Chapter 6.

5.6 CHAPTER SUMMARY

The major concepts that we covered in this chapter relating to multiple linear regression model building are as follows:

Influential points are sample observations that have a disproportionately large influence on the fit of a multiple linear regression model. Such observations may need to be analyzed separately from the other sample observations, or the model might need to be changed in such a way that it can accommodate these observations without them having undue influence on the model fit. Sometimes, identifying influential points can lead to the discovery of mistakes in the data, or finding the influential points can even be the main purpose behind the analysis. Concepts related to influence include the following:

> **Outliers** are sample observations in a multiple linear regression model with a studentized residual less than -3 or greater than $+3$. They represent observations with a particularly unusual response value relative to its predicted value from the model.

> **Leverage** is a numerical measure of the potential influence of a sample observation on a fitted regression model. It enables identification of sample observations that have particularly unusual (combinations of) predictor values (hence their potential for adversely affecting model fit).

> **Cook's distance** is a measure of the potential influence of an observation on a regression model, due to either outlyingness or high leverage. It provides an easy way to quickly identify potentially problematic observations.

Nonconstant variance or heteroscedasticity occurs when the constant variance assumption underlying multiple linear regression is violated. Nonconstant variance can sometimes be detected visually in a residual plot or diagnosed through the use of appropriate tests. Remedies range from applying a variance-stabilizing transformation to the response variable, using weighted least squares or generalized least squares techniques, or applying a generalized linear model.

Autocorrelation occurs when regression model residuals violate the independence assumption because they are highly dependent across time. This can occur when the regression data have been collected over time and the regression model fails to account for any strong time trends. Although this chapter provides an introduction to some of the methods for dealing with this issue, a full discussion lies beyond the scope of this book and can require specialized time series and forecasting methods.

Multicollinearity occurs when there is excessive correlation between quantitative predictor variables that can lead to unstable multiple regression models and inflated standard errors. Potential remedies include collecting less correlated predictor data (if possible), creating new combined predictor variables from the highly correlated predictors (if possible), or removing one of the highly correlated predictors from the model.

Excluding important predictors can sometimes result in multiple linear regression models that provide incorrect, biased conclusions about the remaining predictors. We should strive to include all potentially important predictors in a regression model, and only consider removing a predictor if there are compelling reasons to do so (e.g., if it is causing multicollinearity problems and has a high individual p-value).

Overfitting can occur if the model has been made overly complicated in an attempt to account for every possible pattern in sample data, but the resulting model generalizes poorly to the underlying population. We should always apply a "sanity check" to any regression model we wish to use to make sure that it makes sense from a subject-matter perspective, and its conclusions can be supported by the available data.

Extrapolation occurs when regression model results are used to estimate or predict a response value for an observation with predictor values that are very different from those in the sample. This can be dangerous because it means making a decision about a situation where there are no data values to support our conclusions.

Missing data occurs when particular values in the dataset have not been recorded for particular variables and observations. Dealing with this issue rigorously lies beyond the scope of this book, but there are some simple ideas that we can employ to mitigate some of the major problems—see Section 5.2.7 for more details.

Power and sample size are important considerations in multiple linear regression models insofar as we need sufficient data to be able to derive useful models capable of allowing us to draw reliable conclusions.

Model building guidelines provide a useful framework in which to approach a multiple linear regression analysis, since otherwise the endless possibilities can become overwhelming and the "trial and error" method can become overused. Section 5.3 provides a framework that usually works well, although the multiple linear regression model is sufficiently flexible that other approaches can often be just as successful.

Model selection is challenging when we have a large number of predictor variables available to choose from and endless ways to include a selection of predictors, transformations, and interactions in a model. Computer-automated model selection methods can aid in this process, particularly in the initial stages of a large regression analysis, after which careful, manual methods can be used to fine-tune the model.

Predictor effect plots are line graphs that show how a regression response variable varies with a predictor variable, holding all other predictors constant. They provide a useful graphical method for interpreting the results of a multiple linear regression analysis.

PROBLEMS

- "Computer help" refers to the numbered items in the software information files available from the book website.
- There are *brief* answers to the even-numbered problems in Appendix E.

5.1   This problem is adapted from one in McClave et al. (2005). The **COLLGPA** data file contains data that can be used to determine whether college grade point average (GPA) for 40 students (*Gpa*) can be predicted from:

Verb = verbal score on a college entrance examination (percentile),
Math = mathematics score on a college entrance examination (percentile).

Admission decisions are often based on college entrance examinations (among other things), so this is a common use of regression modeling.

(a) Use statistical software to fit the model

$$E(Gpa) = b_0 + b_1\, Verb + b_2\, Math.$$

Save the studentized residuals and draw a scatterplot with these studentized residuals on the vertical axis and *Verb* on the horizontal axis. Repeat with *Math* on the horizontal axis. Add a "loess" fitted line to each of the plots to help you assess the zero mean regression errors assumption [computer help #35, #15, and #36]. Do either of the plots suggest that this assumption is violated for this model?

(b) Use statistical software to create the interaction *VerbMath*, and the transformations *Verb*^{2} and *Math*^{2}, and fit the full quadratic model:

$$E(Gpa) = b_0 + b_1\, Verb + b_2\, Math + b_3\, VerbMath + b_4\, Verb^2 + b_5\, Math^2.$$

Again save the studentized residuals, and in addition save leverages and Cook's distances [computer help #35, #37, and #38]. Draw a scatterplot of the studentized residuals on the vertical axis versus the (standardized) predicted values on the horizontal axis [computer help #15, #36, and #39]. This residual plot suggests that the zero mean, constant variance, and independence regression error assumptions are satisfied for this quadratic model. Briefly describe what you see (or fail to see) in this residual plot that leads you to this conclusion.
Note: Standardized predicted values are calculated by subtracting their sample mean and dividing by their sample standard deviation (so that they have a mean of 0 and standard deviation of 1). Some software packages can draw residual plots with standardized predicted values on the horizontal axis automatically. Such plots are essentially identical in appearance to residual plots with ordinary (unstandardized) predicted values on the horizontal axis (the only difference is the scale of the horizontal axis of the resulting plot). Plot whatever is easier to do in your particular software. In a real-life regression analysis you would also go on to check the zero mean, constant variance, and independent regression error assumptions in residual plots with Verb on the horizontal axis and also with Math on the horizontal axis. If you were to do that here, you would find that the assumptions are also satisfied in these residual plots.

(c) Draw a histogram and QQ-plot of the studentized residuals for the quadratic model you fit in part (b) [computer help #14 and #22]. What do they suggest about the normality regression error assumption?

(d) Are there any outliers for the quadratic model you fit in part (b)? (Remember to justify your answer.)

(e) Draw a scatterplot with the leverages from the quadratic model you fit in part (b) on the vertical axis and ID on the horizontal axis [computer help #15]. Which student has the highest leverage, and why? *If* you think you should investigate further, do so by seeing what happens if you exclude this student from the analysis [computer help #19].

Hint: Look up this student's predictor values to see why his/her leverage is high, and consult Section 5.1.2 to see if you should investigate further (if so, delete the student, refit the model, and see how much regression parameter estimates change).

(f) Draw a scatterplot with Cook's distances from the quadratic model in part (b) on the vertical axis and ID on the horizontal axis [computer help #15]. Which student has the highest Cook's distance, and why? *If* you think you should investigate further, do so by seeing what happens if you exclude this student from the analysis. *Hint: Look up this student's studentized residual and leverage to see why his/her Cook's distance is high and consult Section 5.1.3 to see if you should investigate further.*

5.2 The **BEVERAGE** data file (kindly provided by Dr. Wolfgang Jank at the University of Maryland) contains the following quarterly data on sales of a popular soft drink for the period 1986–1998:

> *Sales* = quarterly sales (U.S. $ millions)
> *Time* = time period (consecutive numbers from 1 to 52)
> D_1 = indicator variable for quarter 1
> D_2 = indicator variable for quarter 2
> D_3 = indicator variable for quarter 3

The reference level for the indicator variables is quarter 4.

(a) Draw a scatterplot of *Sales* on the vertical axis versus *Time* on the horizontal axis [computer help #15] and connect the points on the plot with lines. Based on the plot, why would it be inappropriate to fit a simple linear regression model with *Sales* as the response variable and *Time* as the predictor variable?

(b) One way to take into account the seasonality of these data is to fit a multiple linear regression model with *Sales* as the response variable and $(D_1, D_2, D_3,$ *Time*) as the predictor variables. Fit this model and draw a scatterplot of the studentized residuals from the model on the vertical axis versus *Time* on the horizontal axis [computer help #31, #35, and #15]. Based on the plot, why is this model also inappropriate?

(c) One possible remedy for models with excessive autocorrelation is to try adding the lag-1 response variable as a predictor variable. Create the lag-1 response variable as follows [computer help #6]:

Sales	LagSales	Time	D_1	D_2	D_3
1735	—	1	1	0	0
2245	1735	2	0	1	0
2534	2245	3	0	0	1
		⋮			

Next remove the first observation of each variable (since there is no value of *LagSales* when *Time* = 1) [computer help #19]. Then fit a multiple linear regression model with *Sales* as the response variable and (D_1, D_2, D_3, *Time*, *LagSales*) as the predictor variables [computer help #31]. Draw a scatterplot of the studentized residuals from this model on the vertical axis versus *Time* on the horizontal axis [computer help #35 and #15]. Comparing this plot with the residual plot from part (b), does including *LagSales* appear to correct any autocorrelation problems?

(d) Use a simple linear regression model with *Sales* as the response variable and *Time* as the predictor variable to predict *Sales* for the four quarters in 1999. Calculate the prediction errors if *Sales* for the four quarters in 1999 were actually 4,428, 5,379, 5,195, and 4,803, respectively [computer help #28]. Also calculate the prediction errors for the models you fit in parts (b) and (c). Which model provides the best predictions overall?

5.3 At the end of Section 5.2.3 we mentioned that models with quadratic transformations or interactions involving quantitative predictors often have large variance inflation factors (VIFs) for the involved predictor terms. *Rescaling* the values of those predictor terms to have a mean close to 0 and a standard deviation close to 1 can mitigate such problems. For example, consider the simulated **RESCALE** data file, which contains 100 observations of a response variable, Y, and a predictor variable X. By working through the following questions, you'll find that a quadratic model fit to the data produces large VIFs for X and X^2, whereas rescaling X first results in a model with considerably reduced VIFs. Nevertheless, results from both models are essentially the same.

(a) Create a squared predictor term, X^2, and fit the quadratic model, $E(Y) = b_0 + b_1 X + b_2 X^2$ [computer help #6 and #31]. Make a note of the t-statistic for b_2, and the values of R^2 and the regression standard error, s, for the model.

(b) One of the regression parameters in the model from part (a) has a large p-value, which might suggest removal of the corresponding predictor term from the model. Which model building principle suggests retaining this predictor term instead, despite its high p-value?

(c) Calculate the variance inflation factors for X and X^2 in the model from part (a) [computer help #41].

(d) Create a rescaled predictor variable, $Z = (X - m_X)/s_X$, where m_X and s_X are the sample mean and standard deviation of X, respectively [computer help #6]. You can check that you've calculated Z correctly by checking that $m_Z = 0$ and $s_Z = 1$. Then create a squared rescaled predictor term, Z^2, and fit the quadratic model, $E(Y) = b_0 + b_1 Z + b_2 Z^2$ [computer help #31]. Make a note of the t-statistic for b_2, and the values of R^2 and the regression standard error, s, for the model. What do you notice?

(e) Calculate the variance inflation factors for Z and Z^2 in the model from part (d) [computer help #41]. What do you notice?

(f) Compare the predicted response values, $\hat{Y}$, from the models in parts (a) and (d) [computer help #28]. What do you notice?

5.4 An issue that often arises in multiple linear regression modeling is whether using a subset of the available predictor variables might be better than using all of them. This can be counterintuitive if we argue that we should use all the available variables, even if some of them aren't particularly helpful. However, not only can some variables be unhelpful, their inclusion can sometimes make things worse. This can happen if the model overfits the sample data so that within-sample predictions are "too specific" and they don't generalize well when the model is used to predict response values for new data. The purpose of this exercise is to demonstrate the phenomenon of overfitting if some "redundant" predictor variables are included in a model. The **SUBSET** data file contains a single response variable, Y, and nine possible predictor variables, $X_1, X_2, X_3, X_4, X_5, X_6, X_7, X_8$, and X_9. The full sample dataset of 100 observations has been split in two, so that the first 50 observations can be used to fit the models and the remaining 50 observations can be used to see how well the models predict the response variable for new data. To facilitate this, Y_1 represents the value of the response for the first 50 observations (and is missing for the rest) and Y_2 represents the value of the response for the last 50 observations (and is missing for the rest).

(a) Fit a multiple linear regression model using all nine X variables as the predictors and Y_1 as the response [computer help #31]. Make a note of the values of the regression standard error, s, the coefficient of determination, R^2, and also adjusted R^2.

(b) You should have noticed that the individual p-values for X_6, X_7, X_8, and X_9 are on the large side, suggesting that they perhaps provide redundant information about Y_1 beyond that provided by X_1, X_2, X_3, X_4, and X_5. Do a nested model F-test to confirm this [computer help #34].

(c) For the reduced model with X_1, X_2, X_3, X_4, and X_5, make a note of the values of the regression standard error, s, the coefficient of determination, R^2, and also adjusted R^2 [computer help #31]. You should find that s has decreased (implying increased predictive ability), R^2 has decreased (which is inevitable—see page 95—so this finding tells us nothing useful), and adjusted R^2 has increased (implying that inclusion of X_6, X_7, X_8, and X_9 was perhaps causing some overfitting).

(d) Calculate predicted response values for the last 50 observations under both models [computer help #28]. Then calculate squared errors (differences between Y_2 and the predicted response values). Finally, calculate the square root of the mean of these squared errors—this is know as the "root mean squared error" or "RMSE." You should find that the RMSE under the first (complete) model is some 5% larger than the RMSE under the second (reduced) model. In other words, we can make more accurate predictions of the response value in a new dataset by using fewer predictors in our model. This confirms that for this dataset, using all nine predictors leads to within-sample overfitting, and using just the first five predictors leads to more accurate out-of-sample predictions.

5.5 This problem is inspired by an example in Cook and Weisberg (1999). UBS AG Wealth Management Research conducts a regular survey of international prices and wages in major cities around the world (UBS, 2009). One variable measured is the price of a Big Mac hamburger, *Bigmac* (measured in the natural logarithm of minutes of working time required by an average worker to buy a Big Mac). The Big Mac is a common commodity that is essentially identical all over the world, and which therefore might be expected to have the same price everywhere. Of course it doesn't, so economists use this so-called "Big Mac parity index" as a measure of inefficiency in currency exchange. The task is to build a multiple regression model to explain the variation in *Bigmac* for 73 cities in 2009 using the following predictor variables available in the **UBS** data file:

Wage = natural logarithm of average net wage, relative to New York = $\log_e(100)$
Bread = natural logarithm of minutes of time required by average worker to buy 1 kg bread
Rice = natural logarithm of minutes of time required by average worker to buy 1 kg rice
Vac = average paid vacation days per year
D_{As} = indicator variable for 14 cities in Asia
D_{Em} = indicator variable for 17 cities in Eastern Europe or the Middle East
D_{Sa} = indicator variable for 10 cities in South America or Africa

The response variable and three of the predictor variables are expressed in natural logarithms to aid modeling since the original variables have highly skewed distributions (a few very high values relative to the rest). The reference region for the indicator variables is "North America, Western Europe, and Oceania" (32 cities).

(a) Draw a scatterplot matrix of *Wage*, *Bread*, *Rice*, *Vac*, and *Bigmac*, and use different plotting symbols for each region [computer help #16 and #17]. Write a couple of sentences on anything of interest that you notice.

(b) Fit a multiple regression model with D_{As}, D_{Em}, D_{Sa}, *Wage*, *Bread*, *Rice*, and *Vac* as predictors, and save the studentized residuals [computer help #31 and #35]. Draw a residual plot with these studentized residuals on the vertical axis and *Bread* on the horizontal axis [computer help #15 and #36]. Write a couple of sentences about how to check three regression assumptions (zero mean, constant variance, independence) using residual plots like this. You should find that the zero mean assumption is most at risk of failing (why?), while the constant variance and independence assumptions probably pass.

(c) To improve the model, consider interactions. In particular, it seems plausible that the effects of *Wage*, *Bread*, *Rice*, and *Vac* on *Bigmac* could vary according to region. So, interactions between the indicator variables and quantitative predictors offer promise for improving the model. Create the 12 interactions: $D_{As}Wage$, $D_{As}Bread$, $D_{As}Rice$, $D_{As}Vac$, $D_{Em}Wage$, $D_{Em}Bread$, $D_{Em}Rice$, $D_{Em}Vac$, $D_{Sa}Wage$, $D_{Sa}Bread$, $D_{Sa}Rice$, and $D_{Sa}Vac$ [computer help #6]. Next, fit the multiple regression model with D_{As}, D_{Em}, D_{Sa}, *Wage*, *Bread*, *Rice*, *Vac*, and these 12 interactions [computer help #31]. Which three interactions have the largest p-values in this model?

(d) Let's see if we can remove these three interactions without significantly reducing the ability of the model to explain *Bigmac*. Do a "nested model F-test" by fitting a reduced multiple regression model with D_{As}, D_{Em}, D_{Sa}, *Wage*, *Bread*, *Rice*, *Vac*, $D_{As}Wage$, $D_{As}Bread$, $D_{As}Rice$, $D_{As}Vac$, $D_{Em}Bread$, $D_{Em}Vac$, $D_{Sa}Bread$, $D_{Sa}Rice$, and $D_{Sa}Vac$ and adding $D_{Em}Wage$, $D_{Em}Rice$, and $D_{Sa}Wage$ to make a complete model [computer help #34]. What are the values of the F-statistic and the p-value for this test? Does this mean that we can remove these three interactions without significantly reducing the ability of the model to explain *Bigmac*? *Hint: See Section 3.3.4.*

(e) Let's see if we can remove three more interactions without significantly reducing the ability of the model to explain *Bigmac*. Do a "nested model F-test" by fitting a reduced multiple regression model with D_{As}, D_{Em}, D_{Sa}, *Wage*, *Bread*, *Rice*, *Vac*, $D_{As}Bread$, $D_{As}Rice$, $D_{As}Vac$, $D_{Em}Bread$, $D_{Em}Vac$, and $D_{Sa}Bread$, and adding $D_{As}Wage$, $D_{Sa}Rice$, and $D_{Sa}Vac$ to make a complete model [computer help #34]. What are the values of the F-statistic and the p-value for this test? Does this mean that we can remove these three interactions without significantly reducing the ability of the model to explain *Bigmac*?

(f) Now fit a multiple regression model with D_{As}, D_{Em}, D_{Sa}, *Wage*, *Bread*, *Rice*, *Vac*, $D_{As}Bread$, $D_{As}Rice$, $D_{As}Vac$, $D_{Em}Bread$, $D_{Em}Vac$, and $D_{Sa}Bread$ [computer help #31]. If we want to have a more parsimonious model that preserves hierarchy, which predictor term can we now consider removing? Do an individual t-test to test this formally.

Hint: See Section 3.3.5.

(g) Our final model has the following predictors: D_{As}, D_{Em}, D_{Sa}, *Wage*, *Bread*, *Rice*, *Vac*, $D_{As}Bread$, $D_{As}Rice$, $D_{As}Vac$, $D_{Em}Vac$, and $D_{Sa}Bread$. Fit this model and save the studentized residuals [computer help #31 and #35]. Draw a residual plot with these studentized residuals on the vertical axis and *Bread* on the horizontal axis [computer help #15 and #36]. Does it appear that the zero mean regression assumption that appeared to be violated in part (b) has now been corrected?

Note: Although the values of the regression standard error, s, and adjusted R^2 suggest that the model from part (f) may be preferable to this model, the individual t-test from part (f) suggests otherwise. For the purposes of this exercise, both models give very similar results and conclusions, and we will proceed with the model from this part as our final model for the remainder of the questions.

(h) Write out the least squares regression equation for this final model [computer help #31]; that is, replace the $\hat{b}$'s with numbers in:

$$\widehat{Bigmac} = \hat{b}_0 + \hat{b}_1 D_{As} + \hat{b}_2 D_{Em} + \hat{b}_3 D_{Sa} + \hat{b}_4 \, Wage + \hat{b}_5 \, Bread$$
$$+ \hat{b}_6 \, Rice + \hat{b}_7 \, Vac + \hat{b}_8 D_{As}Bread + \hat{b}_9 D_{As}Rice$$
$$+ \hat{b}_{10} D_{As}Vac + \hat{b}_{11} D_{Em}Vac + \hat{b}_{12} D_{Sa}Bread.$$

(i) An economist reasons that as net wages increase, the cost of a Big Mac goes down, all else being equal (fewer minutes of labor would be needed to buy a Big Mac since the average wage is higher). According to our final model, is the economist correct?

(j) The economist also reasons that as the cost of bread increases, the cost of Big Macs goes up, all else being equal (food prices tend to fall and rise together). According to our final model, is the economist correct?

Hint: This is trickier to answer than part (i) since the "Bread effect" depends on D_{As} and D_{Sa}. Write this effect out as "$\hat{b}_0 + \hat{b}_1 D_{As} + \hat{b}_2 D_{Em} + \hat{b}_3 D_{Sa} + \hat{b}_4 m_{Wage} + \hat{b}_5 \, Bread + \hat{b}_6 m_{Rice} + \hat{b}_7 m_{Vac} + \hat{b}_8 D_{As}Bread + \hat{b}_9 D_{As}m_{Rice} + \hat{b}_{10} D_{As}m_{Vac} + \hat{b}_{11} D_{Em}m_{Vac} + \hat{b}_{12} D_{Sa}Bread$," replacing the $\hat{b}$'s with numbers, m_{Wage} with the sample mean of Wage, m_{Rice} with the sample mean of Rice, and m_{Vac} with the sample mean of Vac [computer help #10]. Then create this as a variable in the dataset [computer help #6], and draw a predictor effect line graph with the "Bread effect" variable on the vertical axis, Bread on the horizontal axis, and "Region" to mark four separate lines [computer help #42]. This should produce one line for each region; the economist may be correct for some, all, or none of the regions! To ensure that your line graphs are correct, make sure that you can recreate the predictor effect plots in Section 5.5 first, particularly Figure 5.15 on page 230.

(k) The economist also reasons that as the cost of rice increases, the cost of Big Macs goes up, all else being equal (food prices tend to fall and rise together). According to our final model, is the economist correct?

Hint: This is similar to part (j) since the "Rice effect" depends on D_{As}. Write this effect out as "$\hat{b}_0 + \hat{b}_1 D_{As} + \hat{b}_2 D_{Em} + \hat{b}_3 D_{Sa} + \hat{b}_4 m_{Wage} + \hat{b}_5 m_{Bread} + \hat{b}_6 Rice + \hat{b}_7 m_{Vac} + \hat{b}_8 D_{As}m_{Bread} + \hat{b}_9 D_{As}Rice + \hat{b}_{10} D_{As}m_{Vac} + \hat{b}_{11} D_{Em}m_{Vac} + \hat{b}_{12} D_{Sa}m_{Bread}$," replacing the $\hat{b}$'s with numbers, m_{Wage} with the sample mean of Wage, m_{Bread} with the sample mean of Bread, and m_{Vac} with the sample mean of Vac [computer help #10]. Then create this as a variable in the dataset [computer help #6], and draw a predictor effect line graph with the "Rice effect" variable on the vertical axis, Rice on the horizontal axis, and "Region" to mark four separate lines [computer help #42]. Again, the economist may be correct for some, all, or none of the regions.

(l) The economist also reasons that as vacation days increase, the cost of a Big Mac goes up, all else being equal (productivity decreases, leading to an increase in overall production costs). According to our final model, is the economist correct?

Hint: This is similar to part (j) since the "Vac effect" depends on D_{As} and D_{Em}. Write this effect out as "$\hat{b}_0 + \hat{b}_1 D_{As} + \hat{b}_2 D_{Em} + \hat{b}_3 D_{Sa} + \hat{b}_4 m_{Wage} + \hat{b}_5 m_{Bread} + \hat{b}_6 m_{Rice} + \hat{b}_7 Vac + \hat{b}_8 D_{As}m_{Bread} + \hat{b}_9 D_{As}m_{Rice} + \hat{b}_{10} D_{As}Vac + \hat{b}_{11} D_{Em}Vac + \hat{b}_{12} D_{Sa}m_{Bread}$," replacing the $\hat{b}$'s with numbers, m_{Wage} with the sample mean of Wage, m_{Bread} with the sample mean of Bread, and m_{Rice} with the sample mean of Rice [computer help #10]. Then create this as a variable [computer help #6], and draw a predictor effect line graph with "Vac effect" on the vertical axis, Vac on the horizontal axis, and "Region" to mark four separate lines [computer help #42]. Again, the economist may be correct for some, all, or none of the regions.

5.6 The simulated data in the **SIMULATE** data file contain 100 observations of seven predictor variables (an indicator variable, D_1, and six quantitative predictors, $X_2, \ldots, X_7$) and a response variable, Y. We will model this dataset with three different pairs of models, where one of the models in each pair is relatively "good" and the other is relatively "poor." Each of the good models is essentially equally effective at modeling the data, while each of the poor models is deficient in some way. We shall also use predictor effect plots to see that the three good models give very similar predictor effects, while each of the poor models gives at least one predictor effect that is very different from the other models.

Since the data were simulated from one of the models (F below), we know that the good model predictor effects are "correct" (similar to the model effects used to generate the data), while the poor model predictor effects are "incorrect" (very different from the model effects used to generate the data). The goal of this exercise is to demonstrate that good models often give very similar results and conclusions, whereas poor models can be very misleading. There is often no single model that is unambiguously the "best," so the best we can hope in any regression application is that the final model we use is one of the good ones and not one of the poor ones. Fortunately, poor models are usually easy to spot.

(a) Compare the following models using a nested model F-test to decide which is relatively poor and which is relatively good [computer help #34]:
A: $E(Y) = b_0 + b_1 D_1 + b_2 X_2 + b_3 X_3 + b_4 X_4 + b_5 X_5 + b_9 X_3^2$;
B: $E(Y) = b_0 + b_1 D_1 + b_2 X_2 + b_3 X_3 + b_4 X_4 + b_5 X_5 + b_7 X_7 + b_8 D_1 X_2 + b_9 X_3^2$.
Investigate the models to see *why* the good model is superior to the poor one.

(b) Compare the following models using a nested model F-test to decide which is relatively poor and which is relatively good [computer help #34]:

C: $E(Y) = b_0 + b_1 D_1 + b_2 X_2 + b_3 X_3 + b_6 X_6 + b_8 D_1 X_2 + b_9 X_3^2$;

D: $E(Y) = b_0 + b_1 D_1 + b_2 X_2 + b_3 X_3 + b_4 X_4 + b_5 X_5 + b_6 X_6 + b_8 D_1 X_2 + b_9 X_3^2$.

Investigate the models to see *why* the good model is superior to the poor one.

(c) Compare the following models using a nested model F-test to decide which is relatively poor and which is relatively good [computer help #34]:

E: $E(Y) = b_0 + b_1 D_1 + b_2 X_2 + b_3 X_3 + b_4 X_4 + b_8 D_1 X_2$;

F: $E(Y) = b_0 + b_1 D_1 + b_2 X_2 + b_3 X_3 + b_4 X_4 + b_5 X_5 + b_8 D_1 X_2 + b_9 X_3^2$.

Investigate the models to see *why* the good model is superior to the poor one.

(d) Construct predictor effect plots for X_2 for models A, B, C, and F [computer help #42]. Given the presence of the interaction term, $D_1 X_2$, in three of the models, the plots should have two separate lines, one for $D_1 = 0$ and one for $D_1 = 1$. Which of these four models has predictor effects for X_2 that are very different from the other three? (You should find that it is one of the poor models you identified earlier.)

(e) Construct predictor effect plots for X_3 for models B, C, E, and F [computer help #42]. Since there is no interaction term between D_1 and X_3 in any of the models, there is no need to have two separate lines for $D_1 = 0$ and $D_1 = 1$ (although there is no harm in doing so). Which of these four models has predictor effects for X_3 that are very different from the other three? (It should be one of the poor models.)

(f) Construct predictor effect plots for X_4 for models A, B, D, and F [computer help #42]. As in part (e), there is no need to have two separate lines for $D_1 = 0$ and $D_1 = 1$, although there is no harm in doing so. Which of these four models has predictor effects for X_4 that are very different from the other three? (It should be one of the poor models.)

(g) Construct predictor effect plots for X_5 for models A, B, D, and F [computer help #42]. As in part (e), there is no need to have two separate lines for $D_1 = 0$ and $D_1 = 1$, although there is no harm in doing so. Which of these four models has predictor effects for X_5 that are very different from the other three? (It should be one of the poor models.)

(h) Construct predictor effect plots for X_6 for models C and D [computer help #42]. Since there is no interaction term between D_1 and X_6 in either of the models, there is no need to have two separate lines, one for $D_1 = 0$ and one for $D_1 = 1$, in the plots (although there is no harm in doing so). Model F was used to generate the dataset, so X_6 was *not* used. However, variable X_6 is essentially the sum of variables X_4 and X_5 (investigate to see this). Since the only difference between models C and F is that model C includes X_6 instead of X_4 and X_5, the slope parts of the predictor effect for X_6 in model C should be approximately the same as the average of the slope parts of the predictor effects for X_4 and X_5 in model F—see if this is the case. Is the predictor effect for X_6 in model D very different from the predictor effect for X_6 in model C?

5.7 The following problem provides a challenging dataset that you can use to practice multiple linear regression model building. You've been asked to find out how profits for 120 restaurants in a particular restaurant chain are affected by certain characteristics of the restaurants. You would like to build a regression model for predicting $Profit =$ annual profits (in thousands of dollars) from five potential predictor variables:

Cov = number of covers or customers served (in thousands)
Fco = food costs (in thousands of dollars)
Oco = overhead costs (in thousands of dollars)
Lco = labor costs (in thousands of dollars)
$Region$ = geographical location (Mountain, Southwest, or Northwest)

Note that region is a qualitative (categorical) variable with three levels; the **RESTAURANT** data file contains two indicator variables to code the information in region: $D_{Sw} = 1$ for Southwest, 0 otherwise, and $D_{Nw} = 1$ for Northwest, 0 otherwise. Thus, the Mountain region is the reference level with 0 for both D_{Sw} and D_{Nw}. Build a suitable regression model and investigate the role of each of the predictors in the model through the use of predictor effect plots. You may want to consider the following topics in doing so:

- models with both quantitative and qualitative variables;
- polynomial transformations;
- interactions;
- comparing nested models.

You may use the following for terms in your model:

- Cov, Fco, Oco, Lco, D_{Sw}, and D_{Nw};
- interactions between each of the quantitative predictors and the indicator variables, such as $D_{Sw}Cov$, $D_{Nw}Cov$, etc.;
- quadratic terms, such as Cov^2 (do not use terms like D_{Sw}^2, however!);
- use $Profit$ as the response variable [i.e., do not use $\log_e(Profit)$ or any other transformation].

Hint: A "good" model should have a value of R^2 around 0.94 and a regression standard error, s, around 10.

CHAPTER 6

CASE STUDIES

6.1 HOME PRICES

6.1.1 Data description

The **HOMES6** data file contains information on 76 single-family homes in south Eugene, Oregon during 2005. The data were provided by Victoria Whitman, a realtor in Eugene, and were analyzed in Pardoe (2008). We wish to explain and predict the price of a single-family home (*Price*, in thousands of dollars) using the following predictor variables:

$Floor$ = floor size (thousands of square feet)
Lot = lot size category (from 1 to 11—see later)
$Bath$ = number of bathrooms (with half-bathrooms counting as 0.1—see later)
Bed = number of bedrooms (between 2 and 6)
Age = age [standardized: $(\text{year built} - 1970)/10$—see later]
Gar = garage size (0, 1, 2, or 3 cars)
D_{Ac} = indicator for "active listing"
D_{Ed} = indicator for Edison Elementary
D_{Ha} = indicator for Harris Elementary
D_{Ad} = indicator for Adams Elementary
D_{Cr} = indicator for Crest Elementary
D_{Pa} = indicator for Parker Elementary

As discussed in Section 3.2, it is reasonable to assume that homes built on properties with a large amount of land area command higher sale prices than homes with less land, all else being equal. However, it is also reasonable to suppose that an increase in land area of 2,000 square feet from 4,000 to 6,000 would make a larger difference (to sale price) than going from 24,000 to 26,000. Thus, realtors have constructed lot size "categories," which in their experience correspond to approximately equal-sized increases in sale price. The categories (variable *Lot*) used in this dataset are:

Lot size	0-3k	3-5k	5-7k	7-10k	10-15k	15-20k	20k-1ac	1-3ac	3-5ac	5-10ac	10-20ac
Category	1	2	3	4	5	6	7	8	9	10	11

Lot sizes ending in "k" represent thousands of square feet, while "ac" stands for acres—there are 43,560 square feet in an acre.

Realtors have also recognized that "half-bathrooms" (without a shower or bathtub) are not valued by home buyers nearly as highly as "full" bathrooms. In fact, it appears that their value is usually not even one-half of a full bathroom and tends to be closer to one-tenth of their value—this is reflected in the definition of the variable *Bath*, which records half-bathrooms with the value 0.1.

Different housing markets value properties of various ages in different ways. This particular market has a mix of homes that were built from 1905 to 2005, with an average of around 1970. In the realtor's experience, both very old homes and very new homes tend to command a price premium relative to homes of "middle age" in this market. Thus, a quadratic effect (see page 145) might be expected for an age variable in a multiple linear regression model to predict price. As discussed at the end of Section 4.1.2, to facilitate this we calculate a rescaled "age" variable from the "year built" variable by subtracting 1970 (the approximate mean) and dividing by 10. The resulting *Age* variable has a mean close to zero and a standard deviation just over 2. Dividing by 20 instead of 10 in the rescaling would lead to a variable with a standard deviation closer to 1, but dividing by 10 leads to a more intuitive interpretation for *Age*—it represents the number of decades away from 1970.

This dataset includes homes that have recently sold, where *Price* represents the final sale price. However, it also includes homes that are "active listings"—homes offered for sale (at price *Price*) but which have not sold yet. At the time these data were collected, the final sale price of a home could sometimes be considerably less than the price for which it was initially offered. The dataset also includes homes that were "pending sales" for which a sale price had been agreed (*Price*) but paperwork still needed to be completed. To account for possible differences between final sale prices and offer prices, we define an indicator variable, D_{Ac}, to model differences between actively listed homes ($D_{Ac} = 1$) and pending or sold homes ($D_{Ac} = 0$).

This particular housing market comprises a number of different neighborhoods, each with potentially different levels of housing demand. The strongest predictor of demand that is available with this dataset relates to the nearest school for each home. The housing market is contained within the geographic boundaries of a single high school, but there are six different elementary schools within this area. Thus, we define five indicator variables to serve as a proxy for the geographic neighborhood of each home. The most common elementary school in the dataset is Redwood Elementary School, so we select this to be the "reference level" (see Section 4.3.2). The indicator variables D_{Ed} to D_{Pa} then represent differences of the following schools from Redwood: Edison, Harris, Adams, Crest, and Parker.

6.1.2 Exploratory data analysis

Figure 6.1 displays a scatterplot matrix of the quantitative variables in the home prices dataset (see computer help #16 in the software information files available from the book website). Remember that for any plot in the scatterplot matrix, look to the left or right for the label of the vertical axis and up or down for the label of the horizontal axis. This is not so different from what we do in a regular scatterplot (where we look to the left for the label of the vertical axis and down for the label of the horizontal axis). While there are a number of bivariate associations evident, these tell us little about the likely form of a useful multiple linear regression model. The plots do show a number of points that stick out from the dominant patterns. None of these values are so far from the remaining values that they are likely to cause a problem with subsequent analysis, but it is worth making a note of them just in case. Home 76 has a much larger floor size than the rest, while home 74 has a larger lot size than the rest (and is quite expensive). Home 35 is the only one with

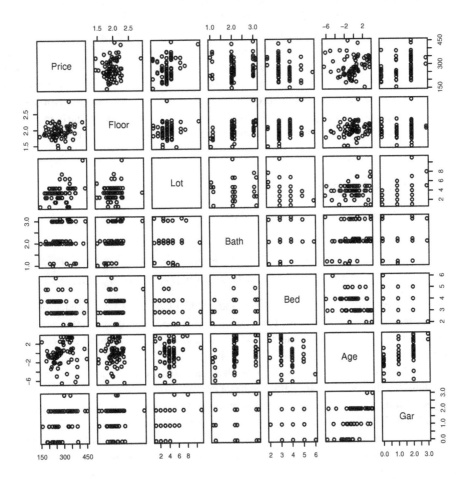

Figure 6.1   Scatterplot matrix of the quantitative variables in the **HOMES6** data file.

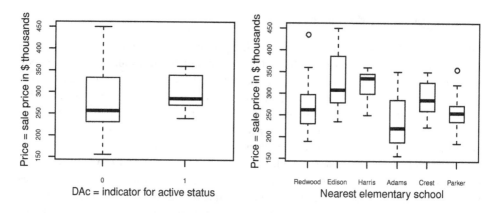

Figure 6.2   Boxplots of *Price* versus status (left) and *Price* versus elementary school (right) for the **HOMES6** data file.

six bedrooms, while home 54 is the oldest home. Homes 21 and 47 are the only ones with three-car garages, while home 2 is the most expensive and home 5 is the cheapest.

Figure 6.2 displays boxplots of *Price* versus the two qualitative variables in the **HOMES6** data file (see computer help #21). Homes prices are less variable and have a higher median for active listings relative to homes that have recently sold (or are pending sales). Prices tend to be higher in neighborhoods near the Harris and Edison schools, and lower for Adams, with the other three schools broadly similar in a "moderate" range. Home 74 is particularly expensive for a typical Redwood home, while home 40 is similarly expensive for a typical Parker home. Keep in mind that these observations do not take into account the quantitative predictors, *Floor*, *Lot*, and so on.

6.1.3   Regression model building

We first try a model with each of the predictor variables "as is" (no transformations or interactions):

$$E(Price) = b_0 + b_1\,Floor + b_2\,Lot + b_3\,Bath + b_4\,Bed + b_5\,Age + b_6\,Gar$$
$$+ b_7\,D_{Ac} + b_8\,D_{Ed} + b_9\,D_{Ha} + b_{10}\,D_{Ad} + b_{11}\,D_{Cr} + b_{12}\,D_{Pa}.$$

This model results in values of $R^2 = 0.5305$ and $s = 45.11$ but isn't very satisfactory for a number of reasons. For example, the residuals from this model fail to satisfy the zero mean assumption in a plot of the residuals versus *Age*, displaying a pronounced curved pattern (plot not shown). To attempt to correct this failing, we will add an Age^2 transformation to the model, which as discussed above was also suggested from the realtor's experience. The finding that the residual plot with *Age* has a curved pattern does not necessarily mean that an Age^2 transformation will correct this problem, but it is certainly worth trying.

In addition, both *Bath* and *Bed* have relatively large individual t-test p-values in this first model, which appears to contradict the notion that home prices should increase with the number of bedrooms and bathrooms. However, the association with bedrooms and bathrooms may be complicated by a possible interaction effect. For example, adding extra bathrooms to homes with just two or three bedrooms might just be considered a waste of space and so have a negative impact on price. Conversely, there is a clearer benefit for homes with four or five bedrooms to have more than one bathroom, so adding bathrooms for these homes probably has a positive impact on price. To model such an association we will add a *BathBed* interaction term to the model.

Therefore, we next try the following model:

$$E(Price) = b_0 + b_1 Floor + b_2 Lot + b_3 Bath + b_4 Bed + b_5 BathBed + b_6 Age + b_7 Age^2$$
$$+ b_8 Gar + b_9 D_{Ac} + b_{10} D_{Ed} + b_{11} D_{Ha} + b_{12} D_{Ad} + b_{13} D_{Cr} + b_{14} D_{Pa}.$$

This model results in values of $R^2 = 0.5989$ and $s = 42.37$ and has residuals that appear to satisfy the four regression model assumptions of zero mean, constant variance, normality, and independence reasonably well (residual plots not shown). However, the model includes some terms with large individual t-test p-values, suggesting that perhaps it is more complicated than it needs to be and includes some redundant terms. In particular, the last three elementary school indicators (for Adams, Crest, and Parker) have p-values of 0.310, 0.683, and 0.389. We conduct a nested model F-test to see whether we can safely remove these three indicators from the model without significantly worsening its fit (at a 5% significance level):

Model Summary

Model	R Squared	Adjusted R Squared	Regression Std. Error	Change Statistics			
				F-stat	df1	df2	Pr(>F)
3 [a]	0.5883	0.5175	41.91				
2 [b]	0.5989	0.5068	42.37	0.5366	3	61	0.659

[a] Predictors: (Intercept), *Floor, Lot, Bath, Bed, BathBed, Age, Age^2, Gar, D_{Ac}, D_{Ed}, D_{Ha}*.
[b] Predictors: (Intercept), *Floor, Lot, Bath, Bed, BathBed, Age, Age^2, Gar, D_{Ac}, D_{Ed}, D_{Ha},*
  D_{Ad}, D_{Cr}, D_{Pa}.

Since the p-value of 0.659 is more than our significance level (0.05), we cannot reject the null hypothesis that the regression parameters for the last three school indicators are all zero. In addition, removing these three indicators improves the values of adjusted R^2 (from 0.5068 to 0.5175) and s (from 42.37 to 41.91). Removing these three school indicators from the model means that the reference level for school now comprises all schools except Edison (D_{Ed}) and Harris (D_{Ha}), in other words, Redwood, Adams, Crest, and Parker (so that we no longer model any systematic differences between these four schools with respect to home prices).

6.1.4 Results and conclusions

Thus, a final model for these data is

$$E(Price) = b_0 + b_1 Floor + b_2 Lot + b_3 Bath + b_4 Bed + b_5 BathBed$$
$$+ b_6 Age + b_7 Age^2 + b_8 Gar + b_9 D_{Ac} + b_{10} D_{Ed} + b_{11} D_{Ha}.$$

Statistical software output for this model (see computer help #31) is:

Model Summary

Model	Sample Size	Multiple R Squared	Adjusted R Squared	Regression Std. Error
3 [a]	76	0.5883	0.5175	41.91

[a] Predictors: (Intercept), *Floor*, *Lot*, *Bath*, *Bed*, *BathBed*, *Age*, Age^2, *Gar*, D_{Ac}, D_{Ed}, D_{Ha}.

Parameters [a]

Model		Estimate	Std. Error	t-stat	Pr(> |t|)
3	(Intercept)	332.471	106.599	3.119	0.003
	Floor	56.721	27.974	2.028	0.047
	Lot	9.916	3.438	2.884	0.005
	Bath	−98.153	42.666	−2.300	0.025
	Bed	−78.909	27.752	−2.843	0.006
	BathBed	30.389	11.878	2.558	0.013
	Age	3.302	3.169	1.042	0.301
	Age^2	1.641	0.733	2.238	0.029
	Gar	13.118	8.285	1.583	0.118
	D_{Ac}	27.426	10.988	2.496	0.015
	D_{Ed}	67.062	16.822	3.987	0.000
	D_{Ha}	47.275	14.844	3.185	0.002

[a] Response variable: *Price*.

The estimated regression equation is therefore

$$\widehat{Price} = 332.47 + 56.72\,Floor + 9.92\,Lot - 98.15\,Bath - 78.91\,Bed + 30.39\,BathBed$$
$$+ 3.30\,Age + 1.64\,Age^2 + 13.12\,Gar + 27.43\,D_{Ac} + 67.06\,D_{Ed} + 47.27\,D_{Ha}.$$

This final model results in residuals that appear to satisfy the four regression model assumptions of zero mean, constant variance, normality, and independence reasonably well (residual plots not shown). Also, each of the individual t-test p-values is below the usual 0.05 threshold (including *Bath*, *Bed*, and the *BathBed* interaction), except *Age* (which is included to preserve hierarchy since Age^2 is included in the model) and *Gar* (which is nonetheless retained since its 0.118 p-value is low enough to suggest a potentially important effect).

The highest leverage in the dataset is for home 76 (with a large floor size), although home 54 (the oldest home) is not far behind. Home 54 has the highest Cook's distance, although it is not above the 0.5 threshold. Neither of these homes dramatically changes the regression results if excluded. None of the studentized residuals are outside the ±3 range, so there are no outliers.

The model can explain 58.8% of the variation in price, and predictions using the model are likely to be accurate to within approximately ±$83,800 (at a 95% confidence level). To put this in context, prices in this dataset range from $155,500 to $450,000. This still leaves more than 40% of the variation in price unexplained by the model, which suggests that the dataset predictors can only go so far in helping to explain and predict home prices in this particular housing market. Variables not measured that could account for the remaining 41.2% of the price variation might include other factors related to the geographical neighborhood, condition of the property, landscaping, and features such as updated kitchens and fireplaces.

A potential use for the model might be to narrow the range of possible values for the asking price of a home about to be put on the market. For example, suppose that a home with the following features is going to be put on the market: 1,879 square feet, lot size category 4, two and a half bathrooms, three bedrooms, built in 1975, two-car garage, and near to Parker Elementary School. A 95% prediction interval ignoring the model (i.e., using the equation on page 26) comes to ($164,800, $406,800). By contrast, a 95% prediction interval using the model results (i.e., using the equation on page 128) comes to ($197,100, $369,000). A realtor could then advise the vendors to price their home somewhere within this range, depending on other factors not included in the model (e.g., toward the upper end of this range if the home is on a nice street, the property is in good condition, and some landscaping has been done to the yard). As is often the case, the regression analysis results are more effective when applied in the context of expert opinion and experience.

A further use for the model might be to utilize the specific findings relating to the effects of each of the predictors on the price. Since $\hat{b}_1 = 56.72$, we expect sale price to increase by $5,672 for each 100-square foot increase in floor size, all else held constant. Similarly, since $\hat{b}_2 = 9.92$, we expect sale price to increase by $9,920 for each one-category increase in lot size, all else held constant. Similarly, since $\hat{b}_8 = 13.12$, we expect sale price to increase by $13,120 for each vehicle increase in garage size, all else held constant.

Interpretation of the parameter estimates for *Bath*, *Bed*, and *Age* are complicated somewhat by their interactions and transformations. In Section 5.5 we showed how to construct predictor effect plots to make such interpretations easier (and to display graphically even the more straightforward interpretations for *Floor*, *Lot*, and *Gar*). First consider how *Price* changes as *Floor* changes. Since *Floor* is not included in any interaction terms, we can isolate this change in *Price* when we hold the remaining predictors constant (at sample mean values for the quantitative predictors, and zero for the indicator variables). Then the "*Floor* effect on *Price*" is

$$Floor \text{ effect} = 134.92 + 56.72\,Floor.$$

The value 56.72 comes directly from the *Floor* part of the estimated regression equation, while the value 134.92 results from plugging in the sample means for *Lot*, *Bath*, *Bed*, *Age*, and *Gar*, and zero for D_{Ac}, D_{Ed}, and D_{Ha} to the rest of the equation. This *Floor* effect then represents how *Price* changes as *Floor* changes for homes with average values for *Lot*, ..., *Gar* that are in the Redwood, Adams, Crest, or Parker neighborhoods.

We can then construct a line plot with this *Floor* effect on the vertical axis and *Floor* on the horizontal axis—the left-hand plot in Figure 6.3 illustrates (see computer help #42). Over the range of values in the dataset, as floor size increases from approximately 1,440 to 2,900 square feet, prices increase from approximately $215k to $300k on average (for homes with average values for *Lot*, ..., *Gar* that are in the Redwood, Adams, Crest, or Parker neighborhoods). Homes with other values for *Lot*, ..., *Gar*, or that are in other neighborhoods, tend to have price differences of a similar magnitude for similar changes in floor size (although the sales prices of individual homes will depend on those values of *Lot*, ..., *Gar* and neighborhood)—the predictor effect plot would simply have different values on the vertical axis, but the slope of the line would be the same.

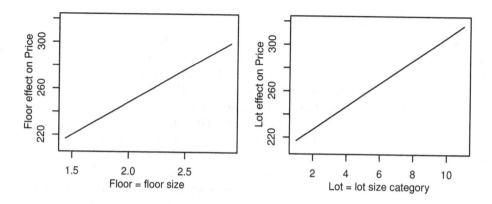

Figure 6.3 Predictor effect plots for *Floor* and *Lot* in the home prices example. In the left plot, the *Floor* effect of $134.92 + 56.72\,Floor$ is on the vertical axis, while *Floor* is on the horizontal axis. In the right plot, the *Lot* effect of $207.15 + 9.92\,Lot$ is on the vertical axis, while *Lot* is on the horizontal axis.

Similarly, the "*Lot* effect on *Price*" is

$$Lot\text{ effect} = 207.15 + 9.92\,Lot.$$

The right-hand plot in Figure 6.3 shows a line plot with this *Lot* effect on the vertical axis and *Lot* on the horizontal axis. Over the range of values in the dataset, as lot size category increases from 1 to 11, prices increase from approximately $215k to $315k on average (for homes with average values for *Floor*, *Bath*, *Bed*, *Age*, and *Gar* that are in the Redwood, Adams, Crest, or Parker neighborhoods). Again, homes with other predictor values tend to have price differences of a similar magnitude for similar changes in lot size.

The "*BathBed* effect on *Price*" involves an interaction:

$$BathBed\text{ effect} = 504.12 - 98.15\,Bath - 78.91\,Bed + 30.39\,BathBed.$$

The left-hand plot in Figure 6.4 shows a line plot with this *BathBed* effect on the vertical axis, *Bath* on the horizontal axis, and lines marked by the value of *Bed* (see computer help #42). Adding bathrooms to homes with just two or three bedrooms (holding all else constant) has a negative impact on price (particularly two-bedroom homes). Conversely, adding bathrooms to homes with four or five bedrooms (holding all else constant) has a positive impact on price (particularly five-bedroom homes). The scale on the plot shows the approximate magnitude of average prices for different numbers of bathrooms and bedrooms (for homes with average values for *Floor*, *Lot*, *Age*, and *Gar* that are in the Redwood, Adams, Crest, or Parker neighborhoods). Again, homes with other predictor values tend to have price differences of a similar magnitude for similar changes in the numbers of bathrooms and bedrooms.

The right-hand plot in Figure 6.4 shows a line plot with the *BathBed* effect on the vertical axis and *Bed* on the horizontal axis, and lines marked by the value of *Bath*. Adding bedrooms to homes with one or two bathrooms (holding all else constant) has a negative impact on price (particularly one-bathroom homes). Conversely, adding bedrooms to homes with three bathrooms (holding all else constant) has a positive impact on price. The scale

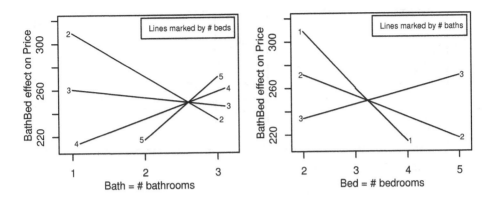

Figure 6.4   Predictor effect plots for *Bath* and *Bed* in the home prices example. In the left plot, the *BathBed* effect of $504.12 - 98.15\,Bath - 78.91\,Bed + 30.39\,BathBed$ is on the vertical axis while *Bath* is on the horizontal axis and the lines are marked by the value of *Bed*. In the right plot, the *BathBed* effect is on the vertical axis while *Bed* is on the horizontal axis and the lines are marked by the value of *Bath*.

on the plot shows the approximate magnitudes of average prices for different numbers of bathrooms and bedrooms (for homes with average values for *Floor*, *Lot*, *Age*, and *Gar* that are in the Redwood, Adams, Crest, or Parker neighborhoods). Again, homes with other predictor values tend to have price differences of a similar magnitude for similar changes in the numbers of bathrooms and bedrooms.

The "*Age* effect on *Price*" is

$$Age \text{ effect} = 246.88 + 3.30\,Age + 1.64\,Age^2.$$

The left-hand plot in Figure 6.5 shows a line plot with this *Age* effect on the vertical axis and year on the horizontal axis (see computer help #42). Over the range of values in the dataset, average prices decrease from a high of approximately $295k to a low of approximately $245k from the early 1900s to 1960, and then increase again up to approximately $280k in 2005 (for homes with average values for *Floor*, *Lot*, *Bath*, *Bed*, and *Gar* that are in the Redwood, Adams, Crest, or Parker neighborhoods). Again, homes with other predictor values tend to have price differences of a similar magnitude for similar changes in age.

The "*Gar* effect on *Price*" is

$$Gar \text{ effect} = 226.15 + 13.12\,Gar.$$

The right-hand plot in Figure 6.5 shows a line plot with this *Gar* effect on the vertical axis and *Gar* on the horizontal axis. Over the range of values in the dataset, as garage size increases from 0 to 3, prices increase from approximately $225k to $265k on average (for homes with average values for *Floor*,...,*Age* that are in the Redwood, Adams, Crest, or Parker neighborhoods). Again, homes with other predictor values tend to have price differences of a similar magnitude for similar changes in garage size.

The indicator variable effects are more easily described in words. An active listing (with all else held constant) tends to add $27,400 to the price of a home (perhaps suggesting that homes tend to be offered for sale substantially overpriced in this housing market). Finally, two elementary schools seem to offer a price premium: Edison (approximately

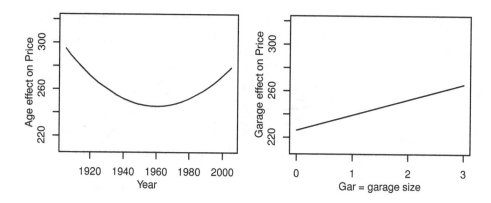

Figure 6.5 Predictor effect plots for *Age* and *Gar* in the home prices example. In the left plot, the *Age* effect of $246.88 + 3.30\,Age + 1.64\,Age^2$ is on the vertical axis while *Age* is on the horizontal axis. In the right plot, the *Gar* effect of $226.15 + 13.12\,Gar$ is on the vertical axis while *Gar* is on the horizontal axis.

$67,100) and Harris (approximately $47,300). This does not necessarily mean that it is just the proximity of these particular schools to a home that is associated with increased sale prices. It is more likely in this case that there are a range of features associated with these neighborhoods that tend to increase home prices (e.g., in this case Edison and Harris elementary schools are both close to the major university in the state, the University of Oregon).

6.1.5   Further questions

1. It is possible that the final model could be improved by considering interactions between the quantitative predictors and the indicator variables, for example, $D_{Ac}Floor$. Investigate whether there are any such interactions that significantly improve the model.

2. Investigate whether an alternative measure of lot size might be more appropriate than the categories used in the dataset. For example, define a new predictor variable that is the natural logarithm of the midpoint of the lot size range (in thousands of square feet) represented by each category [i.e., $\log_e(1.5) = 0.41$ for category 1, $\log_e(4) = 1.39$ for category 2, and so on]. Reanalyze the data with this new predictor in place of *Lot*. Do model results change drastically when you do this?

3. Investigate whether counting half-bathrooms as 0.1 is reasonable. For example, change values ending in .1 in the dataset to end in .5 instead, and reanalyze the data. Do model results change drastically when you do this?

4. Investigate whether there appear to be any systematic differences between pending sale prices and actual sales prices (all else equal). The analysis just described assumes no difference since the only indicator variable for "status" is D_{Ac}, which is 1 for active houses and 0 for both pending sales and sold houses. Add an indicator variable that is 1 for pending sales and 0 for both active and sold houses, and reanalyze the data. Do model results change drastically when you do this?

5. The values for *Price* are slightly skewed in a positive direction, suggesting perhaps that transforming *Price* to $\log_e(Price)$ might result in an improved multiple linear regression model—see Section 4.1.4. Reanalyze the data, but use $\log_e(Price)$ as the response variable instead of *Price*. Interpret results, remembering that regression parameter estimates such as $\hat{b}_1$ will need to be transformed to $\exp(\hat{b}_1)-1$, where they now represent the expected proportional change in *Price* from increasing *Floor* by 1 unit (holding all else constant).

6. Obtain similar data for a housing market near you (e.g., listings of homes for sale are commonly available on the Internet), and perform a regression analysis to explain and predict home prices in that market. Compare and contrast your results with the results presented here.

6.2   VEHICLE FUEL EFFICIENCY

6.2.1   Data description

The **CARS6** data file (obtained from www.fueleconomy.gov) contains information for 351 new U.S. passenger cars for the 2011 model year. We wish to explain and predict *Cmpg*, the city miles per gallon (MPG) using the following predictor variables:

> Eng = engine size (liters)
> Cyl = number of cylinders
> Vol = interior passenger and cargo volume (hundreds of cubic feet)
> D_{Rw} = indicator for rear-wheel drive
> D_{Aw} = indicator for all-wheel drive
> D_{4w} = indicator for four-wheel drive
> D_{St} = indicator for supercharged or turbocharged

We might expect that more powerful, larger vehicles (with high values of *Eng*, *Cyl*, and/or *Vol*) would have relatively low fuel efficiency (low values of *Cmpg*). The vehicles in the data file have also been categorized according to their drive system (127 front-wheel drive, 120 rear-wheel drive, 84 all-wheel drive, and 20 four-wheel drive), and their "air aspiration method" (266 naturally aspirated, 7 supercharged, and 78 turbocharged). Rear-wheel drive, all-wheel drive, and four-wheel drive vehicles are commonly less fuel efficient than similar front-wheel drive vehicles, while supercharged or turbocharged vehicles are commonly less fuel efficient than naturally aspirated vehicles. Thus for modeling purposes, we've defined the indicator variables, D_{Rw}, D_{Aw}, and D_{4w} to allow for differences among the different drive types (relative to front-wheel drive, which is the reference-level category for drive). Similarly, we've defined D_{St} to allow for differences between supercharged or turbocharged vehicles and naturally aspirated vehicles (the reference-level category for aspiration method).

6.2.2   Exploratory data analysis

Figure 6.6 displays a scatterplot matrix of the quantitative variables in the **CARS6** dataset (see computer help #16 in the software information files available from the book website). While there are a number of bivariate associations evident, these tell us little about the

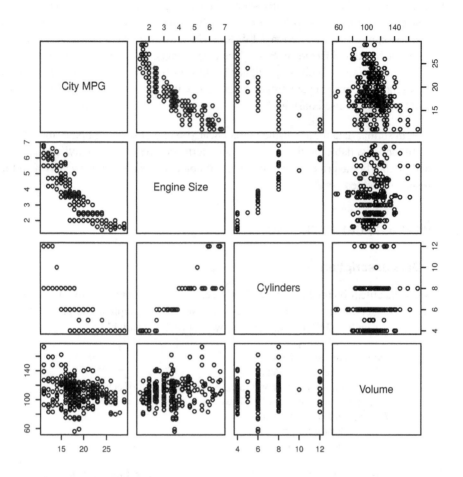

Figure 6.6   Scatterplot matrix of the quantitative variables in the **CARS6** data file.

likely form of a useful multiple linear regression model. The plots do show a number of points that stick out from the dominant patterns. Since some of these observations may cause a problem with subsequent analysis, it is worth making a note of them. Vehicles 202 (Maserati Granturismo Convertible) and 203 (Maserati Quattroporte 4.2L) have low values of *Cmpg* but only medium-high values of *Eng* and *Cyl*. There are a handful of 12-cylinder vehicles with an engine size of 6 liters of less, such as the Bentley Continental GTC FFV (vehicle 28) and BMW 760Li (vehicle 63). By contrast, one of the 8-cylinder vehicles has a 6.8-liter engine (the Bentley Mulsanne, vehicle 29).

Figure 6.7 displays boxplots of *Cmpg* versus the two qualitative variables in the **CARS6** (see computer help #21). Front-wheel drive vehicles appear to be the most fuel efficient overall (although there is quite a lot of variation), with rear-wheel drive, all-wheel drive, and four-wheel drive vehicles having the same median *Cmpg* but slightly different distributions of values. Fuel efficiency tends to be higher for naturally aspirated vehicles relative to supercharged and turbocharged vehicles. Keep in mind that these observations do not take into account the quantitative predictors, *Eng*, *Cyl*, and *Vol*.

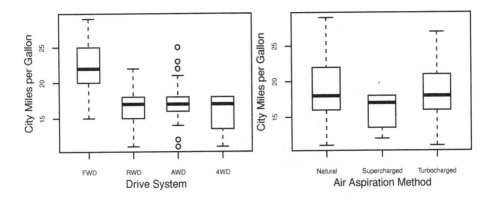

Figure 6.7 Boxplots of *Cmpg* versus drive system (left) and air aspiration method (right) for the **CARS6** data file.

6.2.3 Regression model building

We first try a model with each predictor variable "as is" (no transformations or interactions):

$$E(Cmpg) = b_0 + b_1\,Eng + b_2\,Cyl + b_3\,Vol + b_4\,D_{Rw} + b_5\,D_{Aw} + b_6\,D_{4w} + b_7\,D_{St}.$$

This model results in a value of $R^2 = 0.8274$ but isn't very satisfactory because the residuals from this model fail to satisfy the zero mean assumption in a plot of the residuals versus fitted values, displaying a pronounced curved pattern (plot not shown). As in Sections 4.3.2 and 5.1, we will instead try using $Cgphm = 100/Cmpg$ (city gallons per hundred miles) as our response variable:

$$E(Cgphm) = b_0 + b_1\,Eng + b_2\,Cyl + b_3\,Vol + b_4\,D_{Rw} + b_5\,D_{Aw} + b_6\,D_{4w} + b_7\,D_{St}.$$

This model improves the R^2 value to 0.8535 and now has residuals that appear to satisfy the four regression model assumptions of zero mean, constant variance, normality, and independence reasonably well (residual plots not shown). The regression standard error, s, for this model is 0.4611, while adjusted R^2 is 0.8505.

It might be possible to improve the model further with interactions between the quantitative predictors and the indicator variables, for example, $D_{Rw}\,Eng$. Also, since the vehicles in the sample have different drive systems *and* air aspiration types, we can try interactions between D_{Rw} and D_{St}, and between D_{Aw} and D_{St} (but not between D_{4w} and D_{St} since there are no supercharged or turbocharged four-wheel drive vehicles in the sample). We can then also try interactions such as that between $D_{Rw}D_{St}$ and *Eng*.

Investigations with nested model F-tests (not shown) finds 11 of these interactions to be potentially useful. However, the regression parameter estimates for two of these interactions, $D_{Rw}Eng$ and $D_{Aw}Eng$, are very close to one another. This suggests that perhaps the linear association between *Cgphm* and *Eng* (holding all else constant) is similar for rear-wheel and all-wheel drive vehicles, and we could replace the respective two regression parameters with a single (common) regression parameter. We can test this using a nested model F-test in a similar way to the example on page 182. To do this, we define a new predictor term equal to $D_{RwAw}Eng$, where $D_{RwAw} = D_{Rw} + D_{Aw}$ is an indicator variable for rear-wheel *and* all-wheel drive vehicles, and compare the model with this new predictor term to the model with both $D_{Rw}Eng$ and $D_{Aw}Eng$. The resulting nonsignificant p-value confirms that the simpler (reduced) model with $D_{RwAw}Eng$ is preferred.

6.2.4 Results and conclusions

This final model with 10 interactions is

$$E(Cgphm) = b_0 + b_1\,Eng + b_2\,Cyl + b_3\,Vol + b_4\,D_{Rw} + b_5\,D_{Aw} + b_6\,D_{4w} + b_7\,D_{St}$$
$$+ b_8\,D_{Aw}D_{St} + b_8\,D_{RwAw}Eng + b_{10}\,D_{St}Eng + b_{11}\,D_{Rw}Cyl + b_{12}\,D_{Aw}Cyl$$
$$+ b_{13}\,D_{4w}Cyl + b_{14}\,D_{St}Cyl + b_{15}\,D_{Aw}D_{St}Cyl + b_{16}\,D_{Aw}Vol + b_{17}\,D_{4w}Cyl.$$

Statistical software output for this model (see computer help #31) is:

Model Summary

Model	Sample Size	Multiple R Squared	Adjusted R Squared	Regression Std. Error
3[a]	351	0.8841	0.8782	0.4162

[a] Predictors: (Intercept), Eng, Cyl, Vol, D_{Rw}, D_{Aw}, D_{4w}, D_{St}, 10 interactions.

Parameters[a]

Model		Estimate	Std. Error	t-stat	Pr($>$|t|)
3	(Intercept)	1.896	0.262	7.231	0.000
	Eng	0.879	0.113	7.766	0.000
	Cyl	−0.042	0.086	−0.492	0.623
	Vol	0.619	0.179	3.457	0.001
	D_{Rw}	−0.142	0.262	−0.542	0.588
	D_{Aw}	1.459	0.576	2.534	0.012
	D_{4w}	0.182	0.764	0.238	0.812
	D_{St}	1.384	0.237	5.830	0.000
	$D_{Aw}D_{St}$	−1.499	0.422	−3.556	0.000
	$D_{RwAw}Eng$	−0.675	0.127	−5.327	0.000
	$D_{St}Eng$	0.710	0.116	6.126	0.000
	$D_{Rw}Cyl$	0.452	0.096	4.708	0.000
	$D_{Aw}Cyl$	0.325	0.102	3.175	0.002
	$D_{4w}Cyl$	−0.188	0.156	−1.209	0.228
	$D_{St}Cyl$	−0.570	0.086	−6.632	0.000
	$D_{Aw}D_{St}Cyl$	0.275	0.067	4.121	0.000
	$D_{Aw}Vol$	−0.745	0.411	−1.815	0.070
	$D_{4w}Vol$	1.358	0.468	2.902	0.004

[a] Response variable: $Cgphm$.

This final model results in residuals that appear to satisfy the four regression model assumptions of zero mean, constant variance, normality, and independence reasonably well (residual plots not shown). Also, while not all of the individual t-test p-values are below the usual 0.05 threshold, most are reasonably low (except Cyl, D_{Rw}, and D_{4w}, which are retained for hierarchy reasons). Although the individual p-value for $D_{4w}Cyl$ is high enough to suggest dropping this term from the model, overall results change little if we do and we retain this term for the remainder of this example.

The highest leverage observation in the dataset is vehicle 63 [BMW 760Li, leverage 0.34, which exceeds the $3(k+1)/n = 0.15$ threshold], while the vehicle with the highest Cook's distance is vehicle 28 (Bentley Continental GTC FFV, Cook's distance 0.06, which does not exceed the 0.5 threshold). With 6-liter engines and 12 cylinders, both vehicles are potentially influential, but neither dramatically changes the regression results if excluded.

There are five studentized residuals outside the ±3 range, that is, potential outliers. Each potential outlier has a much higher value of *Cgphm* (i.e., worse fuel efficiency) than expected by the model results: Maserati Granturismo Convertible (vehicle 202), Maserati Quattroporte 4.2L (vehicle 203), Maserati Quattroporte 4.7L (vehicle 204), Mercedes CLS 63 AMG (vehicle 217), and Maserati Granturismo (vehicle 200). With a relatively large sample size like this, we might expect a few studentized residuals outside the ±3 range. Furthermore, none of these five studentized residuals dramatically changes the regression results if excluded. So, there is no need to remove these five vehicles from the dataset—they merely represent the most extreme values of *Cgphm* relative to the model results.

The model can explain 88.4% of the variation in *Cgphm*, and predictions using the model are likely to be accurate to within approximately ±0.832 gallon per hundred miles (at a 95% confidence level). To put this in context, *Cgphm* in this dataset ranges from 3.448 to 9.091 gallons per hundred miles.

A potential use for the model might be to identify the most fuel-efficient vehicles of each vehicle class relative to the model predictions. The model allows prediction of the city MPG for a vehicle with particular values for *Eng*, *Cyl*, and so on. For the sake of illustration, consider two midsize cars: the Buick Lacrosse 2.4L (vehicle 71) and the Kia Optima 2.4L (vehicle 170). Both cars have very similar measured features (2.4-liter engines, 4 cylinders, 116 or 117 cubic feet of interior passenger and cargo volume, naturally aspirated, front-wheel drive), and both are predicted by the model to get a little under 22 city MPG ($\widehat{Cgphm} = 4.55$ for the Buick Lacrosse, equivalent to $100/4.55 = 22.0$ MPG, and $\widehat{Cgphm} = 4.56$ for the Kia Optima, equivalent to $100/4.56 = 21.9$ MPG). However, the Buick Lacrosse actually gets just 19 city MPG, while the Kia Optima gets 24. In other words, to find the most fuel-efficient vehicles, *taking into account* the features represented by *Eng*, *Cyl*, and so on, we should identify those vehicles with the largest negative residuals (since if the actual value of *Cgphm* is much smaller than its prediction from the model, then the corresponding value of *Cmpg*) will be much larger than predicted.

This is *not* the same as just identifying vehicles with the lowest values of *Cgphm* (or, equivalently, the highest values of *Cmpg*). For example, vehicle 92 (Chevrolet Aveo 5 1.6L) gets 25 city MPG, better than the Kia Optima 2.4L, but this is worse than its model prediction ($100/3.74 = 26.7$ city MPG). Vehicles with features like the Kia Optima's can be expected to get just 22 city MPG, but the Kia Optima actually gets 24 city MPG. In other words, we might expect that the different features of the Kia Optima compared with the Chevrolet Aveo (larger engine, roomier) could result in a large trade-off with respect to reduced fuel efficiency (since the model predictions differ greatly from 21.9 for the Kia to 26.7 for the Chevrolet). However, this is not the case since actual fuel efficiency for these two vehicles differs only slightly from 24 for the Kia to 25 for the Chevrolet.

Based on the model results, the "best" vehicles in each class (with respect to high actual city MPG relative to model predictions) are:

- Minicompact and two-seater cars: Aston Martin DB9 5.9L (actual *Cmpg* = 13, predicted 11.9).

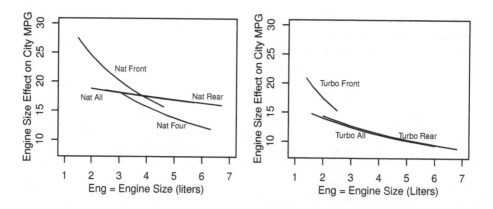

Figure 6.8 Predictor effect plots for *Eng* in the fuel efficiency example. The *Eng* effect of $100/(2.32 + 2.57 D_{Rw} + 2.59 D_{Aw} + 0.54 D_{4w} + 0.25 D_{St} + 0.15 D_{Aw} D_{St} + 0.88 Eng - 0.68 D_{RwAw} Eng + 0.71 D_{St} Eng)$ is on the vertical axis, while *Eng* is on the horizontal axis, and the lines are marked by drive system for naturally aspirated vehicles in the left plot and for supercharged and turbocharged vehicles in the right plot.

- Subcompact cars: Ford Mustang 5L (actual *Cmpg* = 18, predicted 15.0).
- Compact cars: Chevrolet Camaro 6.2L (actual *Cmpg* = 16, predicted 14.4).
- Midsize cars: BMW 528i 3L (actual *Cmpg* = 22, predicted 18.0).
- Large cars: Hyundai Genesis 4.6L (actual *Cmpg* = 17, predicted 14.8).
- Station wagons: Nissan Juke Awd 1.6L (actual *Cmpg* = 25, predicted 22.1).
- Sport utility vehicles: Mercedes ML 550 4matic 5.5L (actual *Cmpg* = 13, predicted 12.5).

A further use for the model might be to utilize the specific findings relating to the effects of each of the predictors on city MPG. Section 5.5 showed how to construct predictor effect plots to facilitate this. First consider how *Cmpg* changes as *Eng* changes. Since *Eng* is included in interaction terms with D_{RwAw} and D_{St}, we can isolate this change in *Cmpg* when we hold the remaining quantitative predictors constant (e.g., at the most common number of cylinders, 6, for *Cyl* and at the sample mean value for *Vol*, 110 cubic feet). Then the "*Eng* effect on *Cmpg*" is

$$Eng \text{ effect} = 100/(2.32 + 2.57 D_{Rw} + 2.59 D_{Aw} + 0.54 D_{4w} + 0.25 D_{St} + 0.15 D_{Aw} D_{St}$$
$$+ 0.88 Eng - 0.68 D_{RwAw} Eng + 0.71 D_{St} Eng).$$

The values 0.88, −0.68, and 0.71 come directly from the estimated regression equation. The other values result from plugging in 6 for *Cyl* and the sample mean for *Vol* into the rest of the equation. This *Eng* effect then represents how *Cmpg* changes as *Eng* changes for vehicles with six cylinders and average interior passenger and cargo volume. We can further plug in values of 0 and 1 for the indicator variables D_{Rw}, D_{Aw}, D_{4w}, D_{St}, and D_{RwAw} to calculate specific equations for different types of vehicle.

This is perhaps clearer to show on a line plot with the *Eng* effect on the vertical axis, *Eng* on the horizontal axis, and lines marked for different types of vehicle—Figure 6.8 illustrates (see computer help #42). Over the range of values in the dataset, as engine size increases

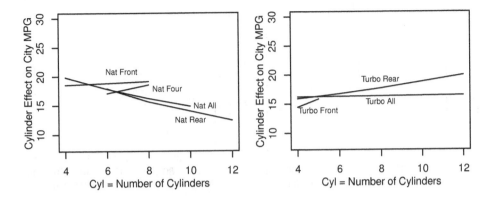

Figure 6.9 Predictor effect plots for *Cyl* in the fuel efficiency example. The *Cyl* effect of $100/(5.57 - 2.44 D_{Rw} - 1.66 D_{Aw} + 1.67 D_{4w} + 3.80 D_{St} - 1.50 D_{Aw} D_{St} - 0.04 Cyl + 0.45 D_{Rw} Cyl + 0.33 D_{Aw} Cyl - 0.19 D_{4w} Cyl - 0.57 D_{St} Cyl + 0.27 D_{Aw} D_{St} Cyl)$ is on the vertical axis, while *Cyl* is on the horizontal axis, and the lines are marked by drive system for naturally aspirated vehicles in the left plot and for supercharged and turbocharged vehicles in the right plot.

from 1.4 to 6.8 liters (holding all else constant), fuel efficiency decreases, first steeply (at small engine sizes) and then becoming more gradual (at larger engine sizes). The rate of decrease depends on the type of vehicle, as shown by the different lines on the plots (which represent vehicles with six cylinders and average interior passenger and cargo volume, 110 cubic feet). Vehicles with other predictor values tend to have fuel efficiency *differences* of a similar magnitude for similar changes in engine size (although the fuel efficiencies of individual vehicles will depend on those predictor values)—the predictor effect plot would simply have different values on the vertical axis, but the general appearance of the lines would be the same.

We can also see from Figure 6.8 that the model successfully captures the tendency for front-wheel drive vehicles to be more fuel efficient than rear-wheel and all-wheel drive vehicles (with similar engine sizes), which in turn tend to be more fuel efficient than four-wheel drive vehicles (with similar engine sizes). We can also confirm that naturally aspirated vehicles tend to be more fuel efficient than supercharged or turbocharged vehicles (with similar engine sizes).

We can also calculate the "*Cyl* effect on *Cmpg*" as

$$Cyl \text{ effect} = 100/(5.57 - 2.44 D_{Rw} - 1.66 D_{Aw} + 1.67 D_{4w} + 3.80 D_{St} - 1.50 D_{Aw} D_{St}$$
$$- 0.04 Cyl + 0.45 D_{Rw} Cyl + 0.33 D_{Aw} Cyl - 0.19 D_{4w} Cyl$$
$$- 0.57 D_{St} Cyl + 0.27 D_{Aw} D_{St} Cyl).$$

Figure 6.9 shows line plots with this *Cyl* effect on the vertical axis, *Cyl* on the horizontal axis, and lines marked for different types of vehicle (see computer help #42).

Over the range of values in the dataset, as the number of cylinders increases from 4 to 12 (holding all else constant), fuel efficiency generally decreases gradually for all-wheel and rear-wheel drive naturally aspirated vehicles, but increases slightly for four-wheel drive naturally aspirated vehicles and front- and rear-wheel drive supercharged or turbocharged vehicles. By contrast, there is little change for front-wheel drive naturally aspirated vehicles and all-wheel drive supercharged or turbocharged vehicles. The rate of change depends on

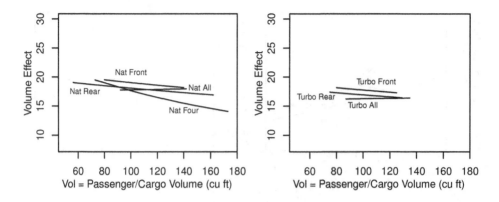

Figure 6.10   Predictor effect plots for *Vol* in the fuel efficiency example. The *Vol* effect of $100/(4.63 + 0.28 D_{Rw} + 1.11 D_{Aw} - 0.95 D_{4w} + 0.38 D_{St} + 0.15 D_{Aw} D_{St} + 0.62 \, Vol - 0.75 D_{Aw} Vol + 1.36 D_{4w} Vol)$ is on the vertical axis, while *Vol* is on the horizontal axis, and the lines are marked by the drive system for naturally aspirated vehicles in the left plot and for supercharged and turbocharged vehicles in the right plot.

the type of vehicle, as shown by the different lines on the plots (which represent vehicles with average engine size, 3.4 liters, and average interior passenger and cargo volume, 110 cubic feet). Vehicles with other predictor values tend to have fuel efficiency differences of a similar magnitude for similar changes in number of cylinders.

These findings are quite complex and it would require careful consideration by a subject-matter expert to make sense of them. What the model appears to be saying is that increasing the number of cylinders in a vehicle (while keeping its engine size and interior passenger and cargo volume fixed) allows for small fuel efficiency gains in some types of vehicle but not others. Of course, since engine size and number of cylinders tend to be closely associated in most vehicles, any such fuel efficiency gains would likely be negated if *both* the engine size and number of cylinders increases.

Finally, we can also calculate the "*Vol* effect on *Cmpg*" as

$$Vol \text{ effect} = 100/(4.63 + 0.28 D_{Rw} + 1.11 D_{Aw} - 0.95 D_{4w} + 0.38 D_{St} + 0.15 D_{Aw} D_{St}$$
$$+ 0.62 \, Vol - 0.75 D_{Aw} Vol + 1.36 D_{4w} Vol).$$

Figure 6.10 shows line plots with this *Vol* effect on the vertical axis, *Vol* on the horizontal axis, and lines marked for different types of vehicle (see computer help #42).

Over the range of values in the dataset, as interior passenger and cargo volume increases from 56 to 173 cubic feet (all else held constant), fuel efficiency generally decreases gradually for front- and rear-wheel drive vehicles, and more steeply for four-wheel drive vehicles. By contrast, there is little change for all-wheel drive vehicles as interior volume increases. These findings hold for both naturally aspirated vehicles and supercharged or turbocharged vehicles. The rate of change depends on the type of vehicle, as shown by the different lines on the plots (which represent vehicles with average engine size, 3.4 liters, and six cylinders). Vehicles with other predictor values tend to have fuel efficiency differences of a similar magnitude for similar changes in interior volume.

6.2.5 Further questions

1. There are many models that appear to fit the sample data as effectively as the final model presented here, but which include different subsets of interactions. Do some model building to find another such model with equally impressive results ($R^2 = 0.8841$ and $s = 0.4162$) and investigate if and how overall model conclusions (as described in Section 6.2.4) change. (You should find that overall conclusions are relatively robust to the precise form of the final model, as long as that model fits as well as the model presented here.)

2. Can you come up with plausible explanations for all the results. For example, why might four-wheel drive vehicles be particularly effected by changes in *Vol* compared with the other types of vehicle (see Figure 6.10)?

3. Instead of using *Cgphm* as the response, try using $\log_e(Cmpg)$ as the response. Compare and contrast your results with the results presented here.

4. The analysis presented here used city MPG. Repeat the analysis using highway MPG instead—this is available in the dataset as the variable "*Hmpg*."

5. Obtain similar data for a more recent vehicle model year, and perform a regression analysis to explain and predict miles per gallon (city or highway). Compare and contrast your results with the results presented here.

6.3 PHARMACEUTICAL PATCHES

6.3.1 Data description

The **PATCHES** data file contains experimental data for a pharmaceutical patch manufactured under varying conditions. The data were provided by Craig Allen, a consultant working in the pharmaceutical industry. The experiment involved the production of 70 patches in different dosage strengths that were formed in a press where the gap height and exposure time were varied. The response variable is *Diff*, the percentage difference from a target value (varying from -25% to 32%), which we wish to explain and predict using the following predictor variables:

> *Dos* = dosage strength (four distinct values: 25, 50, 75, or 100 micrograms)
> *Gap* = gap height (continuous values between 36.45 and 45.72 mm)
> *Exp* = exposure time (five distinct values: 10, 30, 60, 90, or 120 seconds)

Dos and *Exp* can be treated as quantitative predictors or as qualitative predictors using indicator variables.

6.3.2 Exploratory data analysis

Figure 6.11 displays a scatterplot matrix of the quantitative variables in the **PATCHES** data file (see computer help #16 in the software information files available from the book website). Remember that for any plot in the scatterplot matrix, look to the left or right for the label of the vertical axis and up or down for the label of the horizontal axis. There are

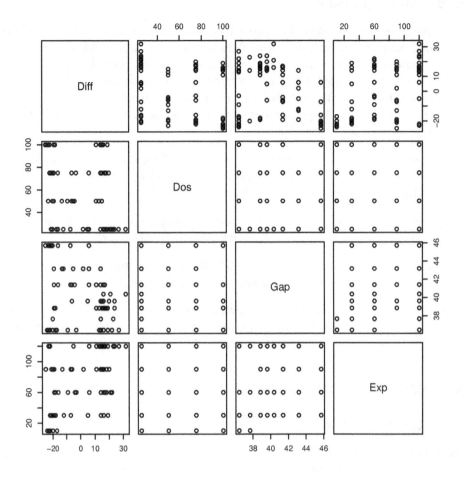

Figure 6.11   Scatterplot matrix of the variables in the **PATCHES** data file.

no obvious points that stand out as anomalous, and none of the variables are particularly skewed. We will therefore proceed using the entire dataset and start off with all variables untransformed. In the second plot in the top row, the values of *Diff* tend to be higher for *Dos* = 25 and fairly consistent for the other three values. This suggests that treating *Dos* as categorical could perhaps be more useful than treating it continuously, with just one indicator variable needed (to distinguish observations with *Dos* = 25 from observations with *Dos* = 50, 75, or 100). However, since a multiple linear regression model takes into account all associations between the predictor and response variables, we cannot rule out the need for the complete set of three indicator variables for *Dos*. For the third plot in the top row, the values of *Diff* tend to be lower for low and high values of *Gap* and higher in the middle, which suggests the possibility that a quadratic or other nonlinear transformation of *Gap* might be needed in the final multiple linear regression model. For the top right plot, there appears to be a linear association between *Diff* and *Exp*, so we might expect to need only an untransformed linear term for *Exp* in a regression model. Again, however, since a multiple linear regression model takes into account all associations between the predictor and response variables, we cannot rule out the need for transforming *Exp* or treating it as categorical and creating indicator variables.

6.3.3 Regression model building

The following table summarizes the model building process.

Predictor Terms	R^2	Adj. R^2	s	Comments
Dos, *Gap*, *Exp*	64.2	62.5	10.48	Variance inflation factors for *Dos*, *Gap*, and *Exp* all low, so probably no multicollinearity problems to worry about.
D_{50}, D_{75}, D_{100}, *Gap*, *Exp*	72.9	70.8	9.26	Indicator variables created for *Dos* (D_{50}, D_{75}, D_{100}) using *Dos* = 25 as a reference level; increase in adjusted R^2 and decrease in s suggest a better fit. However, loess smooths on the residual plots versus *Gap* and *Exp* have "hill shapes," suggesting that quadratic or other nonlinear transformations may be needed.
D_{50}, D_{75}, D_{100}, *Gap*, Gap^2, *Exp*, Exp^2	83.4	81.5	7.37	Gap^2 and Exp^2 added; increase in adjusted R^2 and decrease in s suggests a better fit. However, the estimated regression parameters for D_{50}, D_{75}, and D_{100} are all similar (-20.3, -18.8, and -20.1 with standard errors of 2.4, 2.4, and 2.6, respectively). This suggests that rather than four "groups" we may need to model just two groups, for *Dos* = 25 and for *Dos* = 50, 75, or 100.
D_{25}, *Gap*, Gap^2, *Exp*, Exp^2	83.3	82.0	7.27	Indicator variable created for D_{25} (using *Dos* = 50, 75, or 100 as a reference level); increase in adjusted R^2 and decrease in s suggests a better fit. However, we also need to allow for the possibility of interactions between D_{25} and the other predictor variables.
D_{25}, *Gap*, Gap^2, *Exp*, Exp^2, $D_{25}Gap$, $D_{25}Gap^2$, $D_{25}Exp$, $D_{25}Exp^2$	86.2	84.1	6.83	Interactions added for $D_{25}Gap$, $D_{25}Gap^2$, $D_{25}Exp$, $D_{25}Exp^2$; increase in adjusted R^2 and decrease in s suggests a better fit. However, p-values for $D_{25}Gap$ and $D_{25}Gap^2$ are large (0.56 and 0.55, respectively), suggesting that they could be removed from the model. This is confirmed with a nested model F-test p-value of 0.73.
D_{25}, *Gap*, Gap^2, *Exp*, Exp^2, $D_{25}Exp$, $D_{25}Exp^2$	86.0	84.4	6.76	$D_{25}Gap$ and $D_{25}Gap^2$ removed from the model; increase in adjusted R^2 and decrease in s suggests a better fit. Residual plots and other diagnostic measures suggest that this model fits well (see below), so this is our final model.

6.3.4 Model diagnostics

Residual plots versus each of the predictors for the final model (not shown) suggest that the zero mean, constant variance, and independence regression assumptions seem reasonable. Figure 6.12 displays a plot of the studentized residuals versus fitted values, which also

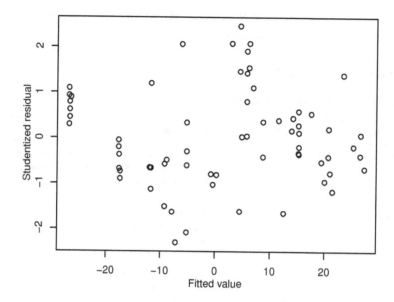

Figure 6.12 Scatterplot of the studentized residuals versus the fitted values from the final model for the pharmaceutical patches example.

show no strong patterns to worry about. Our only slight concern might be the suggestion of nonconstant variance since the residuals appear to be slightly more variable in the middle of the range of fitted values and less variable at the extremes. A histogram and QQ-plot of the studentized residuals (not shown) suggest reasonable normality.

The highest leverage in the dataset is 0.29 for observation 26 (with low values for each of the predictors), although observation 22 at 0.27 is not far behind. The leverage warning threshold is $3(k+1)/n = 0.34$, so neither point is likely to be overly influential. [The $2(k+1)/n$ threshold does not apply since the highest leverage observation is not isolated.] Observations 26 and 22 also have the two highest Cook's distances, although neither is above the 0.5 threshold, and neither dramatically changes the regression results if excluded. Since none of the studentized residuals are outside the ± 3 range, there are no outliers.

6.3.5 Results and conclusions

Thus, a final model for these data is

$$\mathrm{E}(Diff) = b_0 + b_1 D_{25} + b_2 Gap + b_3 Gap^2 + b_4 Exp + b_5 Exp^2 + b_6 D_{25}Exp + b_7 D_{25}Exp^2.$$

Statistical software output for this model (see computer help #31) is:

Model Summary

Model	Sample Size	Multiple R Squared	Adjusted R Squared	Regression Std. Error
6^a	70	0.8601	0.8444	6.757

^{a} Predictors: (Intercept), D_{25}, Gap, Gap^2, Exp, Exp^2, $D_{25}Exp$, $D_{25}Exp^2$.

Parameters[a]

Model		Estimate	Std. Error	t-stat	$\Pr(>|t|)$
6	(Intercept)	−750.499	158.518	−4.734	0.000
	D_{25}	15.373	6.569	2.340	0.023
	Gap	38.257	7.813	4.897	0.000
	Gap^2	−0.508	0.095	−5.352	0.000
	Exp	0.490	0.145	3.385	0.001
	Exp^2	−0.001	0.001	−0.857	0.395
	$D_{25}Exp$	0.384	0.214	1.790	0.078
	$D_{25}Exp^2$	−0.003	0.001	−2.374	0.021

[a] Response variable: *Diff*.

The estimated regression equation is therefore

$$\widehat{Diff} = -750.499 + 15.373\,D_{25} + 38.257\,Gap - 0.508\,Gap^2 + 0.490\,Exp$$
$$- 0.001\,Exp^2 + 0.384\,D_{25}Exp - 0.003\,D_{25}Exp^2.$$

Each of the individual t-test p-values is below the usual 0.05 threshold, except Exp^2 and $D_{25}Exp$ (which have been included to preserve hierarchy since $D_{25}Exp^2$ is included in the model). The model can explain 86.0% of the variation in *Diff*, and predictions using the model are likely to be accurate to within approximately ±13.5% (at a 95% confidence level). To put this in context, *Diff* ranges from −25% to 32% in this dataset.

We next use the methods of Section 5.5 to construct predictor effect plots. First consider how *Diff* changes as *Gap* changes, holding *Exp* constant at its sample mean, 75.29 seconds. The "*Gap* effect on *Diff*" is

$$Gap \text{ effect} = -718.341 + 24.534\,D_{25} + 38.257\,Gap - 0.508\,Gap^2.$$

The values in this expression result from plugging in the sample mean for *Exp* to the estimated regression equation and simplifying as far as possible. This *Gap* effect then represents how *Diff* changes as *Gap* changes for patches with an average value for *Exp*.

The left-hand plot in Figure 6.13 displays a predictor effect plot with this *Gap* effect on the vertical axis and *Gap* on the horizontal axis (see computer help #42). Since the response variable measures the percentage difference from target, finding where the expected response is zero is of particular interest. Thus, a horizontal "zero threshold" line has been added to the plot. The upper curved line is for a dosage strength of 25 micrograms, showing a decreasing nonlinear effect on the response variable as gap height increases, crossing the zero threshold at about 45 mm. The lower line for dosage strengths of 50, 75, and 100 shows a similar decreasing nonlinear effect as gap height increases, but at a lower level overall so that it crosses the zero threshold at about 39 mm.

Similarly, the "*Exp* effect on *Diff*" (holding *Gap* constant at its sample mean, 40.11 mm) is

$$Exp \text{ effect} = -33.992 + 15.373\,D_{25} + 0.490\,Exp - 0.001\,Exp^2$$
$$+ 0.384\,D_{25}Exp - 0.003\,D_{25}Exp^2.$$

The right-hand plot in Figure 6.13 displays a predictor effect plot with this *Exp* effect on the vertical axis and *Exp* on the horizontal axis. The upper line is for a dosage strength of 25

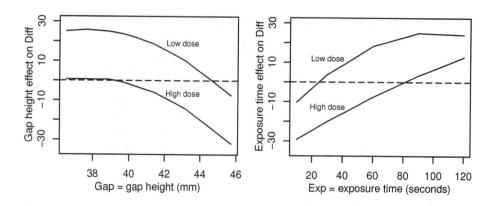

Figure 6.13   Predictor effect plots for *Gap* and *Exp* in the **PATCHES** example. In the left plot, the *Gap* effect, $-718.341 + 24.534 D_{25} + 38.257 Gap - 0.508 Gap^2$, is on the vertical axis, while *Gap* is on the horizontal axis. In the right plot, the *Exp* effect, $-33.992 + 15.373 D_{25} + 0.490 Exp - 0.001 Exp^2 + 0.384 D_{25} Exp - 0.003 D_{25} Exp^2$, is on the vertical axis, while *Exp* is on the horizontal axis.

micrograms, showing an increasing nonlinear effect on the response variable as exposure time increases, crossing the zero threshold at about 25 seconds. The lower line for dosage strengths of 50, 75, and 100 micrograms shows a more linear increasing effect as exposure time increases, but at a lower level overall so that it crosses the zero threshold at about 80 seconds.

An overall conclusion from this analysis might be that the "optimal" manufacturing conditions for the low dose (25 micrograms) pharmaceutical patches are a gap height of 45 mm and an exposure time of 25 seconds, while optimal conditions for the high dose (50, 75, and 100 micrograms) patches are a gap height of 39 mm and an exposure time of 80 seconds. No significant differences were found between the 50-, 75-, and 100-microgram dosage strengths.

6.3.6   Further questions

1. We initially treated *Dos* as a quantitative predictor variable, but then switched to using indicator variables to represent the distinct values of *Dos*. Investigate whether continuing to treat *Dos* as a quantitative predictor variable leads to broadly similar conclusions. Note that you'll probably need to include Dos^2 as an additional predictor term since the values of *Diff* tend to be higher for $Dos = 25$ and fairly consistent for the other three values (in other words, *Dos* seems to affect *Diff* nonlinearly).

2. We treated *Exp* as a quantitative predictor variable throughout the analysis. Investigate whether using indicator variables to represent the distinct values of *Exp* leads to broadly similar conclusions.

CHAPTER 7

EXTENSIONS

In multiple linear regression models, the response variable should be quantitative, having meaningful numerical values where the numbers represent actual quantities of time, money, weight, and so on. By contrast, in this chapter we introduce some extensions of linear regression modeling that are designed for qualitative (categorical) response variables. Rather than give a detailed discussion of these extensions (which lies beyond the scope of this book), we instead give a brief introduction to these more advanced regression modeling techniques through specific applications, and discuss some additional modeling topics.

Section 7.1 includes two examples of generalized linear models: logistic regression for *binary* (two-outcome) responses and Poisson regression for *count data* responses. In Section 7.2 we present a *discrete choice* application in which the response variable observations identify which of a discrete set of alternatives is chosen (e.g., these alternatives might be transit or brand choices, or, for the example considered here, Oscar-winning movie choices). Section 7.3 contains an application of *multilevel* regression modeling, where the data have a hierarchical structure in which the units of observation (homes, say) are grouped (by county, say), and the response of interest (level of radon gas in this example) is expected to vary systematically not only from home to home but also from county to county. Finally, Section 7.4 contains not so much an extension to the regression modeling techniques discussed in this book as an alternative approach to the statistical philosophy that motivates regression modeling. In particular, we briefly introduce the "Bayesian" approach to statistical inference (in contrast to the "frequentist" approach used in the rest of the book).

Applied Regression Modeling, Second Edition. By Iain Pardoe
Copyright © 2012 John Wiley & Sons, Inc.

One regression topic that we do not consider in this book is that of *nonlinear regression*. In linear regression models, the expected value of the response variable can be written as a linear combination of predictor terms (as in the equation at the bottom of page 85). Mathematically, the regression equation is "linear in the regression parameters." By contrast, nonlinear regression is a method for estimating regression models that cannot be written as a linear combination of predictor terms (the regression equation is "nonlinear in the regression parameters"). Such models typically arise in situations where the form of the nonlinear regression equation is theoretically prescribed by the application (e.g., in experimental science contexts). For further details, see Bates and Watts (2007), Seber and Wild (1989), and Ratkowsky (1990).

7.1 GENERALIZED LINEAR MODELS

Generalized linear models (GLMs), first proposed by Nelder and Wedderburn (1972), extend the multiple linear regression model considered in Chapters 3–6 (and which can be motivated using an underlying normal distribution), to include other underlying probability distributions, such as the "binomial" and "Poisson" distributions. The specifics of how this is accomplished lie beyond the scope of this book but are covered rigorously by McCullagh and Nelder (1989). For our purposes, it is sufficient to note that GLMs greatly extend the types of problems we can tackle by allowing not just quantitative response variables (as for multiple linear regression), but also qualitative response variables. For example, in Section 7.1.1 we discuss a medical application in which the response outcome is whether a potentially cancerous tumor is malignant or benign. Such a binary (two-outcome) response can be modeled using "logistic regression," a GLM based on the binomial distribution. Section 7.1.2 contains a marketing example in which the response outcome is a count of the number of bottles of wine sold in a restaurant per week. Such a response can be modeled using "Poisson regression," a GLM based on the Poisson distribution.

The need to use a GLM rather than a linear regression model generally relates to the nature of the *response* variable, not the *predictor* variables. By itself, any nonnormality of predictor variables is not generally relevant to whether a linear regression model or a GLM is more appropriate. If the quantitative predictor variables in a linear regression have approximate normal distributions then the modeling process and results tend to be more straightforward, but there is no requirement for the quantitative predictor variables to have approximate normal distributions. It is possible (even common) to estimate perfectly good linear regression models with nonnormal predictor distributions.

There are many other types of GLM not considered here, probably the most common of which are "log-linear" models. Agresti (2002) provides a comprehensive review of log-linear models and other categorical data analysis methods (or see Agresti, 2007, for an introductory-level book).

7.1.1 Logistic regression

The Wisconsin Breast Cancer Data (Bennett and Mangasarian, 1992), available in the **BCWIS** data file, consist of 683 cases of potentially cancerous tumors, 444 of which turned out to be benign and 239 of which were malignant. Determining whether a tumor is malignant or benign is traditionally accomplished with an invasive surgical biopsy procedure. An alternative, less invasive technique called "fine needle aspiration" (FNA) allows examination of a small amount of tissue from the tumor. For the Wisconsin data, FNA provided nine cell features for each case; a biopsy was then used to determine the tumor status as malignant or benign. Is it possible to determine whether and how these cell features tell us anything about the tumor status? In other words, do the cell features allow us to predict tumor status accurately, so that FNA could be used as an alternative to the biopsy procedure for future patients?

In this situation, the predictors $(X_1, X_2, \ldots)$ are the features of the tissue cells, and the response outcome (Y) is binary; that is, there are two possible values: benign or malignant. We can code this as $Y = 0$ for one outcome (benign, say) and $Y = 1$ for the other outcome (malignant). However, it is inappropriate to then use this as the response variable in a multiple linear regression analysis. In particular, the residuals from such a model would almost certainly not satisfy the four usual linear regression assumptions (zero mean, constant variance, normality, independence). Instead, for binary response outcomes like this, we can use logistic regression to see how the outcome might depend on the values of the predictor variables.

In particular, logistic regression models the *probability* that the response Y is 1 (rather than 0), $\Pr(Y = 1)$. This probability depends on the values of the predictor variables by way of the following equation:

$$\Pr(Y = 1) = \frac{\exp(b_0 + b_1 X_1 + b_2 X_2 + \cdots + b_k X_k)}{1 + \exp(b_0 + b_1 X_1 + b_2 X_2 + \cdots + b_k X_k)}. \tag{7.1}$$

If the value of this equation is high (close to 1), then we predict that the tumor is more likely to be malignant. If the value of this equation is low (close to 0), then we predict that the tumor is more likely to be benign.

We hope that the FNA analysis will be reasonably accurate at predicting the tumor status. However, it is possible that occasionally it will fail to do so, and a benign tumor will be predicted to be malignant (or vice versa). Statistically, we need to be able to account for this possibility using a probability distribution. An appropriate distribution for binary outcomes is the *Bernoulli distribution*, which states that the outcome is either 1 (with a particular probability) or 0 (with 1 minus that probability). We can generalize this to the *binomial distribution*, which considers the probability of obtaining a particular number of "1's" in a series of Bernoulli outcomes.

In linear regression, we estimated the regression parameters (b_0, b_1, etc.) using least squares. That technique does not work here, so instead we use an approach called *maximum likelihood*. Essentially, maximum likelihood estimates are the values of the regression parameters that make the probability of observing the pattern of responses and predictors in the dataset as large as possible. This probability is calculated using formulas (not provided here) based on the binomial distribution just discussed. The important thing for us is that statistical software can perform this calculation to provide us with regression parameter estimates that enable the most accurate predictions of the response outcome Y. Further details on the mechanics of logistic regression are available in Hosmer and Lemeshow (2000).

Returning to the breast cancer example, the dataset consists of the following response and predictor variables:

$Y = 0$ if benign, 1 if malignant
X_1 = clump thickness
X_2 = cell size uniformity
X_3 = cell shape uniformity
X_4 = marginal adhesion
X_5 = single epithelial cell size
X_6 = bare nuclei
X_7 = bland chromatin
X_8 = normal nucleoli
X_9 = mitoses

The predictors, $(X_1, \ldots, X_9)$, are all integer values between 1 and 10 (where 1 represents a "normal" state, and 10 indicates a "most abnormal" state) and were determined by a doctor assessing the tissue cells through a microscope.

Consider fitting a logistic regression model to these data. The probability equation (7.1) can be rewritten

$$\text{logit}(\Pr(Y=1)) \equiv \log\left(\frac{\Pr(Y=1)}{1-\Pr(Y=1)}\right) = b_0 + b_1 X_1 + b_2 X_2 + \cdots + b_k X_k.$$

The ratio $\Pr(Y=1)/(1-\Pr(Y=1))$ is the *odds* of the outcome $Y=1$ occurring. For example, if $\Pr(Y=1)$ is 0.75, then the odds of $Y=1$ occurring is $0.75/0.25 = 3$, that is, three "1's" for every "0."

We first propose the following model:

$$\text{logit}(\Pr(Y=1)) = b_0 + b_1 X_1 + b_2 X_2 + b_3 X_3 + b_4 X_4 + b_5 X_5 + b_6 X_6 + b_7 X_7 + b_8 X_8 + b_9 X_9.$$

Here is part of the output produced by statistical software that displays some results for this model:

Parameters[a]

Model		Estimate	Std. Error	Z-stat	$\Pr(>|Z|)$
1	(Intercept)	−10.104	1.175	−8.600	0.000
	X_1	0.535	0.142	3.767	0.000
	X_2	−0.006	0.209	−0.030	0.976
	X_3	0.323	0.231	1.399	0.162
	X_4	0.331	0.123	2.678	0.007
	X_5	0.097	0.157	0.617	0.537
	X_6	0.383	0.094	4.082	0.000
	X_7	0.447	0.171	2.609	0.009
	X_8	0.213	0.113	1.887	0.059
	X_9	0.535	0.329	1.627	0.104

[a] Response variable: Y.

This output is similar to that for linear regression, the only difference being that the test statistics (parameter estimates divided by their standard errors) use the standard normal distribution rather than t-distributions to determine their significance (hence "Z-stat" rather than "t-stat").

The high individual p-values for X_2 and X_5 suggest that these features add little information to the model beyond that provided by the other cell features. So, we next try the following model:

$$\text{logit}(\Pr(Y=1)) = b_0 + b_1 X_1 + b_3 X_3 + b_4 X_4 + b_6 X_6 + b_7 X_7 + b_8 X_8 + b_9 X_9.$$

Here are the results for this model:

Parameters[a]

Model		Estimate	Std. Error	Z-stat	$\Pr(>|Z|)$
2	(Intercept)	−9.983	1.126	−8.865	0.000
	X_1	0.534	0.141	3.793	0.000
	X_3	0.345	0.172	2.012	0.044
	X_4	0.342	0.119	2.873	0.004
	X_6	0.388	0.094	4.150	0.000
	X_7	0.462	0.168	2.746	0.006
	X_8	0.226	0.111	2.037	0.042
	X_9	0.531	0.324	1.637	0.102

[a] Response variable: Y.

We can formally determine whether X_2 and X_5 can be removed without significant loss of information by conducting a test analogous to the nested model F-test (also called an "analysis of variance" test) of Section 3.3.4. Whereas the nested F-statistic for linear regression compares residual sum of squares (RSS) values for two nested models, the nested model test for GLMs compares the values of an analogous quantity, the *residual deviance*, for two nested models. The resulting test is then called an "analysis of deviance" test (or sometimes a "likelihood ratio" test). For our purposes, the specific details behind calculation of the deviance are less important than knowing how to use it. In particular, we can use statistical software to calculate the residual deviance for each of two nested models, and to calculate a p-value to assess whether the difference in the deviances is significant.

In our example, the first (complete) model (with X_2 and X_5) has a residual deviance of 102.89, while the second (reduced) model (without X_2 and X_5) has a residual deviance of 103.27. The change in deviance (0.38) can be compared with a χ^2 (chi-squared) distribution to find the p-value. Recall from page 204 that the χ^2 distribution, like the F-distribution, is a positively skewed probability distribution that takes only positive values. It is indexed by a single degrees of freedom number, which for this application is equal to the number of predictors we are attempting to remove from the model (i.e., two). We can use computer software to calculate the p-value for a χ^2-distributed random variable (see computer help #9 in the software information files available from the book website). Alternatively, statistical software can provide the p-value directly. In our example the p-value of 0.83 indicates that we cannot reject a null hypothesis stating that the regression parameters for X_2 and X_5 are both zero, and it seems safe to remove X_2 and X_5 from the model. There is also a question mark over X_9, which has an individual p-value of 0.102. However, Pardoe and Cook (2002) demonstrate that removing X_9 from the model is problematic and they prefer to keep it in.

How can we interpret the regression parameter estimates in a logistic regression analysis? Consider the estimate $\hat{b}_1 = 0.534$ for $X_1 =$ clump thickness. This means that if clump thickness increases (becomes more abnormal) by 1 unit (holding all else constant), then the natural logarithm of the odds (of the tumor being malignant rather than benign) will increase by 0.534, and the odds will be *multiplied* by $\exp(0.534) = 1.71$. These exponentiated parameter estimates are known as *odds ratios*. For another example, if $X_3 =$ cell shape uniformity increases by 1 unit (holding all else constant), then the odds (of the tumor being malignant) will be multiplied by $\exp(0.345) = 1.41$.

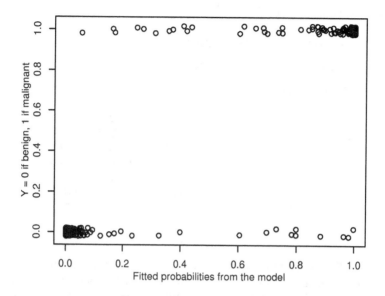

Figure 7.1 Scatterplot of binary response outcome ($Y = 0$ if benign, 1 if malignant) versus fitted probabilities (of malignant status) from the second model for the breast cancer example. The binary outcomes have been "jittered" vertically to enhance the visual appearance of the plot.

Finally, how well does our final fitted logistic regression model predict tumor status in this dataset? We can gain insight into this question by considering the scatterplot in Figure 7.1, which shows the binary response outcome ($Y = 0$ if benign, 1 if malignant) versus fitted probabilities (of malignant status) from this model. Consider drawing a vertical line at probability 0.5 and classifying all tumors to the left of this cutoff (< 0.5) as benign and all tumors to the right of this cutoff (> 0.5) as malignant. There are 11 tumors at the upper left of the plot that would be misclassified as benign when they are actually malignant, and 10 tumors at the lower right of the plot that would be misclassified as malignant when they are actually benign. In this case, the cost of misclassifying a tumor as benign when it is actually malignant is potentially huge, leading to a potentially preventable death. Conversely, the cost of misclassifying a tumor as malignant when it is actually benign is much lower, leading to potentially unnecessary medical intervention, but hopefully not death. We should therefore probably set any cutoff threshold lower than 0.5 in this application. For example, consider moving the cutoff to about 0.2. Now, there are just 3 malignant tumors at the upper left that would be misclassified as benign, and 13 benign tumors at the lower right that would be misclassified as malignant. In practice, a tumor with a fitted probability close to the cutoff probability would probably result in a biopsy being performed to confirm the diagnosis. By contrast, tumors with a fitted probability very close to 0 or 1 probably have no need of an invasive biopsy—diagnosis of these tumors appears to be very accurate based on the model results.

The example in this section was a medical application, but binary response outcomes are common in many fields of application. For example, consider modeling the outcome of a contract bid as a success or failure, or predicting whether a stock price will increase or decrease, or identifying if a current customer will continue a service or cancel it. We have provided just a taste of the power and flexibility of logistic regression modeling here. Many of the model building issues we considered in Chapters 4 and 5 arise in this context, too; a fine resource for exploring these topics in more depth is Hosmer and Lemeshow (2000).

7.1.2 Poisson regression

Durham et al. (2004) analyzed wine demand at a restaurant, using economic hedonic quantity models to evaluate the impact of objective factors (e.g., origin, varietal), sensory descriptors, and price, on the choice of restaurant wines. The data were collected at a high-end restaurant over 19 weeks in 1998. The restaurant offers a wide selection of wines detailed in an extensive menu that describes brand, vintage, origin, price, and sensory qualities. Sensory information includes aroma, flavors, and "mouth feel" (e.g., dry, tannic, smooth, creamy), with typical descriptors for aroma and taste, including fruits (berry, lemon), flowers (apple, rose), and other food associations (herbal, honey).

Wine prices are generally based on expert quality assessments, with adjustments for varietal, origin, and market factors. Only rarely can the price of a wine be said to reflect consumer valuation of its quality. Many wines can appear to be greatly over- or underpriced due to the great variety of wines available, supply variation, and a lack of good information on quality. This study was conducted to explore whether wine demand, measured by quantity sold in each of the 19 weeks, was possibly driven by the objective and sensory descriptors on wine menus rather than simply price.

In this situation, the predictors $(X_1, X_2, \ldots)$ are the wine features, and the response outcome (Y) is a count of the number of bottles of each wine sold in each of the 19 weeks of the study; that is, the only possible values are nonnegative integers $(0, 1, 2, \ldots)$. It is inappropriate to use this as the response variable in a multiple linear regression model, since the resulting residuals would probably not satisfy the four usual linear regression assumptions (zero mean, constant variance, normality, independence). For example, residual plots would tend to look "clumpy" because the response variable is discrete but the predictions are continuous. Instead, for count response outcomes like this, we can use Poisson regression to see how the outcome might depend on the values of the predictors.

Poisson regression models the natural logarithm of the expected value of response Y as

$$\log_e(\mathrm{E}(Y)) = b_0 + b_1 X_1 + b_2 X_2 + \cdots + b_k X_k.$$

The presence of the natural logarithm transformation here ensures that the expected value of the response Y is positive. If the value of this equation is high, then we expect a higher count for Y. If the value of this equation is low, then we expect a lower count for Y.

As with logistic regression, we can use statistical software to estimate the regression parameters $(b_0, b_1,$ etc.) for Poisson regression using maximum likelihood. Recall that maximum likelihood estimates are the values of the regression parameters that make the probability of observing the pattern of responses and predictors in the dataset as large as possible. This probability is calculated using formulas (not provided here) based on the *Poisson distribution* (a probability distribution for nonnegative integers, i.e., count data). Further details are available in Agresti (2002) and Cameron and Trivedi (1998).

The full dataset for the original wine application consists of 47 red and 29 white wines. For the sake of illustration, the dataset considered here, available in the **WINEWHITE** data file, consists of just the white wines, with response variable, $Y =$ quantity sold (in bottles), quantitative predictor variable, $X =$ bottle price (in \$), and the following indicator predictor variables:

D_1 = indicator for having the lowest price in a menu category
D_2 = indicator for being available by the glass
D_3 = indicator for Oregon Chardonnay (ref: California Chardonnay)
D_4 = indicator for Oregon Pinot Gris (ref: California Chardonnay)
D_5 = indicator for California other (ref: California Chardonnay)
D_6 = indicator for Northwest other (ref: California Chardonnay)
D_7 = indicator for French (ref: California Chardonnay)
D_8 = indicator for body
D_9 = indicator for finish
D_{10} = indicator for oak
D_{11} = indicator for rich
D_{12} = indicator for spice
D_{13} = indicator for buttery
D_{14} = indicator for creamy
D_{15} = indicator for dry
D_{16} = indicator for honey
D_{17} = indicator for melon
D_{18} = indicator for citrus
D_{19} = indicator for other tree fruit
D_{20} = indicator for tropical fruit

We first fit a Poisson regression model to these data using all the available predictors. Some high individual p-values in the resulting software output (not shown) indicate predictors that may be redundant (given the presence of the others). So, we next conduct an analysis-of-deviance test to see if we can remove the following eight indicator variables: D_9, D_{11}, D_{12}, D_{13}, D_{16}, D_{17}, D_{19}, and D_{20}. The first (complete) model (with all possible predictors) has a residual deviance of 492.17, while the second (reduced) model (without these eight indicator variables) has a residual deviance of 495.23. We compare the change in deviance (3.05) with a χ^2 distribution with 8 degrees of freedom to find a p-value of 0.93. Thus we cannot reject a null hypothesis of zeros for the eight corresponding regression parameters, and it seems safe to remove those indicator variables from the model.

The reduced model has a reasonable fit to the data. Results are as follows:

Parameters[a]

Model		Estimate	Std. Error	Z-stat	Pr($> |Z|$)
2	(Intercept)	2.204	0.431	5.117	0.000
	X	−0.084	0.015	−5.721	0.000
	D_1	−0.912	0.258	−3.530	0.000
	D_2	1.755	0.185	9.474	0.000
	D_3	−1.309	0.139	−9.442	0.000
	D_4	−0.844	0.126	−6.720	0.000
	D_5	−0.956	0.135	−7.105	0.000
	D_6	−0.910	0.149	−6.087	0.000
	D_7	−1.791	0.440	−4.069	0.000
	D_8	−0.767	0.242	−3.175	0.002
	D_{10}	1.035	0.344	3.007	0.003
	D_{14}	1.105	0.250	4.412	0.000
	D_{15}	0.648	0.195	3.321	0.001
	D_{18}	−0.858	0.272	−3.152	0.002

[a] Response variable: Y.

How can we interpret the regression parameter estimates in a Poisson regression analysis? Consider the estimate $\hat{b}_1 = -0.084$ for $X =$ price. This means that if price increases by 1 dollar (holding all else constant), then the natural logarithm of the expected (average) quantity sold will decrease by 0.084, and the expected quantity sold will be *multiplied* by $\exp(-0.084) = 0.920$ (i.e., demand decreases to 92% of what it was before the price increase). For another example, if a wine has the lowest price in its category on the menu ($D_1 = 1$), then the expected quantity sold will be multiplied by $\exp(-0.912) = 0.402$ (relative to an identical wine that does not have the lowest price in that category).

Looking at the results overall, none of the varietals are as popular as California Chardonnay (the reference varietal), since the parameter estimates for D_3 through D_7 are all negative. The magnitudes of these estimates correspond to the popularity of the other varietals: first Oregon Pinot Gris, then other Northwest, then other California, then Oregon Chardonnay, and finally French whites. There appears to be a recognition of varietals from U.S. regions where those varietals are known for their quality. Favoritism for local wines (the restaurant is located in Oregon) does not appear to extend to varieties in which no local prominence has been achieved (such as Oregon Chardonnay). Rather the local wines of positive reputation, for example, Oregon Pinot Gris, appear to receive favor.

Wines that are available by the glass (D_2) see increased demand beyond that which could be expected from their relative price and origin–varietal information. The price effect (X) is negative (as economic theory expects)—higher prices reduce demand, all else being equal. However, the low price variable effect (D_1) is also negative. This result may be because buying more expensive wines can give the buyer more satisfaction, due to appearing more selective or magnanimous, or at least to avoid giving the opposite impression. The low-price variable could also be considered in the strategic sense for winemakers looking for exposure through restaurant sales. It appears that being the lowest-priced wine in your category is a disadvantage.

In terms of the sensory descriptors, "creamy" (D_{14}) and "oak" (D_{10}) are strongly positive, with "dry" (D_{15}) somewhat positive. By contrast "body" (D_8) and "citrus" (D_{18}) have negative effects. The remaining characteristics—finish, rich, spice, buttery, honey, melon, tree fruit, and tropical fruit—were removed from the final model due to their nonsignificance (and so might be considered mostly neutral). Full interpretation of these results is complicated by whether consumers fully understand the descriptors and their typically strong relation to certain varietals. For example, for white wines, oak is primarily associated with Chardonnay developed in wooden barrels

Although the fit of the Poisson model to these data is reasonable, Durham et al. (2004) found that this model underpredicts the "zero" counts in the dataset (weeks when no one bought a particular wine). They went on to demonstrate the superior fit of some generalizations of the standard Poisson model, in particular, a zero inflated Poisson model (Lambert, 1992) and a negative binomial model (see Agresti, 2002). Overall conclusions were broadly similar to those presented here, however.

7.2 DISCRETE CHOICE MODELS

Pardoe and Simonton (2008) analyzed data on award-winning movies from 1928 to 2006 to determine the extent to which the winner of the top prize in the movie business, the Academy Award (or Oscar), is predictable. Each year, hundreds of millions of people worldwide watch the TV broadcast of the Oscars ceremony, at which the Academy of Motion Picture Arts and Sciences (AMPAS) honors film-making from the previous year. Almost 6,000 members of AMPAS vote for the nominees and final winners of the Oscars, in a wide range of categories for directing, acting, writing, editing, and so on. Oscars have been presented for outstanding achievement in film every year since 1928 and are generally recognized to be the premier awards of their kind.

The research cited focused on the goal of predicting the eventual winners of Oscars from those nominated each year. In terms of data, since the goal was to predict the eventual winner from a list of nominees, any information on the nominees that is available before the announcement of the winner is potentially useful, including other Oscar category nominations, previous nominations and wins, and other (earlier) movie awards—these comprise the predictors $(X_1, X_2, \ldots)$.

There is one winner selected in each category each year from a discrete set of nominees (usually, five), so the response outcome (Y) records which of the nominees is chosen (wins). We can use a discrete choice model to analyze response outcomes such as this, to see how the outcome might depend on the values of the predictor variables. While there are different types of discrete choice model, many derive from the basic *multinomial logit model* of McFadden (1974), which is the model we consider here.

In particular, the multinomial logit model considers the probability of each object in the choice set (nominees in this application) being chosen. Let index j keep track of the objects in the choice set, C. Then the probability of choosing the jth object, $\Pr(Y = j)$, depends on the values of the predictor variables by way of the following equation:

$$\Pr(Y = j) = \frac{\exp(b_1 X_{1j} + b_2 X_{2j} + \cdots + b_k X_{kj})}{\sum_{h \in C} \exp(b_1 X_{1h} + b_2 X_{2h} + \cdots + b_k X_{kh})}. \tag{7.2}$$

The index h in the denominator of this equation runs through each of the objects in the choice set, C. If the value of this equation is high (close to 1), then object j is more likely to be chosen. If the value of this equation is low (close to 0), then object j is less likely to be chosen.

Although this model appears to be quite similar to the logistic regression model considered in Section 7.1.1, here the predictor variables can take different values for different objects in the choice set. This contrasts with logistic regression where each of the two possible response outcomes is associated with the same experimental unit. The multinomial logit model is also general enough to permit the choice set to vary across different choice tasks, which in this case are each of the award categories in each year.

As with logistic and Poisson regression, we can use statistical software to estimate the regression parameters (b_0, b_1, etc.) for the multinomial logit model using maximum likelihood. Recall that maximum likelihood estimates are the values of the regression parameters that make the probability of observing the pattern of responses and predictors in the dataset as large as possible. This probability is calculated using formulas (not provided here) based on the multinomial distribution (a generalization of the binomial distribution for categorical data with more than two outcomes). Further details on the mechanics of the multinomial logit model are available in Ben-Akiva and Lerman (1985), Hensher et al. (2005), Louviere et al. (2000), and Train (2009).

The full dataset for the original Oscars application consisted of data from 1928 to 2006 for the four major awards: best picture, best director, best actor in a leading role, and best actress in a leading role. All data were obtained from a reliable Internet source, namely

"The Internet Movie Database" (www.imdb.com). For the sake of illustration, the dataset considered here, available in the **OSCARS** data file, consists of just the best picture awards, with the following response and predictor variables:

Win = 1 if winning nominee, 0 if losing nominee
Nom = total number of Oscar nominations for the movie
D_{Od} = indicator for also receiving a best director Oscar nomination
D_{Gd} = indicator for winning the best picture (drama) Golden Globe
D_{Gm} = indicator for winning the best picture (musical/comedy) Golden Globe
D_{Pg} = indicator for winning the best picture Producers Guild award (Directors Guild pre-1989)

The Hollywood Foreign Press Association (a group of Southern California-based international journalists) has awarded its Golden Globes every year since 1944 to honor achievements in film during the previous calendar year. Since Oscars are presented some time after Golden Globes (up to two months later), winning a Golden Globe often precedes winning an Oscar. The Directors Guild of America has been awarding its honors for best motion picture director since 1949 (with all but two early awards made before the announcement of the Oscars). Since 1989, the Producers Guild of America has been awarding its honors to the year's most distinguished producing effort (with all but the first awarded before the announcement of the Oscars).

A multinomial logit model using these predictors produces a good fit to the data. Results are as follows:

Parameters[a]

Model		Estimate	Std. Error	Chi-sq	$\Pr(>\chi^2)$
1	*Nom*	0.285	0.085	11.286	0.001
	D_{Od}	2.066	0.654	9.988	0.002
	D_{Gd}	0.709	0.388	3.340	0.068
	D_{Gm}	0.853	0.590	2.087	0.149
	D_{Pg}	1.798	0.367	24.001	0.000

[a] Response variable: *Win*.

This output is similar to that for linear regression, the only difference being that the test statistics are calculated by dividing the parameter estimates by their standard errors and then squaring the resulting ratio. These "χ^2 statistics" then use the χ^2 distribution to determine their significance. Using this criterion, there is perhaps a slight question mark over including D_{Gm} in the model, but we conclude in this case that there is little harm in leaving it in.

To interpret the parameter estimates, consider the following equation derived from the probability equation (7.2):

$$\log \left(\frac{\Pr(Y=a)}{\Pr(Y=b)} \right) = b_1(X_{1a} - X_{1b}) + b_2(X_{2a} - X_{2b}) + \cdots + b_k(X_{ka} - X_{kb}).$$

Here $\Pr(Y=a)/\Pr(Y=b)$ is the odds of choosing nominee a over nominee b. Conditional on the choice being a or b, a predictor variable's effect depends on the difference in the variable's values for those choices. If the values are the same, then the variable has no effect on the choice between a and b.

Consider the estimate $\hat{b}_1 = 0.285$ for *Nom* = total number of Oscar nominations for the movie. This means that if nominee a has one more nomination than nominee b (holding all else constant), then the natural logarithm of the odds (of choosing a over b) will increase by 0.285, and the odds will be *multiplied* by $\exp(0.285) = 1.33$. For another example, if nominee a has a best director Oscar nomination but nominee b does not (holding all else constant), then the odds (of choosing a over b) will be multiplied by $\exp(2.066) = 7.89$. The other odds ratios are $\exp(0.709) = 2.03$ for having won the best picture (drama) Golden Globe, $\exp(0.853) = 2.35$ for having won the best picture (musical or comedy) Golden Globe, and $\exp(1.798) = 6.04$ for having won the Producers Guild award (or Directors Guild award pre-1989).

Pardoe and Simonton (2008) also looked at how patterns of winning versus losing nominees have changed over time. For example, the importance of receiving a best director nomination (for best picture nominees) has tended to increase over time; the Golden Globes have remained useful predictors of future Oscar success since their inception, with musicals and comedies appearing to hold an advantage over dramas in the 1960s and 1970s, dramas dominating in the 1980s and 1990s, and both awards roughly equivalent in recent years; and Guild awards have increasingly enabled quite accurate prediction of best picture winners. Overall, multinomial logit modeling of the four major awards (best picture, best director, best actor in a leading role, and best actress in a leading role) over the 30 years 1977–2006 has been able to correctly predict 95 winners out of the 120 major awards, or 79%. Prediction accuracy has improved further since Pardoe and Simonton published their research, with correct prediction of 98 out of 120 awards (82%) over the 30 years 1982–2011.

The analysis in Pardoe and Simonton (2008) actually used a different method than maximum likelihood to estimate the multinomial model parameters, a method known as Bayesian inference. In this particular application, overall conclusions are broadly similar whichever method is used. In Section 7.4 we discuss how the Bayesian approach to statistical inference contrasts with the "frequentist" approach that lies behind maximum likelihood.

7.3 MULTILEVEL MODELS

Price et al. (1996) analyzed levels of radon gas in homes clustered within counties (see also Lin et al., 1999, and Price and Gelman, 2006). Radon is a carcinogen—a naturally occurring radioactive gas whose decay products are also radioactive—known to cause lung cancer in high concentration and estimated to cause several thousand lung cancer deaths per year in the United States. The distribution of radon levels in U.S. homes varies greatly, with some homes having dangerously high concentrations. In order to identify the areas with high radon exposures, the U.S. Environmental Protection Agency coordinated radon measurements in each of the 50 states.

For the sake of illustration, the dataset considered here, available in the **MNRADON** data file, contains radon measurements for 919 homes in the 85 counties of Minnesota. Radon comes from underground and can enter more easily when a home is built into the ground. So, one potentially useful predictor of the radon level in a home is an indicator for whether the measurement was taken in a basement. We also have an important county-level predictor, a measurement of the average soil uranium content in each county.

Consider analyzing the radon measurements in a simple linear regression model with the basement indicator as the predictor (i.e., ignoring any systematic variation between counties). This seems oversimplistic, particularly as we're specifically interested in identifying counties with high-radon homes (which would be impossible to do using such a model).

One possible way to incorporate county-level information in the analysis is to include indicator variables for each of the counties in a multiple linear regression model. In this case, such a model would have 84 county indicators (the 85th county would be the reference level), together with the basement indicator. This seems overcomplicated, since this model essentially fits a separate regression line to each county, some of which have very small sample sizes. For example, one county has a sample size of just two—the resulting estimate for that county is going to be hard to trust.

Multilevel models, which have a long history that predates this application, provide a compromise between these two extremes of ignoring county variation and modeling each of the counties separately. Such models allow us to fit a regression to the individual measurements while accounting for systematic unexplained variation among the 85 counties. In particular, one possible multilevel model for this dataset starts with the following linear regression equation:

$$E(Y) = b_{0j} + b_1 X,$$

where Y is the logarithm of the radon measurement in a particular home in a particular county, X is the indicator for whether the measurement was in a basement, and j indexes the counties. It is the fact that the regression parameter b_{0j} has a j in its subscript that allows this model to account for variation among the 85 counties.

Consider estimating the regression parameters in this equation. If the 85 regression parameters are constrained to be equal (to a single value b_0), then this equation becomes the simple linear regression model discussed above. On the other hand, if we estimate these 85 regression parameters as a multiple linear regression model using least squares, then this equation becomes the 85 separate models also discussed above. To provide a compromise between these two extremes, we use a second-level model for the b_{0j} regression parameters themselves:

$$b_{0j} \sim \text{Normal}(b_0, s_b^2).$$

In essence, the b_{0j} parameters can vary from county to county by way of a normal distribution with a mean of b_0 and a standard deviation of s_b. If the standard deviation s_b was constrained to be 0, then we would be back in the simple linear regression setting, where each b_{0j} is fixed at b_0. If the standard deviation s_b was constrained to be "infinite," then we would be back in the second setting above, where each b_{0j} is estimated separately. However, multilevel modeling allows the standard deviation s_b to be estimated from the data, thus providing the compromise between these two extremes.

In practice, we can use statistical software to estimate the b_{0j} parameters, and it turns out for the multilevel model that these estimates are between the simple linear regression estimate for b_0 and the estimates obtained from fitting 85 separate models. For counties with a small sample size (i.e., few homes), the multilevel estimate of b_{0j} tends to be close to the simple linear regression estimate for b_0, since there is little information for that county and the model pays more attention to the information available for all the other counties combined. Conversely, for counties with a large sample size, the multilevel estimate of b_{0j} tends to be close to the separate models estimate, since there is a large amount of information for that county and the model pays less attention to the information available for all the other counties combined.

We can extend this basic multilevel model in a number of different ways:

- We can allow other regression parameters in the first-level model to vary by county; for example, the basement indicator parameter, b_1, in our application becomes b_{1j}. This means that there are now two parts to the second-level model—one for b_{0j}, the other for b_{1j}.

- We can introduce county-level predictors to the second-level model, for example, $W =$ the natural logarithm of the county-level uranium measurement.

- We can add further levels to the model if appropriate. For example, if we were to analyze multiple radon measurements within homes, there would be three levels: measurements, homes, and counties.

To illustrate, consider the first two of these extensions. The first-level model is

$$\mathrm{E}(Y) = b_{0j} + b_{1j}X,$$

while the second-level model is

$$\begin{pmatrix} b_{0j} \\ b_{1j} \end{pmatrix} \sim \mathrm{Normal} \left[\begin{pmatrix} c_{00} + c_{01}W_j \\ c_{10} + c_{11}W_j \end{pmatrix}, \begin{pmatrix} s_{b_0}^2 & r_{(b_0,b_1)}s_{b_0}s_{b_1} \\ r_{(b_0,b_1)}s_{b_0}s_{b_1} & s_{b_1}^2 \end{pmatrix} \right].$$

Here, the second-level model includes the county-level predictor, W, and c_{00}, c_{01}, c_{10}, and c_{11} are the second-level model regression parameters. Also, s_{b_0} is the standard deviation of b_{0j}, s_{b_1} is the standard deviation of b_{01}, and $r_{(b_0,b_1)}$ is the correlation between b_{0j} and b_{1j}.

We can use statistical software to estimate all these parameters. It is possible to use maximum likelihood methods to accomplish this, although Bayesian methods—see Section 7.4—have perhaps become more common for multilevel models in recent times. Gelman and Pardoe (2006) used a Bayesian approach to estimate this model. Their analysis confirms that the presence of a basement is associated with an increased level of radon in a home, and that homes in counties with higher levels of uranium are associated with higher levels of radon. This finding is no great surprise, however—the real point of the analysis is getting better estimates for individual counties, especially those with small samples. In particular, the inclusion of the county-level uranium predictor in the model improves the precision of the results, with uranium differences between counties helping to explain not only systematic county differences in radon levels, but also basement effects on those radon levels. Nevertheless, much of the overall variation in radon levels across homes throughout Minnesota remains unexplained by this model, and there are important factors other than basements and underlying uranium levels that affect the radon level. Further information—including a tool to determine whether you should test your home for radon (based on factors that were determined to be important through regression modeling)—is available at www.stat.columbia.edu/~radon/.

The preceding discussion provides just a glimpse of the possibilities offered by multilevel modeling. There are a number of excellent books that describe this area of regression modeling in more depth: for example, Gelman and Hill (2006), Hox (2010), Kreft and De Leeuw (1998), Raudenbush and Bryk (2002), and Snijders and Bosker (2011). When looking for further information, note that multilevel models are also sometimes known as hierarchical models, variance components models, random effects models, or mixed effects models (see also Pinheiro and Bates, 2000).

7.4 BAYESIAN MODELING

In this section we briefly introduce the Bayesian approach to statistical inference (in contrast to the frequentist approach used in the rest of the book). While frequentist methods still dominate in many statistical applications, Bayesian methods are becoming more widely used, particularly in more complex modeling situations. The purpose of this section is to provide some discussion of how Bayesian inference differs from frequentist inference, and why this alternative approach might be preferred in some circumstances.

7.4.1 Frequentist inference

For the purpose of illustration, consider a random sample of n data values, represented by $Y_1, Y_2, \ldots, Y_n$, that comes from a normal population distribution with a mean, $E(Y)$, and a standard deviation, $SD(Y)$. Suppose that we know the value of $SD(Y)$ (let's say that it is S) and we wish to estimate $E(Y)$ based on our sample. Although assuming that we know $SD(Y)$ is uncommon, this will simplify the following discussion without compromising the overall ideas. The normal version of the central limit theorem in Section 1.4.1 on page 12 ensures that the sampling distribution of sample means, M_Y, is a normal distribution with mean $E(Y)$ and standard deviation $S/\sqrt{n}$. Thus, the observed sample mean, m_Y, is an unbiased estimate of $E(Y)$, and the standard deviation of this estimate $(S/\sqrt{n})$ happens to be the smallest out of all possible unbiased estimators of $E(Y)$.

The resulting confidence interval for $E(Y)$ is $m_Y \pm \text{Z-percentile}(S/\sqrt{n})$. (This differs from the confidence interval we saw in Chapter 1 because there we used the sample standard deviation, s_Y, in place of S, and the t-distribution in place of the normal distribution.) For example, using the table on page 8, a 95% confidence interval for $E(Y)$ is $m_Y \pm 1.96(S/\sqrt{n})$. The (frequentist) interpretation of this interval is that if we were to take a large number of random samples of size n from our population and calculate a 95% confidence interval for each, then 95% of those confidence intervals would contain the (unknown) $E(Y)$. This is *not* quite the same as saying there is a 95% probability that $E(Y)$ is in the particular interval calculated for our single sample.

We can also conduct a hypothesis test for $E(Y)$ in this context, let's say an upper-tail test for the sake of argument. To do so, we first calculate the value of a test statistic, in this case Z-statistic $= (m_Y - E(Y))/(S/\sqrt{n})$, where $E(Y)$ is the value in the null hypothesis. (Again, this differs from the t-statistic we saw in Chapter 1 because there we used the sample standard deviation, s_Y, in place of S.) The corresponding p-value is then the probability that in repeated samples we would obtain a test statistic value as large as (or larger than) the calculated value, under the assumption that the null hypothesis is true. A small p-value suggests that the data we have appear unusual under the null hypothesis, which in turn suggests that the alternative hypothesis is more plausible than the null (either that, or the null hypothesis was true and we happened to get an unlucky sample). This is *not* quite the same as saying that the p-value measures the probability that the null hypothesis is true.

7.4.2 Bayesian inference

In general, the Bayesian approach to inference starts with a probability distribution that describes our knowledge about population parameters [like $E(Y)$] before collecting any data, and then updates this distribution using observed data. A Bayesian approach to the example in the preceding section starts off in much the same way. We have a random sample of n data values, represented by $Y_1, Y_2, \ldots, Y_n$, that comes from a normal population distribution with a mean of $E(Y)$ and a standard deviation of $SD(Y)$. Suppose that we know $SD(Y) = S$ and we wish to estimate $E(Y)$ based on our sample. To see how the Bayesian approach tackles this problem we need to introduce some new terminology.

- We'll call the normal distribution assumed for the data $f(Y_1, Y_2, \ldots, Y_n \mid E(Y))$, which in words is the "conditional distribution of the data given $E(Y)$" (also sometimes called the "likelihood function").

- We'll call our knowledge about $E(Y)$ before collecting the data $f(E(Y))$, which in words is the "*prior* distribution of $E(Y)$."

The notation f here represents whichever probability distribution we are using in each case. For the data, this is the normal distribution, while for the prior, a convenient choice is another normal distribution, say, with mean M_0 and variance S_0^2 (we return to this issue later).

To update our knowledge about $E(Y)$ using the observed data, $Y_1, Y_2, \ldots, Y_n$, we use a probability calculation known as Bayes' rule to calculate the "*posterior* distribution of $E(Y)$":

$$f(E(Y) \mid Y_1, Y_2, \ldots, Y_n) = \frac{f(Y_1, Y_2, \ldots, Y_n \mid E(Y)) \, f(E(Y))}{f(Y_1, Y_2, \ldots, Y_n)},$$

where $f(Y_1, Y_2, \ldots, Y_n)$ is the marginal distribution of the data and can be derived from $f(Y_1, Y_2, \ldots, Y_n \mid E(Y))$ and $f(E(Y))$. For our example, the posterior distribution of $E(Y)$ resulting from this calculation is another normal distribution, with the following mean and variance:

$$\text{mean} = M_n = \frac{\frac{n}{S^2} m_Y + \frac{1}{S_0^2} M_0}{\frac{n}{S^2} + \frac{1}{S_0^2}}, \quad \text{variance} = S_n^2 = \frac{1}{\frac{n}{S^2} + \frac{1}{S_0^2}}. \quad (7.3)$$

In other words, the posterior mean for $E(Y)$ is a weighted average of the sample mean, m_Y, and the prior mean, M_0. If our prior information is very vague (large prior variance, S_0^2), then the posterior mean is determined primarily by the data; that is, the posterior mean is very close to the sample mean, m_Y. If, on the other hand, our prior information is very precise (small prior variance, S_0^2), then the posterior mean is heavily influenced by the prior mean, M_0. This provides some insight into selection of the prior distribution. It should have a large variance if we have little information about likely values for $E(Y)$ before collecting the data. Conversely, it should have a small variance if we are more certain about likely values for $E(Y)$ before collecting the data (and the prior mean should also reflect those likely values).

To connect these Bayesian results back to the preceding section, consider a situation in which our prior information is extremely vague, so that the prior variance, S_0^2, is extremely large. Alternatively, consider a situation in which the sample size, n, is extremely large. Either way, the posterior distribution of $E(Y)$ is then essentially normal with mean m_Y and variance S^2/n. This is somewhat similar to the (frequentist) sampling distribution of sample means, M_Y, which is normal with mean $E(Y)$ and variance S^2/n. The distinction is that the roles of M_Y and $E(Y)$ are different in the two approaches. Next we see how this affects the interpretation of confidence intervals and hypothesis testing using the Bayesian approach.

The posterior distribution enables us to calculate intervals, called posterior intervals or credible sets, that contain $E(Y)$ with any particular probability. For example, a 95% posterior interval for $E(Y)$ is $M_n \pm 1.96 \times S_n$, where M_n and S_n^2 are the posterior mean and variance defined in (7.3). With the Bayesian approach, it is permissible to say that for any given dataset, $E(Y)$ lies in the calculated interval with 95% probability. The posterior distribution also enables us to calculate the posterior probability that $E(Y)$ is larger or smaller than some hypothesized value. Decisions regarding the plausibility of different hypotheses are then reasonably straightforward. These probability calculations contrast with the definition of a p-value used in frequentist hypothesis testing.

As alluded to above, a difficulty with applying Bayesian methods in practice is specifying a prior distribution before you start. In large samples, the question is often moot because the data part of the posterior distribution calculation can outweigh the prior part enough that the prior plays little role in the final result. Furthermore, in practice, Bayesian analysts can often select sufficiently vague prior distributions that the final results are reasonably robust to the exact specification of the prior. As discussed earlier, in simple problems (such as estimating $E(Y)$ in a univariate normal population), a Bayesian approach with either a large sample size or a sufficiently vague prior leads to results similar to those using the frequentist approach.

Nevertheless, in more complex situations, the Bayesian approach can offer some distinct advantages over the frequentist approach. In some circumstances, useful prior information is available, and the Bayesian approach provides a rigorous method for updating that information in the light of new data. In other circumstances, the nature of the model to be used lends itself naturally to a Bayesian approach. A good example of this is the multilevel modeling described in Section 7.3. There, the second-level models (for the county parameters) are analogous to Bayesian prior distributions. With more sophisticated modeling like this, the calculation of posterior distributions is usually not as straightforward as it was for the univariate normal example discussed above. Much Bayesian modeling requires computer-intensive simulation techniques, although software to implement this is becoming more widespread and easy to use.

The preceding discussion is based on Stern (1998); Bolstad (2007) and Lee (2004) are good additional introductions to Bayesian inference. Some excellent textbooks on Bayesian modeling include Carlin and Louis (2008), Gelman et al. (2003), and Rossi et al. (2006).

APPENDIX A

COMPUTER SOFTWARE HELP

A recommended way to read the book is to follow along at a computer using statistical software and working with the example datasets (available from the book website at www.iainpardoe.com/arm2e/). Every method that appears in the homework problems at the end of each chapter is first illustrated in the text of the chapter. This allows you to try out a method using your particular statistical software and the book shows you what the results should look like. Then, you know that if you apply the same method for a corresponding homework problem, chances are that the results will be correct (unless you've made some other mistake like applying the wrong method for that problem). To use this approach, you need to get your statistical software to do what you want, which can be slow, painful, and frustrating for many people. To help in this process, there are several software information documents available from the book website. These contain detailed instructions on carrying out the techniques discussed in this book with the major statistical software packages. The software packages included at the time of publication include the following (more may be added in the future):

- Data Desk (a commercial statistical software program for visual data analysis and exploration—see www.datadesk.com)

- EViews (a commercial statistical package used mainly for econometric analysis—see www.eviews.com)

- Minitab (a commercial statistical software package with an easy-to-use graphical user interface—see www.minitab.com)

- R (a free software environment for statistical computing and graphics with a powerful programming interface that is popular among statistics researchers—see `www.r-project.org`)

- S-PLUS (commercial software, which has many features in common with R—see `spotfire.tibco.com/products/s-plus/statistical-analysis-software.aspx`)

- SAS (originally "Statistical Analysis Software": this is probably the leading statistical software package in the world, although its powerful programming interface can be difficult for first-time users to come to terms with—see `www.sas.com`)

- SAS "Analyst Application" (which provides a graphical user interface similar to those in SPSS and Minitab)

- SAS JMP (user-friendly statistical software package with strong graphics—see `www.jmp.com`)

- SPSS (originally "Statistical Package for the Social Sciences": this is a commercial statistical software package with an easy-to-use graphical user interface that is widely used in many fields—see `http://www-01.ibm.com/software/analytics/spss/`)

- Stata (an integrated statistical package with both a point-and-click interface and a powerful programming interface—see `www.stata.com`)

- Statistica (a statistics and analytics software package with a graphical user interface—see `www.statsoft.com`)

Extensive guidance on the use of R and S-PLUS for regression modeling can be found in Fox and Weisberg (2011) and Venables and Ripley (2002). Another package that can be recommended for regression modeling is Arc, free regression software available at `www.stat.umn.edu/arc/`—see also the regression textbook by Cook and Weisberg (1999).

The book website also includes instructions for using Microsoft Excel to carry out *some* of the techniques discussed in this book—further information is available at `office.microsoft.com/excel/`. Although Excel is a commercial spreadsheet package, not dedicated statistical software, it can perform some limited statistical analysis. It is preferable in general to use software such as SPSS, Minitab, SAS, R, or S-PLUS for any serious statistical analysis, but Excel may be able to get you by in a pinch if this is all that is available. There are a number of add-on modules available for Excel that can improve its functionality to something approaching that of a dedicated statistical software package: for example, StatTools (available at `www.palisade.com/stattools/`) and Lumenaut Statistics Package (available at `www.lumenaut.com/statistics.htm`).

PROBLEMS

- "Computer help" refers to numbered items in instructions for various statistical software packages available from the book website. For example, computer help #1–4 describe how to change default options, open a data file, recall previous dialog boxes or commands, and edit results.
- There are *brief* answers to the even-numbered problems in Appendix E.

A.1   The **CARS1** data file (obtained from www.fueleconomy.gov) contains information for 351 new U.S. passenger cars for the 2011 model year. These data come from a larger dataset, which is analyzed more fully in a case study in Section 6.2.

  a) Download and open the **CARS1** data file [computer help #2]. Then make a histogram of highway miles per gallon (variable *Hmpg*) [computer help #14] and copy/paste it into a text document (follow this process for all such questions).

  b) Briefly describe what the histogram says about the distribution of highway miles per gallon for this sample of vehicles.

  c) Compute the mean for *Hmpg* [computer help #10].

  d) Reexpress each vehicle's highway miles per gallon value in "gallons per hundred miles." In other words, create a new variable called *Hgphm* equal to $100/Hmpg$ [computer help #6]. Then compute the mean for the *Hgphm* variable [computer help #10].

  e) Reexpress your answer for part (d) in miles per gallon. In other words, calculate $100/m_{Hgphm}$, where m_{Hgphm} is the sample mean of *Hgphm*.

  f) Are your answers to parts (c) and (e) the same or different? Explain.

A.2   Consider interior passenger and cargo volume in hundreds of cubic feet (variable *Vol*).

  a) Make a histogram of *Vol* [computer help #14].

  b) Briefly describe what the histogram says about the distribution of interior passenger and cargo volumes for this sample of vehicles.

  c) Compute the mean and the median for *Vol* [computer help #10].

  d) Which of the two statistics in part (c) best summarizes the "center" of the distribution of interior passenger and cargo volumes?

  e) Compute the quartiles for *Vol* [computer help #10].

  f) Within which values do the middle 50% of the interior passenger and cargo volumes fall?

A.3   Consider mean city miles per gallon (variable *Cmpg*) for different combinations of vehicle class (variable *Class*) and drive system (variable *Drive*). The codes for *Class* are MC: minicompact and two-seater car; SC: subcompact car; CO: compact car; MI: midsize car; LA: large car; SW: station wagon; SU: sport utility vehicle. The codes for *Drive* are F: front-wheel drive; R: rear-wheel drive; A: all-wheel drive; 4: four-wheel drive.

  a) Find the mean *Cmpg* for each vehicle class/drive combination (e.g., for front-wheel drive minicompact and two-seater cars, front-wheel drive subcompact cars, . . . , four-wheel drive sport utility vehicles). Use grouping variables *Class* and *Drive* [computer help #12].

  b) Make a clustered bar chart of mean *Cmpg* by *Class* (category axis) and *Drive* (clusters) [computer help #20].

c) Are differences clearer in the table in part (a) or the bar chart in part (b)? Why?

d) Compare fuel efficiency for front-wheel, rear-wheel, all-wheel, and four-wheel drive vehicles across different vehicle classes in a couple of sentences.

e) Suggest some possible reasons for the differences in fuel efficiency for different vehicle classes and drive systems. Be specific.

A.4 Consider the frequencies (counts) of vehicles in different combinations of class (variable *Class*) and drive system (variable *Drive*).

a) Make a cross-tabulation table in which *Class* is the row variable and *Drive* is the column variable, and calculate the row and column percentages [computer help #11].

b) How many front-wheel drive midsize cars are there?

c) What percentage of midsize cars are front-wheel drive?

d) What percentage of front-wheel drive vehicles are midsize cars?

e) Write a short paragraph summarizing all the results in the table.

A.5 Consider city miles per gallon (variable *Cmpg*) for different vehicle classes (variable *Class*).

a) Create boxplots of *Cmpg* for each vehicle class [computer help #21].

b) Write a short paragraph summarizing what the boxplots show.

Hint: Statistical software packages differ in the way they construct boxplots, so use the "Help" facility in your software to find out exactly what the boxplots are displaying [computer help #5].

A.6 Consider the associations between city miles per gallon (variable *Cmpg*), engine size in liters (variable *Eng*), number of cylinders (variable *Cyl*), and interior passenger and cargo volume in hundreds of cubic feet (variable *Vol*).

a) Make a scatterplot matrix of *Cmpg*, *Eng*, *Cyl*, and *Vol* [computer help #16].

b) Comment on the association between each pair of variables in the scatterplot matrix.

Hint: One key to understanding a particular graph in a scatterplot matrix is to look left or right to see the label for the variable plotted on the vertical axis and to look up or down to see the label for the variable plotted on the horizontal axis. In this respect, scatterplot matrices are similar to scatterplots where the vertical axis label is typically to the left of the graph and the horizontal axis label is typically below the graph.

A.7 Consider the association between highway miles per gallon (variable *Hmpg*) and city miles per gallon (variable *Cmpg*).

a) Make a scatterplot with *Hmpg* on the vertical (Y) axis and *Cmpg* on the horizontal (X) axis [computer help #15].

b) Briefly describe the association between the two variables.

c) Redraw the plot so that each axis has the same range and add a "Y=X" line to the plot (by hand if you cannot get the computer to do it).

d) Describe anything of interest that the "Y=X" line helps you see. Be specific.

APPENDIX B

CRITICAL VALUES FOR t-DISTRIBUTIONS

Table B.1 contains critical values or percentiles for t-distributions; a description of how to use the table precedes it. Figure B.1 illustrates how to use the table to find bounds for an upper-tail p-value. Bounds for a lower-tail p-value involve a similar procedure for the negative (left-hand) side of the density curve. To find bounds for a two-tail p-value, multiply each bound for the corresponding upper-tail p-value by 2; for example, the two-tail p-value for the situation in Figure B.1 lies between 0.05 and 0.10.

Use Table B.1 and Figure B.2 to find critical values or percentiles for t-distributions; each row of the table corresponds to a t-distribution with the degrees of freedom shown in the left-hand column. The critical values in the body of the table represent values along the horizontal axis of the figure. Each upper-tail significance level in bold at the top of the table represents the area under the curve to the right of a critical value. For example, if the curve in the figure represents a t-distribution with 60 degrees of freedom, the right-hand shaded area under the curve to the right of the critical value 2.000 represents an upper-tail significance level of 0.025. Each two-tail significance level in bold at the bottom of the table represents the sum of the areas to the right of a critical value and to the left of the negative of that critical value. For example, for a t-distribution with 60 degrees of freedom, the sum of the shaded areas under the curve to the right of the critical value 2.000 and to the left of −2.000 represents a two-tail significance level of 0.05.

For t-distributions with degrees of freedom *not* in the table (e.g., 45), to be conservative you should use the table row corresponding to the next *lowest* number (i.e., 40 for 45 degrees of freedom), although you will lose some accuracy when you do this. Alternatively, use

Applied Regression Modeling, Second Edition. By Iain Pardoe
Copyright © 2012 John Wiley & Sons, Inc.

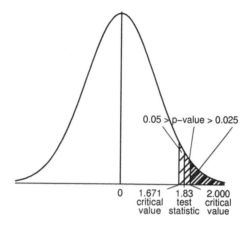

Figure B.1   Density curve for a t-distribution showing two critical values from Table B.1 to the left and to the right of a calculated test statistic. The upper-tail p-value is between the corresponding upper-tail significance levels at the top of the table, in this case 0.025 and 0.05.

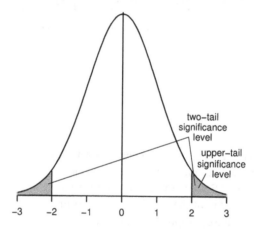

Figure B.2   Density curve for a t-distribution showing critical values (or percentiles or t-statistics) along the horizontal axis and significance levels (or probabilities or p-values) as areas under the curve.

computer help #8 in the software information files available from the book website to find exact percentiles (or critical values). For example, computer software will show that the 97.5th percentile of the t-distribution with 40 degrees of freedom is 2.021, while the 97.5th percentile of the t-distribution with 45 degrees of freedom is 2.014. Be careful to input the correct significance level when using software to find t-percentiles. For example, if you enter "0.05," software that expects a one-tail significance level will return the 95th percentile, whereas software that expects a two-tail significance will return the 97.5th percentile. Use computer help #9 to turn these calculations around and find tail areas (or p-values). Again, be careful about whether the software is working with one- or two-tail areas. For example, the upper-tail area corresponding to the test statistic 2.021 for a t-distribution with 40 degrees of freedom is 0.025, while the two-tail area corresponding to the test statistic 2.021 for a t-distribution with 40 degrees of freedom is 0.05.

Table B.1 Percentiles or critical values for t-distributions. The final row of the table labeled Z represents the standard normal distribution (equivalent to a t-distribution with infinite degrees of freedom).

df	\multicolumn{6}{c}{t-distribution upper-tail significance level}
	0.1	**0.05**	**0.025**	**0.01**	**0.005**	**0.001**
2	1.886	2.920	4.303	6.965	9.925	22.327
3	1.638	2.353	3.182	4.541	5.841	10.215
4	1.533	2.132	2.776	3.747	4.604	7.173
5	1.476	2.015	2.571	3.365	4.032	5.893
6	1.440	1.943	2.447	3.143	3.707	5.208
7	1.415	1.895	2.365	2.998	3.499	4.785
8	1.397	1.860	2.306	2.896	3.355	4.501
9	1.383	1.833	2.262	2.821	3.250	4.297
10	1.372	1.812	2.228	2.764	3.169	4.144
11	1.363	1.796	2.201	2.718	3.106	4.025
12	1.356	1.782	2.179	2.681	3.055	3.930
13	1.350	1.771	2.160	2.650	3.012	3.852
14	1.345	1.761	2.145	2.624	2.977	3.787
15	1.341	1.753	2.131	2.602	2.947	3.733
16	1.337	1.746	2.120	2.583	2.921	3.686
17	1.333	1.740	2.110	2.567	2.898	3.646
18	1.330	1.734	2.101	2.552	2.878	3.610
19	1.328	1.729	2.093	2.539	2.861	3.579
20	1.325	1.725	2.086	2.528	2.845	3.552
21	1.323	1.721	2.080	2.518	2.831	3.527
22	1.321	1.717	2.074	2.508	2.819	3.505
23	1.319	1.714	2.069	2.500	2.807	3.485
24	1.318	1.711	2.064	2.492	2.797	3.467
25	1.316	1.708	2.060	2.485	2.787	3.450
26	1.315	1.706	2.056	2.479	2.779	3.435
27	1.314	1.703	2.052	2.473	2.771	3.421
28	1.313	1.701	2.048	2.467	2.763	3.408
29	1.311	1.699	2.045	2.462	2.756	3.396
30	1.310	1.697	2.042	2.457	2.750	3.385
40	1.303	1.684	2.021	2.423	2.704	3.307
50	1.299	1.676	2.009	2.403	2.678	3.261
60	1.296	1.671	2.000	2.390	2.660	3.232
70	1.294	1.667	1.994	2.381	2.648	3.211
80	1.292	1.664	1.990	2.374	2.639	3.195
90	1.291	1.662	1.987	2.368	2.632	3.183
100	1.290	1.660	1.984	2.364	2.626	3.174
200	1.286	1.653	1.972	2.345	2.601	3.131
500	1.283	1.648	1.965	2.334	2.586	3.107
1000	1.282	1.646	1.962	2.330	2.581	3.098
Z	1.282	1.645	1.960	2.326	2.576	3.090
df	**0.2**	**0.1**	**0.05**	**0.02**	**0.01**	**0.002**
	\multicolumn{6}{c}{t-distribution two-tail significance level}

APPENDIX C

NOTATION AND FORMULAS

C.1 UNIVARIATE DATA

Applied Regression Modeling, Second Edition. By Iain Pardoe
Copyright © 2012 John Wiley & Sons, Inc.

Notation and Formulas	Page
Upper-tail critical value for testing $\mathrm{E}(Y)$: t-percentile from t_{n-1} (significance level = area to the right)	20
Lower-tail critical value for testing $\mathrm{E}(Y)$: t-percentile from t_{n-1} (significance level = area to the left)	20
Two-tail critical value for testing $\mathrm{E}(Y)$: t-percentile from t_{n-1} (significance level = sum of tail areas)	20
Upper-tail p-value for testing $\mathrm{E}(Y)$: area under t_{n-1} curve to right of t-statistic	21
Lower-tail p-value for testing $\mathrm{E}(Y)$: area under t_{n-1} curve to left of t-statistic	21
Two-tail p-value for testing $\mathrm{E}(Y)$: $2\times$ area under t_{n-1} curve beyond t-statistic	21
Model for univariate data: $Y = \mathrm{E}(Y) + e$	26
Point estimate for $\mathrm{E}(Y)$: m_Y	15
Confidence interval for $\mathrm{E}(Y)$: $m_Y \pm$ (t-percentile from t_{n-1})$(s_Y/\sqrt{n})$	17
Point estimate for Y^* (prediction): m_Y	25
Prediction interval for Y^*: $m_Y \pm$ (t-percentile from t_{n-1})$(s_Y\sqrt{1+1/n})$	26

C.2 SIMPLE LINEAR REGRESSION

Notation and Formulas	Page
Response values: Y; predictor values: X; sample size: n	35
Simple linear regression model: $Y = \mathrm{E}(Y) + e = b_0 + b_1 X + e$	40
Fitted regression model for $\mathrm{E}(Y)$: $\hat{Y} = \hat{b}_0 + \hat{b}_1 X$	42
Estimated errors or residuals: $\hat{e} = Y - \hat{Y}$	42
Residual sum of squares: $\mathrm{RSS} = \sum_{i=1}^{n} \hat{e}_i^2$	42
Regression standard error: $s = \sqrt{\dfrac{\mathrm{RSS}}{n-2}}$	46
(with 95% confidence, we can expect to predict Y to within approx. $\pm 2s$)	47
Total sum of squares: $\mathrm{TSS} = \sum_{i=1}^{n}(Y_i - m_Y)^2$	48
Coefficient of determination: $\mathrm{R}^2 = 1 - \dfrac{\mathrm{RSS}}{\mathrm{TSS}}$	48
(linear association between Y and X explains R^2 of the variation in Y)	49
Coefficient of correlation: $r = \dfrac{\sum_{i=1}^{n}(Y_i - m_Y)(X_i - m_X)}{\sqrt{\sum_{i=1}^{n}(Y_i - m_Y)^2}\sqrt{\sum_{i=1}^{n}(X_i - m_X)^2}}$	50
(r tells us strength and direction of any linear association between Y and X)	50
t-statistic for testing b_1: $\dfrac{\hat{b}_1 - b_1}{s_{\hat{b}_1}}$ (the test value, b_1, is usually 0)	53
Upper-tail critical value for testing b_1: t-percentile from t_{n-2} (significance level = area to the right)	54
Lower-tail critical value for testing b_1: t-percentile from t_{n-2} (significance level = area to the left)	54
Two-tail critical value for testing b_1: t-percentile from t_{n-2} (significance level = sum of tail areas)	55
Upper-tail p-value for testing b_1: area under t_{n-2} curve to right of t-statistic	54

Notation and Formulas	Page
Lower-tail p-value for testing b_1: area under t_{n-2} curve to left of t-statistic	54
Two-tail p-value for testing b_1: $2\times$ area under t_{n-2} curve beyond t-statistic	55
Confidence interval for b_1: $\hat{b}_1 \pm (\text{t-percentile from } t_{n-2})(s_{\hat{b}_1})$	58
Point estimate for $E(Y)$ (estimation) at X_p: $\hat{Y} = \hat{b}_0 + \hat{b}_1 X_p$	68
Confidence interval for $E(Y)$ at X_p: $\hat{Y} \pm (\text{t-percentile from } t_{n-2})(s_{\hat{Y}})$	68
Standard error of estimation: $s_{\hat{Y}} = s\sqrt{\dfrac{1}{n} + \dfrac{(X_p - m_X)^2}{\sum_{i=1}^{n}(X_i - m_X)^2}}$	68
Point estimate for Y^* (prediction) at X_p: $\hat{Y} = \hat{b}_0 + \hat{b}_1 X_p$	69
Prediction interval for Y^* at X_p: $\hat{Y} \pm (\text{t-percentile from } t_{n-2})(s_{\hat{Y}*})$	69
Standard error of prediction: $s_{\hat{Y}*} = s\sqrt{1 + \dfrac{1}{n} + \dfrac{(X_p - m_X)^2}{\sum_{i=1}^{n}(X_i - m_X)^2}}$	70

C.3 MULTIPLE LINEAR REGRESSION

Notation and Formulas	Page
Response values: Y; predictor values: $X_1, X_2, \ldots, X_k$; sample size: n	83
Multiple linear regression model: $Y = E(Y) + e = b_0 + b_1 X_1 + b_2 X_2 + \cdots + b_k X_k + e$	85
Interpreting regression parameters in models such as $E(Y) = b_0 + b_1 X_1 + b_2 X_2$: b_1 = expected change in Y when X_1 increases by one unit (and X_2 stays fixed)	85
Fitted regression model for $E(Y)$: $\hat{Y} = \hat{b}_0 + \hat{b}_1 X_1 + \hat{b}_2 X_2 + \cdots + \hat{b}_k X_k$	87
Estimated errors or residuals: $\hat{e} = Y - \hat{Y}$	88
Residual sum of squares: $\text{RSS} = \sum_{i=1}^{n} \hat{e}_i^2$	88
Regression parameter estimates: $(\mathbf{X}^T\mathbf{X})^{-1}\mathbf{X}^T\mathbf{Y}$	92
Regression standard error: $s = \sqrt{\dfrac{\text{RSS}}{n-k-1}}$	93
(with 95% confidence, we can expect to predict Y to within approx. $\pm 2s$)	93
Total sum of squares: $\text{TSS} = \sum_{i=1}^{n}(Y_i - m_Y)^2$	94
Coefficient of determination: $R^2 = 1 - \dfrac{\text{RSS}}{\text{TSS}}$	94
[the linear regression model for $(X_1, \ldots, X_k)$ explains R^2 of the variation in Y]	94
Adjusted $R^2 = 1 - \left(\dfrac{n-1}{n-k-1}\right)(1 - R^2)$	96
Multiple $R = \sqrt{R^2}$	100
(the correlation between the observed Y-values and the fitted $\hat{Y}$-values)	100
Global F-statistic for testing $b_1 = b_2 = \cdots = b_k = 0$: $\dfrac{(\text{TSS} - \text{RSS})/k}{\text{RSS}/(n-k-1)} = \dfrac{R^2/k}{(1-R^2)/(n-k-1)}$	101
Critical value: F-percentile from $F_{k,n-k-1}$ (significance level = area to the right)	102
p-value: area under the $F_{k,n-k-1}$ curve to the right of the F-statistic	102

Notation and Formulas	Page
Nested F-statistic for testing $b_{r+1} = b_{r+2} = \cdots = b_k = 0$: $$\frac{(\text{RSS}_R - \text{RSS}_C)/(k-r)}{\text{RSS}_C/(n-k-1)}$$	105
Critical value: F-percentile from $F_{k-r,n-k-1}$ (significance level = area to the right)	105
p-value: area under the $F_{k-r,n-k-1}$ curve to the right of the F-statistic	105
t-statistic for testing b_p: $\dfrac{\hat{b}_p - b_p}{s_{\hat{b}_p}}$ (the test value, b_p, is usually 0)	109
Upper-tail critical value for testing b_p: t-percentile from t_{n-k-1} (significance level = area to the right)	111
Lower-tail critical value for testing b_p: t-percentile from t_{n-k-1} (significance level = area to the left)	111
Two-tail critical value for testing b_p: t-percentile from t_{n-k-1} (significance level = sum of tail areas)	110
Upper-tail p-value for testing b_p: area under t_{n-k-1} curve to right of t-statistic	111
Lower-tail p-value for testing b_p: area under t_{n-k-1} curve to left of t-statistic	111
Two-tail p-value for testing b_p: $2\times$ area under t_{n-k-1} curve beyond t-statistic	111
Confidence interval for b_p: $\hat{b}_p \pm (\text{t-percentile from } t_{n-k-1})(s_{\hat{b}_p})$	113
Regression parameter standard errors: square roots of the diagonal entries of $s^2(\mathbf{X}^T\mathbf{X})^{-1}$	118
Point estimate for E(Y) (estimation) at $(X_1, X_2, \ldots, X_k)$: $\hat{Y} = \hat{b}_0 + \hat{b}_1 X_1 + \hat{b}_2 X_2 + \cdots + \hat{b}_k X_k$	126
Confidence interval for E(Y) at $(X_1, X_2, \ldots, X_k)$: $\hat{Y} \pm (\text{t-percentile from } t_{n-k-1})(s_{\hat{Y}})$	126
Point estimate for Y^* (prediction) at $(X_1, X_2, \ldots, X_k)$: $\hat{Y} = \hat{b}_0 + \hat{b}_1 X_1 + \hat{b}_2 X_2 + \cdots + \hat{b}_k X_k$	128
Prediction interval for Y^* at $(X_1, X_2, \ldots, X_k)$: $\hat{Y} \pm (\text{t-percentile from } t_{n-k-1})(s_{\hat{Y}*})$	128
Standard error of estimation: $s_{\hat{Y}} = s\sqrt{\mathbf{x}^T(\mathbf{X}^T\mathbf{X})^{-1}\mathbf{x}}$	129
Standard error of prediction: $s_{\hat{Y}*} = s\sqrt{1 + \mathbf{x}^T(\mathbf{X}^T\mathbf{X})^{-1}\mathbf{x}}$	129
Models with $\log_e(Y)$ as the response, for example, E$(\log_e(Y)) = b_0 + b_1 X_1 + b_2 X_2$: $\exp(b_1) - 1 = $ proportional change in Y when X_1 increases by 1 unit (and X_2 stays fixed)	155
Standardized residual: $r_i = \dfrac{\hat{e}_i}{s\sqrt{1-h_i}}$	193
Studentized residual: $t_i = r_i \sqrt{\dfrac{n-k-2}{n-k-1-r_i^2}}$	193
Leverages: diagonal entries of $\mathbf{H} = \mathbf{X}(\mathbf{X}^T\mathbf{X})^{-1}\mathbf{X}^T$	196
Leverage (alternate formula): $h_i = \dfrac{(k+1)D_i}{(k+1)D_i + r_i^2}$	199
Cook's distance: $D_i = \dfrac{r_i^2 h_i}{(k+1)(1-h_i)}$	199
Variance inflation factor for X_p: $1/(1-R_p^2)$	209

APPENDIX D

MATHEMATICS REFRESHER

While having a certain comfort with mathematical reasoning will help with following the material in this book, very few technical mathematical methods are used (e.g., there is no calculus). The only requirement is the ability to use a standard calculator capable of adding, subtracting, multiplying, dividing, and performing basic transformations including squares, reciprocals, natural logarithms, and exponentials.

D.1 THE NATURAL LOGARITHM AND EXPONENTIAL FUNCTIONS

The natural logarithm transformation (also known as "log to base-e" or by the symbols "$\log_e$" or "ln") is a way to transform (rescale) skewed positive values to make them more symmetric and normal. To back-transform a variable in natural logarithms to its original scale, use the exponentiation function on a calculator [denoted $\exp(X)$ or e^X, where X is the variable expressed in natural logarithms]. This is because $\exp(\log_e(Y)) = Y$.

Other useful properties of these functions derive from the following mathematical relationships:

$$\log_e(ab) = \log_e(a) + \log_e(b)$$

and

$$\exp(c+d) = \exp(c)\exp(d),$$

where a and b are positive numbers, and c and d are any numbers—try some examples out on a calculator to practice working with these transformations.

Applied Regression Modeling, Second Edition. By Iain Pardoe
Copyright © 2012 John Wiley & Sons, Inc.

D.2   ROUNDING AND ACCURACY

This book emphasizes the practical interpretation of statistical analyses, and it is usually more practical to employ some rounding when reporting final results rather than report exact numbers in all their gory detail. In particular, rounding to three significant digits seems a reasonable compromise to balance accuracy with practicality. For example, the following table contrasts calculated values with how these values should be reported:

Calculated	Reported
0.012345	0.0123
0.12345	0.123
1.2345	1.23
12.345	12.3
123.45	123
1,234.5	1,230
12,345	12,300

However, be careful *not* to round numbers used in intermediate calculations since this can adversely affect the accuracy of a final answer. For example, the following table illustrates the impact on the simple addition of 1.5739 and 2.6048:

	Correct	Incorrect
Intermediate	1.5739	1.57
	2.6048	2.60
Calculation	4.1787	4.17
Rounded	4.18	4.17

While the difference between the correct rounded answer (4.18) and the incorrect rounded answer (4.17) may not seem particularly large, this type of sloppiness with intermediate calculations can lead to very serious problems in complex calculations involving many steps. For example, with just one additional step (exponentiating the sum of 1.5739 and 2.6048), the difference in accuracy becomes very pronounced:

	Correct	Incorrect
Intermediate	$\exp(4.1787)$	$\exp(4.17)$
Calculation	65.2809	64.7155
Rounded	65.3	64.7

APPENDIX E

ANSWERS FOR SELECTED PROBLEMS

This appendix contains *brief* answers for even-numbered problems—these are intended to help you review material. The odd-numbered problems tend to be more challenging and normally require more extensive solutions than those given here.

Chapter 1

1.2   (a) $\Pr(Z < 1.282) = 0.90 \Leftrightarrow \Pr((Y - 70)/10 < 1.282) = 0.90$
$\Leftrightarrow \Pr(Y < 70 + 1.282(10)) = 0.90 \Leftrightarrow \Pr(Y < 82.8) = 0.90.$

   (b) $\Pr(Z < 2.326) = 0.99 \Leftrightarrow \Pr((Y - 70)/10 < 2.326) = 0.99$
$\Leftrightarrow \Pr(Y < 70 + 2.326(10)) = 0.99 \Leftrightarrow \Pr(Y < 93.3) = 0.99.$

   (c) $\Pr(Z < -1.645) = 0.05 \Leftrightarrow \Pr((Y - 70)/10 < -1.645) = 0.05$
$\Leftrightarrow \Pr(Y < 70 - 1.645(10)) = 0.05 \Leftrightarrow \Pr(Y < 53.6) = 0.05.$

1.4   (a) $\Pr(Y > 34.1) = \Pr((Y - 3)/10 > (34.1 - 3)/10)$
$= \Pr(Z > (34.1 - 3)/10) = \Pr(Z > 3.11) \approx 0.001$ (from last row of Table B.1).

   (b) $\Pr(Y > 15.7) = \Pr((Y - 3)/10 > (15.7 - 3)/10)$
$= \Pr(Z > (15.7 - 3)/10) = \Pr(Z > 1.27) \approx 0.1$ (from last row of Table B.1).

   (c) $\Pr(Y < -13.3) = \Pr((Y - 3)/10 < (-13.3 - 3)/10)$
$= \Pr(Z < (-13.3 - 3)/10) = \Pr(Z < -1.63) \approx 0.05$ (from last row of Table B.1).

Applied Regression Modeling, Second Edition. By Iain Pardoe
Copyright © 2012 John Wiley & Sons, Inc.

(d) $\Pr(M_Y > 7.4) = \Pr((M_Y - 3)/(10/\sqrt{50}) > (7.4 - 3)/(10/\sqrt{50}))$
$= \Pr(Z > (7.4 - 3)/(10/\sqrt{50})) = \Pr(Z > 3.11) \approx 0.001$ (from last row of Table B.1).

(e) $\Pr(M_Y > 4.8) = \Pr((M_Y - 3)/(10/\sqrt{50}) > (4.8 - 3)/(10/\sqrt{50}))$
$= \Pr(Z > (4.8 - 3)/(10/\sqrt{50})) = \Pr(Z > 1.27) \approx 0.1$ (from last row of Table B.1).

(f) $\Pr(M_Y < 0.7) = \Pr((M_Y - 3)/(10/\sqrt{50}) < (0.7 - 3)/(10/\sqrt{50}))$
$= \Pr(Z < (0.7 - 3)/(10/\sqrt{50})) = \Pr(Z < -1.63) \approx 0.05$ (from last row of Table B.1).

1.6 (a) $\Pr(Price < 215) = \Pr((Price - 280)/50 < (215 - 280)/50) = \Pr(Z < -1.30)$
$=$ slightly less than 10% (since $\Pr(Z < -1.282) = 10\%$).

(b) $m_Y \pm$ 95th percentile $(s_Y/\sqrt{n}) = 278.6033 \pm 1.699(53.8656/\sqrt{30})$
$= 278.6033 \pm 16.7088 = (262, 295)$.

(c) (i) Since the t-statistic of $(278.6033 - 265)/(53.8656/\sqrt{30}) = 1.38$ is less than 1.699 (the 95th percentile of the t-distribution with $n-1 = 29$ degrees of freedom), fail to reject NH in favor of AH;
(ii) Since the t-statistic of $(278.6033 - 300)/(53.8656/\sqrt{30}) = -2.18$ is less than -1.699 (the 5th percentile of the t-distribution with $n-1 = 29$ degrees of freedom), reject NH in favor of AH;
(iii) Since the t-statistic of $(278.6033 - 290)/(53.8656/\sqrt{30}) = -1.16$ is more than -1.699 (the 5th percentile of the t-distribution with $n-1 = 29$ degrees of freedom), fail to reject NH in favor of AH;
(iv) Since the t-statistic of $(278.6033 - 265)/(53.8656/\sqrt{30}) = 1.38$ is less than 2.045 (the 97.5th percentile of the t-distribution with $n-1 = 29$ degrees of freedom), fail to reject NH in favor of AH.

(d) $m_Y \pm$ 95th percentile $(s_Y\sqrt{1+1/n}) = 278.6033 \pm 1.699(53.8656\sqrt{1+1/30})$
$= 278.6033 \pm 93.0304 = (186, 372)$.

1.8 (a) $m_Y \pm$ 95th percentile $(s_Y/\sqrt{n}) = 2.9554 \pm 2.037(1.48104/\sqrt{33})$
$= 2.9554 \pm 0.5252 = (2.43, 3.48)$.

(b) $m_Y \pm$ 95th percentile $(s_Y\sqrt{1+1/n}) = 2.9554 \pm 2.037(1.48104\sqrt{1+1/33})$
$= 2.9554 \pm 3.0622 = (-0.107, 6.018)$.
Since the data range is 1.642 to 7.787, this does not seem very reasonable.

(c) $0.3956 \pm 2.037(0.12764/\sqrt{33}) = 0.3956 \pm 0.0453 = (0.3503, 0.4409)$.

(d) $(1/0.4409, 1/0.3503) = (2.27, 2.85)$.

(e) $0.3956 \pm 2.037(0.12764\sqrt{1+1/33}) = 0.3956 \pm 0.2639 = (0.1317, 0.6595)$.
In original units this corresponds to $(1.52, 7.59)$, a much more reasonable interval based on the range of Y-values in the data. Although not as obvious, the confidence interval in part (d) is also more reasonable than the confidence interval in part (a)—if you look at a histogram of the Y-values, it seems very unlikely that the "center" could be as high as 3.5, as suggested by the confidence interval in part (a). The data in original units were far from normally distributed, whereas taking the reciprocal transformation made the sample values look more normal—confidence intervals and (particularly) prediction intervals tend to be more effective the closer to normal the data look.

Chapter 2

2.2 (a) The slope should be positive since a higher batting average should result in more wins, all other things being equal.

(b) The points in the scatterplot (not shown) with *Win* on the vertical axis and *Bat* on the horizontal axis have a general upward trend from left to right, so this does agree with the answer in part (a).

(c) $\widehat{Win} = -127 + 813\,Bat$.

(d) The line seems to represent the linear trend in the data reasonably well and is not overly influenced by any isolated points in the scatterplot.

(e) There is some kind of linear association, but it would probably not be described as "strong." The variation of the data points about the least squares line is somewhat less than the overall variation in the number of wins, but there still remains quite a lot of unexplained variation—winning baseball games depends on more than just a team's batting average.

(f) The estimated intercept of -127 has little practical interpretation since it corresponds to the expected number of wins for a team that has a zero batting average—a nonsensical value in this context. The estimated slope of 813 corresponds to the expected change in the number of wins as a team's batting average increases by 1 unit. It might be clearer to say that we expect the number of wins to increase by 8.13 games on average when a team's batting average increases by 0.01 unit (e.g., from 0.250 to 0.260).

(g) The linear association between the number of wins and the batting average is not as strong for the American League teams as it is for the National League teams.

2.4 (a) $\hat{b}_1 \pm \text{95th percentile}\,(s_{\hat{b}_1}) = 40.800 \pm 2.353 \times 5.684 = 40.800 \pm 13.374 = (27.4, 54.2)$.

(b) Conduct this upper-tail test: NH: $b_1 = 20$ versus AH: $b_1 > 20$. Since the t-statistic of $(40.800 - 20)/5.684 = 3.66$ is greater than 2.353 (the 95th percentile of the t-distribution with $n-2=3$ degrees of freedom), reject NH in favor of AH. Thus, the sample data suggest that the population slope is greater than 20 (at a 5% significance level). In other words, putting a 500-square foot addition onto a house could be expected to increase its sale price by $10,000 or more.

2.6 (a) Our best point estimate for E(*Price*) at *Floor*=2 is
$\widehat{Price} = 190.318 + 40.800 \times 2 = 271.918$.
Then, $\widehat{Price} \pm \text{95th percentile}\,(s_{\hat{Y}}) = 271.918 \pm 2.353 \times 1.313 = 271.918 \pm 3.089 = (268.8, 275.0)$.

(b) Our best point estimate for *Price** at *Floor*=2 is
$\widehat{Price} = 190.318 + 40.800 \times 2 = 271.918$.
Then, $\widehat{Price}^* \pm \text{95th percentile}\,(s_{\hat{Y}*}) = 271.918 \pm 2.353 \times 3.080 = 271.918 \pm 7.247 = (264.7, 279.2)$.

2.8 You should find that the model with X_1 has a higher s, a lower R^2, and a lower absolute t-statistic for b_1 than the other three models (which all have essentially the same values of s, R^2, and $|t_{b_1}|$). However, the model with X_1 is *more* appropriate than each of the other three models:

- The model with X_2 has a clear problem with increasing variance of the residuals as X_2 increases, which violates the constant variance regression assumption.
- The model with X_3 fails to pick up the clear curved trend in the data points.
- The model with X_4 seems overly influenced by the way the data seem to be in two clusters, one at the bottom left and one at the top right rather than in a more dispersed straight-line pattern.

Concluding message: measures of regression model fit like the regression standard error, s, the coefficient of determination, R^2, and the absolute t-statistic for b_1 are meaningful only when the assumptions of the model are broadly satisfied.

2.10   (a) NH: $b_1 = 0$ vs. AH : $b_1 > 0$.

(b) Yes, since the p-value of the test (0.0005) is less than the significance level (0.05).

(c) For every additional customer, we estimate costs to increase by \$9.87.

(d) For a day with no customers (presumably a day the restaurant is closed), we estimate costs (presumably fixed costs) to be \$1,192.

(e) We expect about 95% of the observed cost values to lie within \$448 of their least squares predicted values.

(f) About 90.5% of the total variation in the sample cost values (about their mean) can be explained by (or attributed to) the linear association between cost and number of customers.

Chapter 3

3.2   (a) $\hat{b}_1 \pm 95\text{th percentile}\,(s_{\hat{b}_1}) = 6.074 \pm 1.753 \times 2.662 = 6.074 \pm 4.666 = (1.41, 10.74)$.

(b) $\hat{b}_1 \pm 95\text{th percentile}\,(s_{\hat{b}_1}) = 5.001 \pm 1.740 \times 2.261 = 5.001 \pm 3.934 = (1.07, 8.94)$. This interval is narrower (more precise) than the one in (a) because the two-predictor model is more accurate than the four-predictor model (which contains two unimportant predictors).

3.4   (a) Our best point estimate for E(*Lab*) at *Tws*=6 and *Asw*=20 is
$\widehat{Lab} = 110.431 + 5.001 \times 6 - 2.012 \times 20 = 100.2$.
Then, $\widehat{Lab} \pm 95\text{th percentile}\,(s_{\hat{y}}) = 100.2 \pm 1.740 \times 2.293 = 100.2 \pm 3.990 = (96.2, 104.2)$.

(b) Our best point estimate for *Lab** at *Tws*=6 and *Asw*=20 is
$\widehat{Lab} = 110.431 + 5.001 \times 6 - 2.012 \times 20 = 100.2$.
Then, $\widehat{Lab}^* \pm 95\text{th percentile}\,(s_{\hat{y}*}) = 100.2 \pm 1.740 \times 9.109 = 100.2 \pm 15.850 = (84.4, 116.1)$.

3.6   (a) The least squares equation is $\widehat{Mort} = 1{,}006.244 - 15.346\,Edu + 4.214\,Nwt - 2.150\,Jant + 1.624\,Rain + 18.548\,Nox + 0.537\,Hum - 0.345\,Inc$.

(b) NH: $b_6 = b_7 = 0$ versus AH: at least one of b_6 or b_7 is not equal to zero.

$$\text{Nested F-statistic} = \frac{(RSS_R - RSS_C)/(k-r)}{RSS_C/(n-k-1)}$$
$$= \frac{(60{,}948 - 60{,}417)/(7-5)}{60{,}417/(56-7-1)}$$
$$= 0.211.$$

Since this is less than the 95th percentile of the F-distribution with 2 numerator degrees of freedom and 48 denominator degrees of freedom (3.19), do not reject the null hypothesis in favor of the alternative. In other words, *Hum* and *Inc* do not provide significant information about the response, *Mort*, beyond the information provided by the other predictor variables, and the reduced model is preferable.

(c) NH: $b_p = 0$ versus AH: $b_p \neq 0$. Individual t-statistics (from statistical software output) are $-2.418, 6.334, -3.347, 2.942$, and 3.479. Since the absolute values of these statistics are all greater than the 97.5th percentile of the t-distribution with 50 degrees of freedom (2.01), we reject the null hypothesis (in each case) in favor of the alternative. In other words, each of the variables has a significant linear association with *Mort*, controlling for the effects of all the others.

(d) Looking at residual plots and a histogram and QQ-plot of the residuals (not shown), the four assumptions seem reasonable.

(e) $\widehat{Mort} = 1{,}028.232 - 15.589\,Edu + 4.181\,Nwt - 2.131\,Jant + 1.633\,Rain + 18.413\,Nox$. This model shows positive associations between mortality and each of percentage nonwhite, rainfall, and nitrous oxide, and negative associations between mortality and each of education and temperature. All of these associations might have been expected.

(f) Using statistical software, the 95% confidence interval is (946, 979).

(g) Using statistical software, the 95% prediction interval is (891, 1,035).

Chapter 4

4.2 95% prediction intervals for model 1 at $100k, $150k, and $200k are ($466, $1,032), ($808, $1,377), and ($1,146, $1,727), while the equivalent intervals for model 2 are ($534, $1,018), ($794, $1,520), and ($1,049, $2,026). The model 2 intervals seem to be more appropriate than the model 1 intervals based on visual inspection of a scatterplot of *Tax* versus *Price*.

4.4 When the prevailing interest rate is 3%, we expect to increase sales by $1.836 - 0.126(3) = \$1.46m$ for each additional $1m we spend on advertising.

4.6 (a) NH: $b_1 = b_2 = \cdots = b_5 = 0$ versus AH: at least one of $b_1, b_2, \ldots, b_5$ is not equal to zero.

(b) The test statistic calculation is as follows:

$$\text{global F-statistic} = \frac{(TSS - RSS)/k}{RSS/(n-k-1)} = \frac{(733.520 - 97.194)/5}{97.194/(200-5-1)}$$
$$= 254.022.$$

Since this is greater than the 95th percentile of the F-distribution with 5 numerator degrees of freedom and 194 denominator degrees of freedom (2.26), reject the null hypothesis in favor of the alternative. In other words, at least one of the predictor terms—*Freq, Amt, FreqAmt, Freq²*, and *Amt²*—is linearly associated with *Score*.

(c) NH: $b_3 = b_4 = b_5 = 0$ versus AH: at least one of b_3, b_4, or b_5 is not equal to zero.

(d) The test statistic calculation is as follows:

$$\text{nested F-statistic} = \frac{(\text{RSS}_R - \text{RSS}_C)/(k-r)}{\text{RSS}_C/(n-k-1)}$$
$$= \frac{(124.483 - 97.194)/(5-2)}{97.194/(200-5-1)}$$
$$= 18.156.$$

Since this is greater than the 95th percentile of the F-distribution with 3 numerator degrees of freedom and 194 denominator degrees of freedom (2.65), reject the null hypothesis in favor of the alternative. In other words, *FreqAmt, Freq²*, and *Amt²* provide useful information about *Score* beyond the information provided by *Freq* and *Amt* alone.

(e) Use the complete model to predict *Score* (assuming that it passes the regression assumption checks) since it provides significantly more predictive power than the reduced model.

4.8 (a) $E(Y) = b_0 + b_1 X$.

(b) $E(Y) = b_0 + b_1 X + b_2 D_1 + b_3 D_2$, where D_1 and D_2 are indicator variables for two of the levels relative to the third (reference) level.

(c) $E(Y) = b_0 + b_1 X + b_2 D_1 + b_3 D_2 + b_4 D_1 X + b_5 D_2 X$.

(d) When $b_4 = b_5 = 0$.

(e) When $b_2 = b_3 = b_4 = b_5 = 0$.

Chapter 5

5.2 (a) It would be inappropriate to fit a simple linear regression model with *Sales* as the response variable and *Time* as the predictor variable because this model cannot account for the strong seasonal pattern in sales that is evident in the scatterplot (not shown).

(b) A multiple linear regression model with *Sales* as the response variable and (D_1, D_2, D_3, *Time*) as the predictor variables is also inappropriate because there is a strong autocorrelation pattern evident in the residual plot (not shown).

(c) Compared with the residual plot from part (b), the residual plot for the model including *LagSales* shows no strong autocorrelation patterns, so including *LagSales* does appear to correct the autocorrelation problem.

(d) The prediction errors for the four quarters in 1999 are $(-717, 165, -88, -549)$ for model (a), $(-383, -160, -245, -380)$ for model (b), and $(6, 110, -139,$

-211) for model (c), indicating that model (c) provides the best predictions overall.

5.4   (a) $s = 0.7656$, $R^2 = 0.8835$, and adjusted $R^2 = 0.8573$.

    (b) Nested model F-test p-value is 0.7995.

    (c) $s = 0.7448$, $R^2 = 0.8788$, and adjusted $R^2 = 0.8650$.

    (d) RMSE under the first model is 5.46, while RMSE under the second model is 5.19.

5.6   (a) The nested F-statistic $= 42.294$ and p-value $= 0.000$ suggest that model A is relatively poor and model B is relatively good. Model A is poor because it doesn't include the $D_1 X_2$ interaction, which is very significant in model B (individual p-value $= 0.000$).

    (b) The nested F-statistic $= 0.9473$ and p-value $= 0.3916$ suggest that model D is relatively poor and model C is relatively good. Model D is poor because it includes X_4 and X_5 in addition to X_6, and these three predictors are collinear (individual p-values for X_4, X_5, and X_6 in model D are $0.174, 0.183$, and 0.374, respectively, but the individual p-value for X_6 in model C is 0.000)—see also part (h).

    (c) The nested F-statistic $= 26.435$ and p-value $= 0.000$ suggest that model E is relatively poor and model F is relatively good. Model E is poor because it doesn't include the X_3^2 transformation, which is very significant in model F (individual p-value $= 0.000$).

    (d) A: X_2 effect $= 8.19 - 0.05X_2$ (for $D_1 = 0$), $5.50 - 0.05X_2$ (for $D_1 = 1$).
       B: X_2 effect $= 5.98 + 0.79X_2$ (for $D_1 = 0$), $8.18 - 1.09X_2$ (for $D_1 = 1$).
       C: X_2 effect $= 5.98 + 0.78X_2$ (for $D_1 = 0$), $8.20 - 1.10X_2$ (for $D_1 = 1$).
       F: X_2 effect $= 5.94 + 0.80X_2$ (for $D_1 = 0$), $8.13 - 1.08X_2$ (for $D_1 = 1$).
       The X_2 predictor effects are very different for model A.

    (e) B: X_3 effect $= 6.76 + 2.78X_3 - 1.28X_3^2$ ($D_1 = 0$), $4.02 + 2.78X_3 - 1.28X_3^2$ ($D_1 = 1$).
       C: X_3 effect $= 6.76 + 2.76X_3 - 1.26X_3^2$ ($D_1 = 0$), $4.03 + 2.76X_3 - 1.26X_3^2$ ($D_1 = 1$).
       E: X_3 effect $= 9.44 - 1.30X_3$ (for $D_1 = 0$), $6.87 - 1.30X_3$ (for $D_1 = 1$).
       F: X_3 effect $= 6.77 + 2.74X_3 - 1.26X_3^2$ ($D_1 = 0$), $4.03 + 2.74X_3 - 1.26X_3^2$ ($D_1 = 1$).
       The X_3 predictor effects are very different for model E.

    (f) A: X_4 effect $= 6.80 + 0.43X_4$ (for $D_1 = 0$), $4.11 + 0.43X_4$ (for $D_1 = 1$).
       B: X_4 effect $= 6.58 + 0.50X_4$ (for $D_1 = 0$), $3.84 + 0.50X_4$ (for $D_1 = 1$).
       D: X_4 effect $= 3.77 + 1.46X_4$ (for $D_1 = 0$), $1.03 + 1.46X_4$ (for $D_1 = 1$).
       F: X_4 effect $= 6.54 + 0.51X_4$ (for $D_1 = 0$), $3.80 + 0.51X_4$ (for $D_1 = 1$).
       The X_4 predictor effects are very different for model D.

    (g) A: X_5 effect $= 6.95 + 0.44X_5$ (for $D_1 = 0$), $4.26 + 0.44X_5$ (for $D_1 = 1$).
       B: X_5 effect $= 6.87 + 0.47X_5$ (for $D_1 = 0$), $4.13 + 0.47X_5$ (for $D_1 = 1$).
       D: X_5 effect $= 4.53 + 1.39X_5$ (for $D_1 = 0$), $1.79 + 1.39X_5$ (for $D_1 = 1$).
       F: X_5 effect $= 6.86 + 0.47X_5$ (for $D_1 = 0$), $4.11 + 0.47X_5$ (for $D_1 = 1$).
       The X_5 predictor effects are very different for model D.

    (h) C: X_6 effect $= 6.46 + 0.49X_6$ (for $D_1 = 0$), $3.72 + 0.49X_6$ (for $D_1 = 1$).
       D: X_6 effect $= 11.09 - 0.94X_6$ (for $D_1 = 0$), $8.35 - 0.94X_6$ (for $D_1 = 1$).

The slope part of the predictor effect for X_6 in model C (0.49) is approximately the same as the average of the slope parts of the predictor effects for X_4 and X_5 in model F $((0.51+0.47)/2=0.49)$. The X_6 predictor effect is different in models C and D.

Appendix A

A.2 (a) Histogram of *Vol* (not shown).

(b) The histogram shows interior passenger and cargo volumes ranging between 50 and 180 cubic feet, with many values tending to cluster around 90–130 cubic feet rather than at the extremes. The distribution seems relatively symmetric.

(c) Mean $=1.10$ (110 cubic feet), median $=1.11$ (111 cubic feet).

(d) For symmetric data such as these, the mean has nice technical properties that make it a more appropriate measure of the center. By contrast, for highly skewed data, the median is a better summary than the mean of the central tendency of the data. For example, when a dataset has a few very large values, this can cause the mean to be relatively large in comparison to the median (which is not affected to the same extent by these large values). Since the mean and median are quite close together in this case, any skewness in these data is practically nonexistent.

(e) 25th percentile (first quartile) $=1.00$ (100 cubic feet);
50th percentile (second quartile, or median) $=1.11$ (111 cubic feet);
75th percentile (third quartile) $=1.20$ (120 cubic feet).

(f) The middle 50% of the volumes fall between 1.00 (100 cubic feet) and 1.20 (120 cubic feet).

A.4 (a) Cross-tabulation (not shown).

(b) There are 36 front-wheel drive midsize cars.

(c) 43% of midsize cars are front-wheel drive.

(d) 28% of front-wheel drive vehicles are midsize cars.

(e) The most common vehicle class is midsize car (83 vehicles), mostly front-wheel drive (43%) or all-wheel drive (30%). The next most common classes are compact, large, and subcompact cars, with compact cars mostly front-wheel drive (54%), and large and subcompact cars mostly rear-wheel drive (50% and 51%, respectively). Station wagons and minicompacts/two-seater cars are the next most frequent, with station wagons mostly front-wheel drive (50%), and minicompacts/two-seater cars mostly rear-wheel drive (57%). Finally, there are just 20 sport utility vehicles, mostly all-wheel drive (40%) and four-wheel drive (30%). The most common front-wheel drive vehicles are compact and midsize cars (28% each), while the most common rear-wheel drive vehicles are subcompact and large cars (25% each). The most common all-wheel drive vehicles are midsize cars (30%), while the most common four-wheel drive vehicles are minicompacts/two-seater cars and sport utility vehicles (30% each).

A.6   (a) Scatterplot matrix (not shown).

    (b) There are fairly strong negative associations between *Cmpg* and each of *Eng* and *Cyl*, but the associations appear curved—steeper for low values of *Eng* and *Cyl* and becoming shallower as *Eng* and *Cyl* each increase. There doesn't appear to be much of an association between *Cmpg* and *Vol* other than a slight tendency for the few vehicles with very large volumes to have lower fuel efficiencies. There is a positive, reasonably linear association between *Eng* and *Cyl*, but no clear association between *Eng* and *Vol* or between *Cyl* and *Vol*. A few vehicles "stick out" from the dominant patterns in the plots. For example, vehicle 29 (Bentley Mulsanne 6.8L) with 8 cylinders has a particularly high value of *Eng*.

REFERENCES

Agresti, A. *Categorical Data Analysis*. Wiley, Hoboken, NJ, 2nd edition, 2002.

Agresti, A. *An Introduction to Categorical Data Analysis*. Wiley, Hoboken, NJ, 2nd edition, 2007.

Aiken, L. S. and West, S. G. *Multiple Regression: Testing and Interpreting Interactions*. Sage, Thousand Oaks, CA, 1991.

Andersen, R. *Modern Methods for Robust Regression*. Sage, Thousand Oaks, CA, 2007.

Bates, D. M. and Watts, D. G. *Nonlinear Regression Analysis and Its Applications*. Wiley, Hoboken, NJ, 2007.

Beach, C. and MacKinnon, J. A maximum likelihood procedure for regression with autocorrelated errors. *Econometrica*, 46:51–58, 1978.

Belsley, D. A., Kuh, E., and Welsch, R. E. *Regression Diagnostics: Identifying Influential Data and Sources of Collinearity*. Wiley, Hoboken, NJ, 2004.

Ben-Akiva, M. and Lerman, S. *Discrete Choice Analysis: Theory and Application to Travel Demand (Transportation Studies)*. The MIT Press, Cambridge, MA, 1985.

Bennett, K. P. and Mangasarian, O. L. Robust linear programming discrimination of two linearly inseparable sets. In *Optimization Methods and Software*, volume 1, pages 23–34. Gordon and Breach Science Publishers, New York, 1992.

Bolstad, W. M. *Introduction to Bayesian Statistics*. Wiley, Hoboken, NJ, 2nd edition, 2007.

Box, G. E. P. and Cox, D. R. An analysis of transformations. *Journal of the Royal Statistical Society, Series B*, 26:211–246, 1964.

Box, G. E. P. and Pierce. D. Distribution of residual autocorrelations in autoregressive moving average time series models. *Journal of the American Statistical Association*, 65:1509–1526, 1970.

Box, G. E. P. and Tidwell, P. W. Transformation of the independent variables. *Technometrics*, 4:531–550, 1962.

Breusch, T. S. Testing for autocorrelation in dynamic linear models. *Australian Economic Papers*, 17:334–355, 1978.

Breusch, T. S. and Pagan, A. R. A simple test for heteroscedasticity and random coefficient variation. *Econometrica*, 47:1287–1294, 1979.

Cameron, A. C. and Trivedi, P. K. *Regression Analysis of Count Data*. Cambridge University Press, Cambridge, UK, 1998.

Carlin, B. P. and Louis, T. A. *Bayesian Methods for Data Analysis*. Chapman & Hall/CRC, Boca Raton, FL, 3rd edition, 2008.

Chow, G. C. Tests of equality between sets of coefficients in two linear regressions. *Econometrica*, 28:591–605, 1960.

Cochrane, D. and Orcutt. G. Application of least squares regression to relationships containing autocorrelated error terms. *Journal of the American Statistical Association*, 44:32–61, 1949.

Cook, R. D. Detection of influential observations in linear regression. *Technometrics*, 19:15–18, 1977.

Cook, R. D. Exploring partial residual plots. *Technometrics*, 35:351–362, 1993.

Cook, R. D. Added variable plots and curvature in linear regression. *Technometrics*, 38:a–b, 1996.

Cook, R. D. and Weisberg, S. *Residuals and Influence in Regression*. Chapman & Hall/CRC, Boca Raton, FL, 1982. Out of print: available at www.stat.umn.edu/rir/.

Cook, R. D. and Weisberg, S. Diagnostics for heteroscedasticity in regression. *Biometrika*, 70:1–10, 1983.

Cook, R. D. and Weisberg, S. *Applied Regression Including Computing and Graphics*. Wiley, Hoboken, NJ, 1999.

Dallal, G. E. and Wilkinson, L. An analytic approximation to the distribution of Lilliefors' test for normality. *The American Statistician*, 40:294–296, 1986.

Data and Story Library. Available at lib.stat.cmu.edu/DASL/.

De Rose, D. and Galarza, R. Major League Soccer: predicting attendance for future expansion teams. *Stats*, 29:8–12, 2000.

Dielman, T. E. *Applied Regression Analysis: A Second Course in Business and Economic Statistics*. Brooks/Cole, Stamford, CT, 4th edition, 2004.

Draper, N. R. and Smith, H. *Applied Regression Analysis*. Wiley, Hoboken, NJ, 3rd edition, 1998.

Durbin, J. Testing for serial correlation in least squares regression when some of the regressors are lagged dependent variables. *Econometrica*, 38:410–421, 1970.

Durbin, J. and Watson, G. S. Testing for serial correlation in least squares regression I. *Biometrika*, 37:409–428, 1950.

Durbin, J. and Watson, G. S. Testing for serial correlation in least squares regression II. *Biometrika*, 38:159–178, 1951.

Durbin, J. and Watson, G. S. Testing for serial correlation in least squares regression III. *Biometrika*, 58:1–19, 1971.

Durham, C. A., Pardoe, I., and Vega, E. A methodology for evaluating how product characteristics impact choice in retail settings with many zero observations: an application to restaurant wine purchase. *Journal of Agricultural and Resource Economics*, 29:112–131, 2004.

Fox, J. *Applied Regression Analysis and Generalized Linear Models*. Sage, Thousand Oaks, CA, 2008.

Fox, J. and Weisberg, S. *An R Companion to Applied Regression*. Sage, Thousand Oaks, CA, 2nd edition, 2011.

Freedman, D., Pisani, R., and Purves, R. *Statistics*. W. W. Norton, New York, 4th edition, 2007.

Frees, E. W. Estimating densities of functions of observations. *Journal of the American Statistical Association*, 89:517–526, 1994.

Frees, E. W. *Data Analysis Using Regression Models: The Business Perspective*. Prentice Hall, Upper Saddle River, NJ, 1995.

Gelman, A., Carlin, J. B., Stern, H. S., and Rubin, D. B. *Bayesian Data Analysis*. Chapman & Hall/CRC, Boca Raton, FL, 2nd edition, 2003.

Gelman, A. and Hill, J. *Data Analysis Using Regression and Multilevel/Hierarchical Models*. Cambridge University Press, Cambridge, UK, 2006.

Gelman, A. and Pardoe, I. Bayesian measures of explained variance and pooling in multilevel (hierarchical) models. *Technometrics*, 48:241–251, 2006.

Godfrey, L. G. Testing against general autoregressive and moving average error models when the regressors include lagged dependent variables. *Econometrica*, 46:1293–1302, 1978.

Goldfeld, S. and Quandt, R. Some tests for homoscedasticity. *Journal of the American Statistical Association*, 60:539–547, 1965.

Greene, W. H. *Econometric Analysis*. Prentice Hall, Upper Saddle River, NJ, 7th edition, 2011.

Hastie, T. and Tibshirani, R. J. *Generalized Additive Models*. Chapman & Hall/CRC, Boca Raton, FL, 1990.

Hensher, D. A., Rose, J. M., and Greene, W. H. *Applied Choice Analysis: A Primer*. Cambridge University Press, Cambridge, UK, 2005.

Hildreth, C. and Lu, J. Demand relations with autocorrelated disturbances. Technical Report 276, Michigan State University Agricultural Experiment Station, 1960.

Horngren, C. T., Foster, G., and Datar, S. M. *Cost Accounting*. Prentice Hall, Upper Saddle River, NJ, 1994.

Hosmer, D. W. and Lemeshow, S. *Applied Logistic Regression*. Wiley, Hoboken, NJ, 2nd edition, 2000.

Hox, J. *Multilevel Analysis: Techniques and Applications*. Routledge, London, 2nd edition, 2010.

Huber, P. and Ronchetti, E. M. *Robust Statistics*. Wiley, Hoboken, NJ, 2nd edition, 2009.

Koenker, R. *Quantile Regression*. Cambridge University Press, Cambridge, UK, 2005.

Kreft, I. and De Leeuw, J. *Introducing Multilevel Modeling*. Sage, Thousand Oaks, CA, 1998.

Kutner, M. H., Nachtsheim, C. J., and Neter, J. *Applied Linear Regression Models*. McGraw-Hill/Irwin, New York, 4th edition, 2004.

Lambert, D. Zero-inflated Poisson regression, with an application to defects in manufacturing. *Technometrics*, 34:1–14, 1992.

Lee, P. M. *Bayesian Statistics: An Introduction*. Wiley, Hoboken, NJ, 3rd edition, 2004.

Lewis, M. *Moneyball: The Art of Winning an Unfair Game*. W. W. Norton, New York, 2003.

Lin, C. Y., Gelman, A., Price, P. N., and Krantz, D. H. Analysis of local decisions using hierarchical modeling, applied to home radon measurement and remediation (with disussion and rejoinder). *Statistical Science*, 14:305–337, 1999.

Little, R. J. A. and Rubin, D. B. *Statistical Analysis with Missing Data*. Wiley, Hoboken, NJ, 2nd edition, 2002.

Ljung, G. and Box, G. E. P. On a measure of lack of fit in time series models. *Biometrika*, 66:265–270, 1979.

Long, J. S. and Ervin, L. H. Using heteroscedasity consistent standard errors in the linear regression model. *The American Statistician*, 54:217–224, 2000.

Louviere, J. J., Hensher, D. A., and Swait, J. D. *Stated Choice Methods: Analysis and Applications.* Cambridge University Press, Cambridge, UK, 2000.

Mallows, C. Some comments on C_p. *Technometrics*, 15:661–676, 1973.

Marquardt, D. W. Generalized inverses, ridge regression and biased linear estimation. *Technometrics*, 12:591–612, 1970.

McClave, J. T., Benson, P. G., and Sincich, T. *Statistics for Business and Economics.* Prentice Hall, Upper Saddle River, NJ, 9th edition, 2005.

McCullagh, P. and Nelder, J. A. *Generalized Linear Models.* Chapman & Hall/CRC, Boca Raton, FL, 2nd edition, 1989.

McFadden, D. Conditional logit analysis of qualitative choice behavior. In Zarembka, P., editor, *Frontiers in Econometrics*, pages 105–142. Academic Press, New York, 1974.

Mendenhall, W. and Sincich, T. *A Second Course in Statistics: Regression Analysis.* Pearson, Upper Saddle River, NJ, 7th edition, 2011.

Moore, D. S. Tests of the chi-squared type. In D'Agostino, R. B. and Stephens, M. A., editors, *Goodness-of-Fit Techniques.* Marcel Dekker, New York, 1986.

Moore, D. S., Notz, W. I., and Fligner, M. A. *The Basic Practice of Statistics.* W. H. Freeman, New York, 6th edition, 2011.

Nelder, J. A. and Wedderburn, R. W. M. Generalized linear models. *Journal of the Royal Statistical Society, Series A*, 135:370–384, 1972.

Newey, W. and West, K. A simple positive semi-definite, heteroscedasticity and autocorrelation consistent covariance matrix. *Econometrica*, 55:703–708, 1987.

Pardoe, I. Modeling home prices using realtor data. *Journal of Statistics Education*, 16, 2008. Available at www.amstat.org/publications/jse/.

Pardoe, I. and Cook, R. D. A graphical method for assessing the fit of a logistic regression model. *The American Statistician*, 56:263–272, 2002.

Pardoe, I. and Simonton, D. K. Applying discrete choice models to predict Academy Award winners. *Journal of the Royal Statistical Society, Series A*, 171:375–394, 2008.

Pinheiro, J. C. and Bates, D. M. *Mixed Effects Models in S and S-PLUS.* Springer-Verlag, New York, 2000.

Prais, S. and Winsten, C. Trend estimation and serial correlation. Technical Report 383, Cowles Commission Discussion Paper, Chicago, 1954.

Price, P. N. and Gelman, A. Should you measure the concentration in your home? In *Statistics: A Guide to the Unknown.* Duxbury, Belmont, CA, 4th edition, 2006.

Price, P. N., Nero, A. V., and Gelman, A. Bayesian prediction of mean indoor radon concentrations for Minnesota counties. *Health Physics*, 71:922–936, 1996.

Ratkowsky, D. A. *Handbook of Nonlinear Regression Models.* Marcel Dekker, New York, 1990.

Raudenbush, S. W. and Bryk, A. S. *Hierarchical Linear Models: Applications and Data Analysis Methods.* Sage, Thousand Oaks, CA, 2nd edition, 2002.

Rea, J. D. The explanatory power of alternative theories of inflation and unemployment. *Review of Economics and Statistics*, 65:183–195, 1983.

Rossi, P., Allenby, G., and McCulloch, R. *Bayesian Statistics and Marketing.* Wiley, Hoboken, NJ, 2006.

Royston, J. P. An extension of Shapiro and Wilk's W test for normality to large samples. *Applied Statistics*, 31:115–124, 1982a.

Royston, J. P. Algorithm AS 181: The W test for normality. *Applied Statistics*, 31:176–180, 1982b.

Royston, J. P. A pocket-calculator algorithm for the Shapiro-Francia test for non-normality: an application to medicine. *Statistics in Medicine*, 12:181–184, 1993.

Royston, J. P. A remark on Algorithm AS 181: The *W* test for normality. *Applied Statistics*, 44:547–551, 1995.

Ryan, T. P. *Modern Regression Methods*. Wiley, Hoboken, NJ, 2nd edition, 2008.

Sakamoto, Y., Ishiguro, M., and Kitagawa, G. *Akaike Information Criterion Statistics*. Reidel, Dordrecht, Netherlands, 1987.

Schwarz, G. Estimating the dimension of a model. *The Annals of Statistics*, 6:461–464, 1978.

Seber, G. and Wild, C. *Nonlinear Regression*. Wiley, Hoboken, NJ, 2003.

Shapiro, S. S. and Wilk, M. B. An analysis of variance test for normality (complete samples). *Biometrika*, 52:591–611, 1965.

Simpson, E. H. The interpretation of interaction in contingency tables. *Journal of the Royal Statistical Society, Series B*, 13:238–241, 1951.

Snijders, T. A. B. and Bosker, R. J. *Multilevel Analysis: An Introduction to Basic and Advanced Multilevel Modeling*. Sage, Thousand Oaks, CA, 2nd edition, 2011.

Stephens, M. A. EDF statistics for goodness of fit and some comparisons. *Journal of the American Statistical Association*, 69:730–737, 1974.

Stephens, M. A. Tests based on EDF statistics. In D'Agostino, R. B. and Stephens, M. A., editors, *Goodness-of-Fit Techniques*. Marcel Dekker, New York, 1986.

Stern, H. A primer on the Bayesian approach to statistical inference. *Stats*, 23:3–9, 1998.

Thisted, R. A. *Elements of Statistical Computing: Numerical Computation*. Chapman & Hall/CRC, Boca Raton, FL, 1988.

Thode Jr., H. C. *Testing for Normality*. Chapman & Hall/CRC, Boca Raton, FL, 2002.

Train, K. E. *Discrete Choice Methods with Simulation*. Cambridge University Press, Cambridge, UK, 2009 edition, 2009.

Tukey, J. W. *Exploratory Data Analysis*. Addison-Wesley, Reading, MA, 1977.

UBS. *Prices and Earnings: A Comparison of Purchasing Power Around the Globe*. UBS AG, Wealth Management Research, Zurich, 2009.

U.S. Central Intelligence Agency. *The World Factbook 2010*. Central Intelligence Agency, Washington, DC, 2010. Available at `https://www.cia.gov/library/publications/the-world-factbook/`.

U.S. Environmental Protection Agency and U.S. Department of Energy. *Fuel Economy Guide, Model Year 2011*. The Office of Energy Efficiency and Renewable Energy Information Center, Washington, DC, 2011. Available at `www.fueleconomy.gov`.

Velilla, S. A note on the multivariate Box-Cox transformation to normality. *Statistics and Probability Letters*, 17:259–263, 1993.

Venables, W. N. and Ripley, B. D. *Modern Applied Statistics with S*. Springer-Verlag, New York, 4th edition, 2002.

Wald, A. and Wolfowitz, J. On a test whether two samples are from the same population. *Annals of Mathematical Statistics*, 11:147–162, 1940.

Weisberg, S. *Applied Linear Regression*. Wiley, Hoboken, NJ, 3rd edition, 2005.

White, H. A heteroscedasticity-consistent covariance matrix estimator and a direct test for heteroscedasticity. *Econometrica*, 48:817–838, 1980.

Yule, G. U. Notes on the theory of association of attributes in statistics. *Biometrika*, 2:121–134, 1903.

GLOSSARY

ANOVA test   *See* Global usefulness test *and* Nested model test.

Autocorrelation   Data collected over time can result in regression model residuals that violate the independence assumption because they are highly dependent across time (p. 202). Also called *serial correlation.*

Average   *See* Mean.

Bivariate   Datasets with two variables measured on a sample of observations (p. 35).

Categorical   *See* Qualitative.

Collinearity   *See* Multicollinearity.

Confidence interval   A range of values that we are reasonably confident (e.g., 95%) contains an unknown population parameter such as a population mean or a regression parameter (p. 16). Also called a *mean confidence interval.*

Cook's distance   A measure of the potential influence of an observation on a regression model, due to either outlyingness or high leverage (p. 196).

Correlation   A measure of linear association between two quantitative variables (p. 50).

Covariate(s)   *See* Predictor variable(s).

Critical value   A percentile from a probability distribution (e.g., t or F) that defines the rejection region in a hypothesis test (p. 20).

Degrees of freedom   Whole numbers for t, F, and χ^2 distributions that determine the shape of the density function, and therefore also critical values and p-values (p. 14).

Density curve   Theoretical smoothed histogram for a probability distribution that shows the relative frequency of particular values for a random variable (p. 6).

Dependent variable   *See* Response variable.

Distribution   Theoretical model that describes how a random variable varies, that is, which values it can take and their associated probabilities (p. 5).

Dummy variables   *See* Indicator variables.

Expected value   The population mean of a variable.

Extrapolation   Using regression model results to estimate or predict a response value for an observation with predictor values that are very different from those in our sample (p. 213).

Fitted value   The estimated expected value, $\hat{Y}$, of the response variable in a regression model (p. 88). Also called an *(unstandardized) predicted value*.

Global usefulness test   Hypothesis test to see whether any of the predictors in a multiple linear regression model are significant (p. 101). An example of an *ANOVA test*.

Hierarchy   A modeling guideline that suggests including lower-order predictor terms when also using higher-order terms, for example, keep X_1 when using X_1^2, keep X_1 and X_2 when using $X_1 X_2$, and keep X_2 when using DX_2 (p. 145).

Histogram   A bar chart showing relative counts (frequencies) within consecutive ranges (bins) of a variable (p. 3).

Hypothesis test   A method for deciding which of two competing hypotheses about a population parameter seems more reasonable (p. 19).

Imputation   One method for dealing with missing data by replacing the missing values with imputed numbers, which might be sample means, model predictions, and so on (p. 215).

Independent variable(s)   *See* Predictor variable(s).

Indicator variables   Variables derived from qualitative variables that have values of 1 for one category and 0 for all other categories (p. 167). Also called *dummy variables*.

Individual prediction interval   *See* Prediction interval.

Input variable(s)   *See* Predictor variable(s).

Interaction   When the effect of one predictor variable on a response variable depends on the value of another predictor variable (p. 159).

Least squares   The computational criterion used to derive regression parameter estimates by minimizing the residual sum of squares, where the residuals are the differences between observed Y-values and fitted $\hat{Y}$-values (p. 88).

Leverage   A measure of the potential influence of a sample observation on a fitted regression model (p. 194).

Loess fitted line   A smooth line for a scatterplot that fits a general nonlinear curve representing the association between the variables on the two axes (p. 120).

Mean   A measure of the central tendency of a variable, also known as the *average* (p. 4).

Median   An alternative measure of the central tendency of a variable, which is greater than half the sample values and less than the other half (p. 4).

Multicollinearity   When there is excessive correlation between quantitative predictor variables that can lead to unstable multiple regression models and inflated standard errors (p. 206). Also called *collinearity*.

Multiple R The correlation between the observed Y-values and the fitted $\hat{Y}$-values from a regression model (p. 100).

Multivariate Datasets with two or more variables measured on a sample of observations (p. 83).

Natural logarithm transformation A mathematical transformation for positive-valued quantitative variables which spreads out low values and pulls in high values; that is, it makes positively skewed data look more normal (p. 142).

Nested model test Hypothesis test to see whether a subset of the predictors in a multiple linear regression model is significant (p. 104). An example of an *ANOVA test*. Also called an *R-squared change test*.

Nominal *See* Qualitative.

Normal probability plot *See* QQ-plot.

Observed significance level *See* p-value.

Ordinal *See* Qualitative.

Outcome variable *See* Response variable.

Outlier A sample observation in a linear regression model with a studentized residual less than -3 or greater than $+3$ (p. 190).

Output variable *See* Response variable.

p-value The probability of observing a test statistic as extreme as the one observed or even more extreme (in the direction that favors the alternative hypothesis) (p. 21).

Parameter A numerical summary measure for a population such as a population mean or a regression parameter (p. 11).

Percentile A number that is greater than a certain percentage (say, 95%) of the sample values and less than the remainder (5% in this case) (p. 4). Also called a *quantile*.

Point estimate A single number used as an estimate of a population parameter. For example, the sample mean is a point estimate of the population mean (p. 15).

Polynomial transformation A mathematical transformation involving increasing powers of a quantitative variable, for example, X, X^2, and X^3 (p. 144).

Population The entire collection of objects of interest about which we would like to make statistical inferences (p. 5).

Predicted value *See* Fitted value.

Prediction interval A range of values that we are reasonably confident (e.g., 95%) contains an unknown data value (such as for univariate data or for a regression response variable) (p. 25). Also called an *individual prediction interval*.

Predictor effect plot A line graph that shows how a regression response variable varies with a predictor variable holding all other predictors constant (p. 224).

Predictor variable(s) Variable(s) in a regression model that we use to help estimate or predict the response variable; also known as independent or input variable(s), or covariate(s) (p. 83).

Probability Mathematical method for quantifying the likelihood of particular events occurring (p. 9).

QQ-plot A scatterplot used to assess the normality of some sample values (p. 8).

Quadratic   A particular type of polynomial transformation that uses a variable and its square, for example, X and X^2 (p. 145).

Qualitative   Data variable that contains labels for categories to which each sample observation belongs (p. 166). Also called *categorical*, *nominal* (if there is no natural order to the categories, e.g., male/female), or *ordinal* (if there is a natural order to the categories, e.g., small/medim/large).

Quantile   *See* Percentile.

Quantitative   Data variable that contains meaningful numerical values that measure some characteristic for each sample observation. Also called a scale measure (p. 35).

R-squared (R^2)   The proportion of variation in a regression response variable (about its mean) explained by the model (p. 94).

R-squared change test   *See* Nested model test.

Reciprocal transformation   A mathematical transformation that divides a quantitative variable into 1, for example, $1/X$ (p. 147).

Reference level   One of the categories of a qualitative variable selected to be the comparison level for all the other categories. It takes the value zero for each of the indicator variables used (p. 174).

Regression coefficients   *See* Regression parameters.

Regression parameters   The numbers multiplying the predictor values in a multiple linear regression model, that is, $(b_1, b_2, \ldots)$ in $E(Y) = b_0 + b_1 X_1 + b_2 X_2 + \cdots$. Also called (unstandardized) regression coefficients (p. 86).

Regression standard error (s)   An estimate of the standard deviation of the random errors in a multiple linear regression model (p. 93). Also called *standard error of the estimate* in SPSS, *root mean squared error* in SAS, and *residual standard error* in R.

Rejection region   The range of values for a probability distribution that leads to rejection of a null hypothesis if the test statistic falls in this range (p. 20).

Residual   The difference, $\hat{e}$, between a response Y-value and a fitted $\hat{Y}$-value in a regression model (p. 119).

Residual standard error   R terminology for regression standard error.

Response variable   Variable, Y, in a regression model that we would like to estimate or predict (p. 83). Also known as a dependent, outcome, or output variable.

Root mean squared error   SAS terminology for regression standard error.

Sample   A (random) subset of the population for which we have data values (p. 11).

Sampling distribution   The probability distribution of a test statistic under (hypothetical) repeated sampling (p. 12).

Scatterplot   A graph representing bivariate data with one variable on the vertical axis and the other on the horizontal axis (p. 37).

Scatterplot matrix   A matrix of scatterplots representing all bivariate associations in a set of variables (p. 89).

Serial correlation   *See* Autocorrelation.

Significance level   The probability of falsely rejecting a null hypothesis when it is true— used as a threshold for determining significance when a p-value is less than this (p. 20).

Standardize   Rescale a variable by subtracting a sample mean value and dividing by a sample standard deviation value. The resulting Z-value has a mean equal to 0 and a standard deviation equal to 1 (p. 4).

Standard deviation   A measure of the spread of a variable, with most of the range of a normal random variable contained within 3 standard deviations of the mean (p. 4).

Standard error   An estimate of a population standard deviation, often used to quantify the sampling variability of a test statistic or model estimate (p. 26).

Standard error of a regression parameter   A standard deviation estimate used in hypothesis tests and confidence intervals for regression parameters (p. 111).

Standard error of estimation   A standard deviation estimate used in hypothesis tests and confidence intervals for a univariate population mean (p. 26).

Standard error of estimation for regression   A standard deviation estimate used in confidence intervals for the population mean in a regression model (p. 126).

Standard error of prediction   A standard deviation estimate used in prediction intervals for a univariate prediction (p. 26).

Standard error of prediction for regression   A standard deviation estimate used in prediction intervals for an individual response value in a regression model (p. 128).

Standard error of the estimate   SPSS terminology for regression standard error.

Statistic   A numerical summary measure for a sample such as a sample mean or an estimated regression parameter (p. 11).

Stem-and-leaf plot   A variant on a histogram where numbers in the plot represent actual sample values or rounded sample values (p. 2).

Test statistic   A rescaled numerical summary measure for a sample that has a known sampling distribution under a null hypothesis, for example, a t-statistic for a univariate mean or a t-statistic for a regression parameter (p. 19).

Unbiased   When a statistic is known to estimate the value of the population parameter correctly on average under repeated sampling (p. 11).

Univariate   Datasets with a single variable measured on a sample of observations (p. 1).

Variance   The square of the standard deviation (p. 10).

Variance inflation factor (VIF)   An estimate of how much larger the variance of a regression parameter estimate becomes when the corresponding predictor is included in the model (p. 206).

Z-value   *See* Standardize.

INDEX

χ^2 distribution, *see* Chi-squared distribution

A

Added variable plot, 117
Adjusted R^2, 95–99
AIC, *see* Akaike Information Criterion
Akaike Information Criterion, 222
Alternative hypothesis, 19
Analysis of deviance, 271
Analysis of variance, 103
Anderson-Darling normality test, 124
ANOVA, *see* Analysis of variance
AR(1), *see* First-order autoregressive errors
Assumptions
    multiple linear regression, 118–124
    simple linear regression, 59–66
Autocorrelation, 202–206

B

Bayes Information Criterion, 222
Bayesian inference, 278, 280–283
Bernoulli distribution, 269
BIC, *see* Bayes Information Criterion
Binary, 269
Binning, 198
Binomial distribution, 269
Bivariate data, 35
Box-Cox transformations, 157
Box-Pierce autocorrelation Q test, 204
Box-Tidwell transformations, 158

Boxplot, 5, 123, 246, 254–255
Breusch-Godfrey autocorrelation test, 124, 204
Breusch-Pagan nonconstant variance test, 124, 200

C

Categorical variable, 166
Causality, 36, 44, 84, 91, 224
Central limit theorem, 12, 15
Ceres plot, 158
Chi-squared distribution, 204
Chow test, 174
Coefficient
    regression, *see* Regression parameter
Coefficient of correlation, 50–52
Coefficient of determination
    multiple linear regression, 94–95
    simple linear regression, 48–50
Collinearity, *see* Multicollinearity
Component plus residual plot, 158
Confidence interval, 16
    interpretation, 18, 59, 69, 114, 127
    multiple linear regression mean, 126–127
    multiple linear regression parameters, 113–114
    simple linear regression mean, 68–69
    simple linear regression slope, 58–59
    univariate population mean, 15–19
Confounding variable, 212
Constant variance, 60, 118
Convenience sample, 5
Cook's distance, 196–199

Applied Regression Modeling, Second Edition. By Iain Pardoe
Copyright © 2012 John Wiley & Sons, Inc.